Interfacial Science

Interfacial Science

A 'Chemistry for the 21st Century' monograph

EDITED BY

M.W. ROBERTS PhD, DSc, CChem, FRSC

Department of Chemistry
University of Wales, Cardiff, UK

Blackwell
Science

© 1997 by International Union of Pure and
Applied Chemistry and published for them by
Blackwell Science Ltd
Editorial Offices:
Osney Mead, Oxford OX2 0EL
25 John Street, London WC1N 2BL
23 Ainslie Place, Edinburgh EH3 6AJ
350 Main Street, Malden
  MA 02148 5018, USA
54 University Street, Carlton
  Victoria 3053, Australia

Other Editorial Offices:
Blackwell Wissenschafts-Verlag GmbH
Kurfürstendamm 57
10707 Berlin, Germany

Blackwell Science KK
MG Kodenmacho Building
7–10 Kodenmacho Nihombashi
Chuo-ku, Tokyo 104, Japan

First published 1997

Set by Semantic Graphics, Singapore
Printed and bound in Great Britain
by Hartnolls Ltd, Bodmin, Cornwall

The Blackwell Science logo is a
trade mark of Blackwell Science Ltd,
registered at the United Kingdom
Trade Marks Registry

DISTRIBUTORS

Marston Book Services Ltd
PO Box 269
Abingdon
Oxon OX14 4YN
(*Orders*: Tel: 01235 465500
         Fax: 01235 465555)

USA
Blackwell Science, Inc.
Commerce Place
350 Main Street
Malden, MA 02148 5018
(*Orders*: Tel: 800 759 6102
           617 388 8250
       Fax: 617 388 8255)

Canada
Copp Clark Professional
200 Adelaide St West, 3rd Floor
Toronto, Ontario M5H 1W7
(*Orders*: Tel: 416 597-1616
           800 815-9417
       Fax: 416 597-1617)

Australia
Blackwell Science Pty Ltd
54 University Street
Carlton, Victoria 3053
(*Orders*: Tel: 3 9347 0300
       Fax: 3 9347 5001)

A catalogue record for this title
is available from the British Library

ISBN 0-632-04219-2

Library of Congress
Cataloging-in-publication Data

Interfacial science/
edited by M.W. Roberts.
    p.   cm. —
  (A "chemistry for the 21st century" monograph)
  Includes bibliographical references and index.
  ISBN 0-632-04219-2
  1 Surface chemistry.
  2. Interfaces (Physical science)
I Roberts, M. W. (Meirion Wyn)
II. Series.
QD508.I58   1997
541.3′3—dc21   97-7683
                                    CIP

Contents

Contributors

H.N. AIYER MSc, *Materials Research Centre and CSIR Centre of Excellence in Chemistry, Indian Institute of Science, Bangalore 560 012, India* [1]

J.P.S. BADYAL MA, PhD, *Department of Chemistry, Science Laboratories, University of Durham, Durham DH1 3LE, UK* [237]

V.I. BUKHTIYAROV PhD, *Boreskov Institute of Catalysis, Prospekt Akademika Lavrentieva, 5, Novosibirsk, 630090 Russia* [109]

A.F. CARLEY MA, PhD, *Department of Chemistry, University of Cardiff, PO Box 912, Cardiff CF1 3TB, UK* [77]

C.R.A. CATLOW *The Royal Institution of Great Britain, 21 Albemarle Street, London W1X 4BS, UK* [195]

P.R. DAVIES BSc, PhD, *Department of Chemistry, University of Cardiff, PO Box 912, Cardiff CF1 3TB, UK* [77]

D. DIESING Dr rer nat, *Lehrstuhl für Oberflächenwissenschaft (IPkM), Heinrich-Heine-Universität, Düsseldorf, D-40225 Düsseldorf, Germany* [163]

M. FANFONI *Dipartimento di Fisica, Università di Roma 'Tor Vergata' and Istituto Nazionale di Fisica della Materia, Via della Ricerca Scientifica, 00133 Rome, Italy* [129]

J.D. GALE *Department of Chemistry, Imperial College of Science, Technology and Medicine, Exhibition Road, London SW7 2AZ, UK* [195]

D.H. GAY *The Royal Institution of Great Britain, 21 Albemarle Street, London W1X 4BS, UK* [195]

M. HÄNISCH Dr rer nat, *Lehrstuhl für Oberflächenwissenschaft (IPkM), Heinrich-Heine-Universität, Düsseldorf, D-40225 Düsseldorf, Germany* [163]

K.D.M. HARRIS BSc, PhD, *School of Chemistry, University of Birmingham, Edgbaston, Birmingham B15 2TT, UK* [21]

Y. IWASAWA DSc, *Department of Chemistry, Graduate School of Science, University of Tokyo, Hongo, Bunkyo-ku, Tokyo 113, Japan* [57]

H. JANSSEN Dr rer nat, *Lehrstuhl für Oberflächenwissenschaft (IPkM), Heinrich-Heine-Universität, Düsseldorf, D-40225 Düsseldorf, Germany* [163]

D. KÖRWER Dr rer nat, *Lehrstuhl für Oberflächenwissenschaft (IPkM), Heinrich-Heine-Universität, Düsseldorf, D-40225 Düsseldorf, Germany* [163]

G. KRITZLER Dipl phys, *Lehrstuhl für Oberflächenwissenschaft (IPkM), Heinrich-Heine-Universität, Düsseldorf, D-40225 Düsseldorf, Germany* [163]

M.M. LOHRENGEL Dr rer nat, *AGEF-Institut an der Heinrich-Heine-Universität, Düsseldorf, D-40225 Düsseldorf, Germany* [163]

N. MAUNG PhD, *Advanced Materials Research Laboratory, Multidisciplinary Research and Innovation Centre, North East Wales Institute, Plas Coch, Mold Road, Wrexham, LL11 2AW, UK* [257]

M.A. NYGREN *The Royal Institution of Great Britain, 21 Albemarle Street, London W1X 4BS, UK* [195]

H. ONISHI DSc, *Department of Chemistry, Graduate School of Science, University of Tokyo, Hongo, Bunkyo-ku, Tokyo 113, Japan* [57]

A. OTTO Prof Dr rer nat, *Lehrstuhl für Oberflächenwissenschaft (IPkM), Heinrich-Heine-Universität, Düsseldorf, D-40225 Düsseldorf, Germany* [163]

M.E. PEMBLE BSc, PhD, CChem, FRSC, CPhys, MInstP, *Division of Chemical Sciences, Science Research Institute, University of Salford, Salford M5 4WT, UK* [217]

A.R. RAJU MPhil, PhD, *Materials Research Centre and CSIR Centre of Excellence in Chemistry, Indian Institute of Science, Bangalore 560 012, India and Jawaharlal Nehru Centre for Advanced Scientific Research, Jakkur, Bangalore 560 064, India* [1]

C.N.R. RAO PhD, DSc, *Materials Research Centre and CSIR Centre of Excellence in Chemistry, Indian Institute of Science, Bangalore 560 012, India and Jawaharlal Nehru Centre for Advanced Scientific Research, Jakkur, Bangalore 560 064, India* [1]

S. RÜßE Dr rer nat, *AGEF-Institut an der Heinrich-Heine-Universität, Düsseldorf, D-40225 Düsseldorf, Germany* [163]

D.C. SAYLE *The Royal Institution of Great Britain, 21 Albemarle Street, London W1X 4BS, UK* [195]

A. SCHAAK Dipl phys, *Lehrstuhl für Oberflächenwissenschaft (IPkM), Heinrich-Heine-Universität, Düsseldorf, D-40225 Düsseldorf, Germany* [163]

S. SCHATTEBURG Dipl phys, *Lehrstuhl für Oberflächenwissenschaft (IPkM), Heinrich-Heine-Universität, Düsseldorf, D-40225 Düsseldorf, Germany* [163]

M. TOMELLINI *Dipartimento di Scienze e Tecnologie Chimiche, Università di Roma 'Tor Vergata', Via della Ricerca Scientifica, 00133 Rome, Italy* [129]

S.H. WHEALE BSc, *Department of Chemistry, Science Laboratories, University of Durham, Durham DH1 3LE, UK* [237]

J.O. WILLIAMS PhD, DSc, *Advanced Materials Research Laboratory, Multidisciplinary Research and Innovation Centre, North East Wales Institute, Plas Coch, Mold Road, Wrexham, LL11 2AW, UK* [257]

H. WINKES Dr rer nat, *Lehrstuhl für Oberflächenwissenschaft (IPkM), Heinrich-Heine-Universität, Düsseldorf, D-40225 Düsseldorf, Germany* [163]

A.C. WRIGHT PhD, *Advanced Materials Research Laboratory, Multidisciplinary Research and Innovation Centre, North East Wales Institute, Plas Coch, Mold Road, Wrexham, LL11 2AW, UK* [257]

Foreword

The instigation of the series of books of which this one forms an integral part, owes a great deal to Kirill Ilyich Zamaraev, former President of the International Union of Pure and Applied Chemistry (1993–95). News of his death in June 1996 caused grief in many corners of the world. But the memory of his gifts and talents lives on, not least in books such as these — there were many others — that were stimulated by his concerns, enthusiasms and perspicacity.

Many tributes have already been paid to the life and work of Kirill Zamaraev (see, for example, *Mendeleev Communications*, 1996, 213 and *Catalysis Letters*, 1996, Vol. 41, No. 1,2, p(i)). Doubtless many more will be composed. What needs to be emphasised about Kirill Zamaraev, however, is that he combined in a remarkable way the ability to be forward-looking, with a deep awareness and appreciation of the past. Whilst he loved the past, he did not live in it. But the finest tradition bequeathed to us by great human beings was something of which he was acutely and sensitively aware. Even while lying ill in a Moscow hospital in the autumn of 1994 he wrote a beautiful booklet that chronicles the achievements of Russian scientists in catalysis from the days of Lomonosov (1711–65) onwards.

To some thinkers, tradition is, by definition, an irrational inheritance from the dead past, a mere residue of obsolete beliefs and customs which an enlightened society, especially scientists, should discard as an impediment to reason and progress. Kirill rejected such arguments: we owe as much to the traditions of our scientific predecessors as we do to their discoveries and inventions. This ethos always shone through whenever one had a philosophical talk with Kirill. I shall always treasure his quotation from Pushkin: 'How many and varied are the discoveries prepared for us by the spirit of the enlightenment, by experiment, the child of error and effort, by genius, the friend of paradox, and by that divine inventor, chance'.

He would be proud of this book.

Sir John Meurig Thomas
Royal Institution of Great Britain, London,
and Peterhouse, University of Cambridge

Preface

It was Kirill Zamaraev who, on a visit to Cardiff to deliver the Royal Society of Chemistry Centenary Lecture in 1995, invited me to edit this monograph on Interfacial Science. I had first met Professor Zamaraev at the Indian National Science Academy's Golden Jubilee Symposium held in New Delhi in 1984. This led to a most fruitful collaboration with two of his colleagues at the Institute of Catalysis in Novosibirsk, A. Boronin and V. Bukhtiyarov. We agreed that the monograph should not dwell exclusively on surface science, heterogeneous catalysis or materials science; rather that it should reflect the wider perspective of interfacial science. It was, therefore, with this very much in mind that I approached the authors of these eleven articles.

The ability to design a solid surface with specific properties, whether as an efficient and selective catalyst or for the creation of a new optical device, has been a central theme in both fundamental and applied research over the last two decades. It has not only been driven by technological demands but also through academic curiosity and the emergence of sophisticated surface-sensitive techniques which have provided precise answers to previously unanswerable questions. In addition, and especially over the last decade, computational methods have been developed which have played an important role in constructing models of surfaces and the associated chemistry of a wide range of solid materials; these methods have already opened up new avenues for application and for fundamental thinking. The largely unpredictable nature of molecular events at surfaces does, however, emphasise the need to ensure that the computational results need always to be assessed against the relevant experimental data.

In this monograph we see a blurring of the traditional approaches to condensed phase and surface chemistry:
- surface modelling as a development of bulk computational methods (Catlow *et al.*);
- whether oxygen states present at metal surfaces mimic those at bulk oxide surfaces (Carley and Davies; Onishi and Iwasawa);
- the chemistry and model of growth of small metal clusters compared with the characteristics of the corresponding bulk metal surfaces (Bukhtiyarov; Tomellini and Fanfoni);
- growing single-crystal films of complex metal oxides by epitaxial or vapour deposition methods (Rao *et al.*; Williams *et al.*);
- studying semiconductor surfaces at 'high' pressures using new experimental techniques (Pemble);
- controlling the properties of polymer surfaces by plasma etching (Wheale and Badyal);
- the technological implications of novel inclusion compounds (Harris); and
- the role of hot electrons in surface reactions (Otto *et al.*).

The authors provide both authoritative and critical accounts of the present status of their research topics which will appeal to a wide spectrum of surface scientists.

M.W. Roberts

1 Solid–Solid Interfaces in the Epitaxial Films of Complex Oxides Deposited by Chemical Methods

A.R. RAJU, H.N. AIYER* and C.N.R. RAO

Materials Research Centre and CSIR Centre of Excellence in Chemistry, Indian Institute of Science, Bangalore 560 012, India and Jawaharlal Nehru Centre for Advanced Scientific Research, Jakkur, Bangalore 560 064, India

* Materials Research Centre and CSIR Centre of Excellence in Chemistry, Indian Institute of Science, Bangalore 560 012, India

1 Introduction

The term *epitaxy* literally means 'arrangement upon'; it describes the oriented growth of one crystalline material (guest) on the surface of a single crystal of a different material (substrate). The adjective *epitaxial* implies not only that one particular crystal plane of the deposited guest crystal (deposited film) comes into contact with the surface of the host crystal, but also that one particular crystallographic direction in the contact plane of the guest crystal is parallel to a specific crystallographic direction in the contact plane of the host crystal. This is usually described in terms of Miller indices of the crystal planes and directions. For instance, (001)F [100]F//(001)S [100]S means that the (001) plane of the deposited film is in contact or aligned with the (001) plane of the substrate surface and that the [100] film growth direction coincides with the [100] direction of the substrate surface. For epitaxial growth to occur, it is desirable to have good lattice matching between the material of the film and the substrate, which would alleviate the problems of stress and relaxation through the formation of misfit dislocations. Another requirement for epitaxial nucleation is that the surface energy of the nucleus–substrate interface should be lower for the epitaxial orientation than for other orientations, so that the nucleation rate for the epitaxial orientation is greater. However, epitaxy can also occur due to different growth processes.

Preparation of single-crystalline films of complex oxides by means of epitaxial growth on suitable single-crystal substrates is an important area of research today because of the extensive, sophisticated technological applications of the epitaxial films in various types of devices. The single-crystalline or epitaxial nature of a film is established by X-ray diffraction techniques and more particularly by direct observation of the interface between the substrate and the deposited film by means of high-resolution transmission electron microscopy (HRTEM). By studying HRTEM images, one can determine how good the lattice match is between the lattice of the deposited film and that of the substrate. One can estimate the strain at the interface based on the mismatch, besides observing structural defects at the interface. In recent years, excellent single-crystalline (epitaxial) films of ferroelectric oxides such as $PbTiO_3$ and $PbZr_xTi_{1-x}O_3$ (PZT), superconducting $YBa_2Cu_3O_{7-x}$ and other complex oxide systems have been prepared on different single-crystal substrates by several techniques [1–8]. A few important deposition techniques are: (i) pulsed laser deposition (PLD); (ii) molecular beam epitaxy (MBE); and (iii) metal–organic chemical vapour deposition (MOCVD). Of these three methods, MOCVD is truly a chemical method involving reactions of precursors, while PLD involves the physical process of transferring the material from the target to the substrate. MBE involves vaporisation and monolayer deposition, followed by chemical reaction of the layers at the atomic scale. All three methods have yielded high-quality films, but they are all expensive and need sophisticated instrumentation. This is especially true of MBE.

In our laboratory, we have established inexpensive but powerful techniques for preparing oriented films of complex oxides by nebulised spray pyrolysis. In this paper, we shall briefly describe the PLD, MBE and MOCVD methods and illustrate results obtained by them with respect to epitaxial films of complex oxides on single-crystal substrates. The single-crystal substrates are generally oxides such as MgO, $SrTiO_3$ (STO) and $LaAlO_3$ (LAO), giving rise to oxide–oxide (solid–solid) interfaces. We provide a detailed description of the results on thin films of complex oxides obtained by nebulised spray pyrolysis and demonstrate how this inexpensive technique gives excellent-quality epitaxial films of complex oxides of technological importance, which include ferroelectric $PbZr_xTi_{1-x}O_3$ ($x = 0.5$), metallic $LaNiO_3$, and $LaMnO_3$ which exhibits giant magnetoresistance (GMR). In so doing, we describe the method in some detail and provide the HRTEM evidence for the epitaxial nature of the films obtained by this technique, showing the presence of an excellent lattice match between the deposited film and the substrate.

2 Pulsed laser deposition

In PLD, the material is evaporated from a target composed of the stoichiometric solid by means of an intense pulsed laser beam [9–12]. An excimer laser (KrF type, with a wavelength of 248 nm and a pulsewidth of c. 30 ns, operated at 100–1000 mJ pulse energy and a typical repetition rate of 1–10 Hz) is generally employed. A lower-power HeNe laser is employed to align and focus the excimer laser on to the target. The PLD process can be classified into three regions.

1 Absorption of the laser radiation by the target material resulting in the ablation (evaporation) of the surface layer.

2 Interaction of the evaporated material with the laser beam during which the evaporated ions and atoms absorb the photon energy.

3 Anisotropic adiabatic expansion of the plasma resulting in a forward-directed plume.

The epitaxial growth of the oxide superconductors on single-crystal substrates is required for microwave and other applications. LAO is a strong candidate as a substrate for superconducting thin films for microwave applications due to its low dielectric constant and loss tangent [13]. Although the actual structure of $LaAlO_3$ at room temperature is rhombohedral, its unit cell can be described as a slightly distorted pseudocubic perovskite providing a close lattice match with the unit cell of YBCO. Highly epitaxial thin films of YBCO have been obtained on (001) single-crystal LAO by PLD [14]. An HRTEM image of the interface between the YBCO film and the LAO substrate is presented in Fig. 1.1. The image shows near-perfect epitaxial c perpendicular growth of YBCO on (001) LAO. This interface microstructure also reveals the presence of regions of strain which extend into both the film and the substrate. These strained regions correspond to interfacial dislocations. The presence of interfacial dislocations is due to the slight mismatch of the lattice parameters of YBCO and LAO.

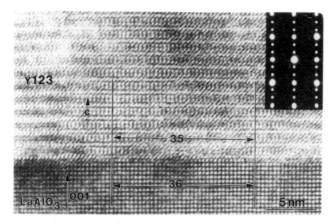

Figure 1.1. HRTEM image of the interface between the YBCO film deposited by PLD and the LaAlO$_3$(001) substrate. The inset SAED pattern taken from an area including both the film and the substrate confirms the epitaxial growth. The presence of an interfacial dislocation is seen in form of an extra half-plane in the substrate within the marked regions. From [14].

The epitaxial c perpendicular growth of the YBCO film is further confirmed by the selected area electron diffraction (SAED) pattern which is taken from an area at the interface including both the film and the substrate (see inset Fig. 1.1). The coincidence of the diffraction spots of the film and the substrate shows the epitaxial nature of the film. The epitaxial relationship is (001)YBa$_2$Cu$_3$O$_{7-x}$//(001)LaAlO$_3$. Another example of an epitaxial film obtained by PLD is illustrated in Fig. 1.2 [15]. The figure shows an HRTEM image of a PbTiO$_3$ film on an (001)STO substrate. This image reveals a perfect in-plane matching and a complete absence of a-axis-oriented domains establishing the c-axis orientation of the film. The SAED pattern shown in the inset also confirms the c-axis orientation of the film. The orientation relationship of the film and the substrate is (001)PbTiO$_3$//(001)SrTiO$_3$.

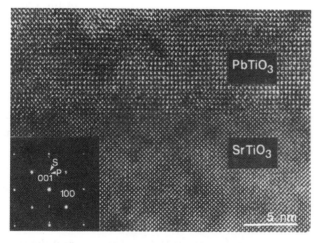

Figure 1.2. HRTEM image of the interface between a PbTiO$_3$(P) film deposited on the SrTiO$_3$(001) substrate by PLD showing perfect in-plane matching. The SAED pattern shown in the inset also evidences the c orientation of the PbTiO$_3$ film. From [15].

3 Molecular beam epitaxy

MBE is a versatile technique for growing epitaxial films of metals, semiconductors, insulators and oxide materials [16,17]. In this technique, molecular beams of the constituent atoms formed from Knudsen effusion cells are allowed to impinge on a heated substrate under ultra-high vacuum (c. 10^{-10} torr). The heated substrate provides the impinging molecules with sufficient energy to be mobile on the substrate so that these atoms migrate on the surface until they encounter an appropriate site on which to condense. Under satisfactory conditions, one obtains the growth of a single-crystal epitaxial film on a single-crystal substrate. Typical rates used for deposition in MBE are about one atomic layer per second. The rate of deposition is controlled by the intensity of the molecular beam. Therefore, the uniformity of the film depends on the uniformity of the beams. Thus, the MBE process offers a high degree of uniformity, controllability and reproducibility. In the MBE system, several beams of different substances can be operated in the same chamber and directed on to the substrate surface. Given the independent control of the intensities of the beams and the control of the substrate temperature, structures of the prescribed composition with a variety of layers can be deposited in the form of a film. MBE has been used to prepare high-quality epitaxial films of a variety of oxides as well as superlattices on different single-crystal substrates. One of the important applications of the ferroelectric perovskite $BaTiO_3$ is in electro-optic devices. The large electro-optic coefficient of $BaTiO_3$, coupled with the small microwave dielectric constant of a substrate such as MgO, could substantially improve the device speed and efficiency. Devices based on these materials in thin-film form require long-range structural coherence at the interface, which can be realised in epitaxial thin films. Epitaxial $BaTiO_3$ films have been prepared on a MgO substrate by using MBE [18]. Figure 1.3 shows a scanning electron micrograph of $BaTiO_3$ on MgO(001) in cross-sectional view. We see that the film is 0.6 µm thick, crack-free, dense, single-phase and optically transparent. Figure 1.4 shows an HRTEM image of the $BaTiO_3$–MgO interface. The cube-on-cube epitaxy can be readily observed in this image. At a larger scale, the HRTEM images showed dislocations nucleated to relieve the interfacial strain.

Figure 1.3. Cross-sectional scanning electron micrograph of $BaTiO_3$ film deposited on MgO(001) by MBE. Film thickness is 0.6 µm. From [18].

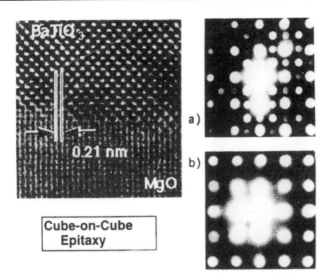

Figure 1.4. HRTEM image of the BaTiO$_3$–MgO(001) interface (in a film obtained by MBE) with the SAED pattern inserts from: (a) BaTiO$_3$; (b) MgO. From [18].

4 Metal–organic chemical vapour deposition

In MOCVD, a metal–organic precursor compound or combination of precursor compounds from the gas phase condenses on a substrate surface where a chemical reaction occurs, leading to the formation of a solid film [19]. If the (source) material to be deposited is not in the vapour state, it can be converted into the vapour state by volatilising from either the solid or the liquid source. In its most common form, the CVD process relies upon elevated substrate temperatures to pyrolyse the gaseous precursor for film growth. However, when high growth temperatures are incompatible with the substrate material, or when metastable materials are to be deposited, it is necessary to activate the chemical reaction near or on the surface to facilitate film growth at reduced temperatures. This is accomplished by the application of a radio-frequency (RF) field (plasma excitation), light (photoexcitation) or by direct heating of the gaseous source (thermal excitation). MOCVD has been widely employed for the deposition of epitaxial ferroelectric films. An epitaxial ferroelectric BaTiO$_3$ film is obtained on a MgO(100) substrate by MOCVD at 875 K by using titanium isopropoxide and bis(2,2,6,6,-tetramethyl 3,5 heptanedionato)barium, i.e. Ba(thd)$_2$, as precursors [20]. An HRTEM image of this BaTiO$_3$ film, showing the BaTiO$_3$–MgO interface, is shown in Fig. 1.5(a). We see that the (200) lattice planes are continuous across the interface, demonstrating the epitaxial nature of the film. The image also shows that the film–substrate interface is sharp and contains no second phases. Misfit dislocations in the BaTiO$_3$ film were observed at the interface. One such dislocation is indicated by an arrow in the micrograph. Formation of the dislocations probably occurred during the deposition process in order to accommodate the 5.4% lattice mismatch between the substrate and the growing film at the growth temperature. The SAED pattern of the BaTiO$_3$–MgO interface shown in Fig. 1.5(b) shows a cube-on-cube orientation between BaTiO$_3$ and MgO (i.e. alignment of both normal and in-plane lattice vectors of the film with that of the substrate) confirming that the film is epitaxial. Higher-order diffraction spots showed some arcing, indicating that the different regions of the film were slightly

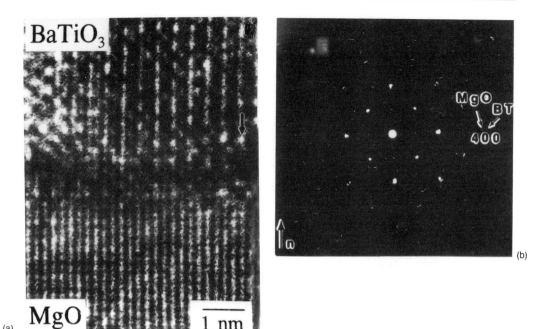

Figure 1.5. (a) HRTEM image of the interface between the BaTiO$_3$ film deposited on MgO(100) by MOCVD. (b) SAED pattern from the BaTiO$_3$–MgO(100) interface showing the 'cube-on-cube' orientation relationship between the film and the substrate. The arrow indicates the direction of the normal (n) to the film–substrate interface. From [20].

misoriented relative to the ⟨100⟩ direction in the MgO substrate. An upper limit of 1% for the misorientation angle was estimated from the extent of arcing. Lattice parameters calculated from the pattern confirmed that the film was epitaxial and oriented with the *a*-axis normal to the substrate surface.

Epitaxial thin films of PbTiO$_3$ have recently been deposited on SrRuO$_3$-buffered SrTiO$_3$(001) substrates by MOCVD [21]. Tetraethyl-lead and titanium isopropoxide were used as precursors in this study. The SrRuO$_3$ buffer layer was prepared by RF sputtering. In Fig. 1.6, the cross-sectional TEM image of the PbTiO$_3$–SrRuO$_3$–SrTiO$_3$

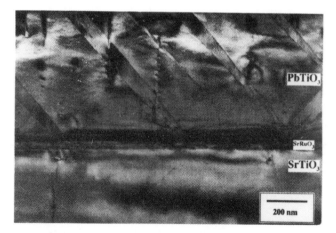

Figure 1.6. Cross-sectional TEM image of a PbTiO$_3$ film prepared by MOCVD deposited on a SrTiO$_3$ substrate with a SrRuO$_3$(001) epitaxial buffer layer. The image shows that the dominant defects in the film are 90° domains and threading dislocations. From [21].

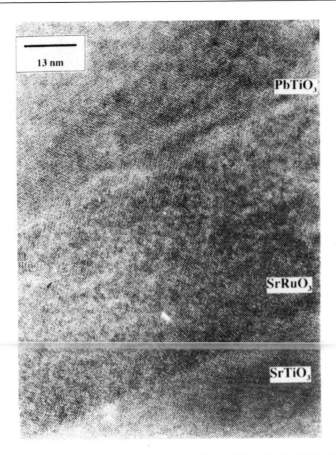

Figure 1.7. HRTEM image of $PbTiO_3(001)$–$SrRuO_3(001)$–$SrTiO_3(001)$ interfaces showing atomically sharp interfaces between individual layers. From [21].

film is shown. The image shows that the film is epitaxial and c-axis oriented, with 90° domains, with threading dislocations as the primary structural defects. The thickness of the $SrRuO_3$ buffer layer is about 33 nm. The 90° domains visible in the image nucleate at the structural defects in the substrate. The strain contrast associated with the substrate defect sites appears to propagate directly through the $SrRuO_3$ layer into the $PbTiO_3$ layer. The threading dislocations appear normal to the substrate–film interface and propagate through the 90° domains. This is indicative of the formation of dislocations prior to the ferroelectric phase transition, while the film is in the cubic state. A cross-sectional HRTEM image of the $PbTiO_3(001)$–$SrRuO_3(001)$–$SrTiO_3(001)$ interfaces is shown in Fig. 1.7. The image shows atomically sharp interfaces. The $PbTiO_3(001)$–$SrRuO_3(001)$ interface appears to be cleaner than the $SrRuO_3(001)$–$SrTiO_3(001)$ interface, indicating that the deposition of the buffer layer appears to improve the quality of the substrate surface, resulting in an improved ferroelectric film.

5   Nebulised spray pyrolysis

The technique of nebulised spray pyrolysis developed in our laboratory does not employ a vacuum. It is a simpler, low-cost alternative for the deposition of high-quality thin films of novel complex oxides [22–24]. Pyrolysis of sprays is a well-known method for depositing films. A novel improvement in this technique is the so-called pyrosol process or nebulised spray pyrolysis of a spray generated by an ultrasonic atomiser. In nebulised spray pyrolysis, a solution containing the organometallic derivatives of the relevant metals in a suitable solvent (source liquid) is nebulised by making use of a PZT transducer operated at a frequency of 1.72 MHz. The nebulised spray is slowly deposited on a solid substrate at a relatively low temperature with sufficient control of the rate of deposition to yield the oxide film of the desired stoichiometry. The rate of deposition is controlled by means of the flow rate of the carrier gas used, the temperature of the substrate and the concentration of the organometallic precursors. The film thickness can vary from few hundred nanometres (*c.* 200 nm) to a few micrometres, depending upon the solution concentration, deposition time, etc.

Figure 1.8 is a schematic diagram showing the different parts of our apparatus for ultrasonically nebulised spray pyrolysis. It consists of two independent zones: the atomisation chamber and the pyrolysis reactor. The source liquid is kept in the atomisation chamber. The atomisation chamber is designed so that the bottom of the chamber has a cylindrical opening to fit the PZT transducer of 20 mm diameter (1 mm thick). The special feature of this design is that the upper portion of the PZT transducer is in direct contact with the source liquid. This gives the highest energy transfer to the liquid when compared to the non-contact method. Its disadvantage is that the pH of the liquid has to be maintained at around 7. The liquid level in the atomisation chamber is maintained by using a constant level burette which allows the measurement of the

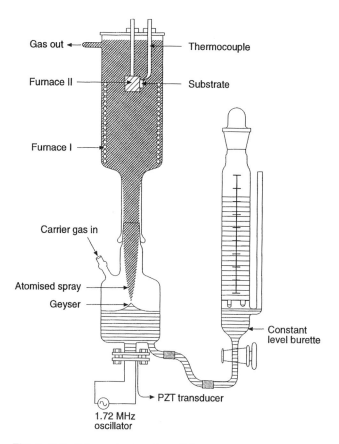

Figure 1.8. Schematic drawing of the apparatus used for nebulised spray pyrolysis.

volume of the liquid nebulised. When a high-frequency ultrasonic beam is directed at the gas–source-liquid interface through the PZT transducer, a geyser forms at the surface, the height of which is proportional to the acoustic intensity and the physical properties of the liquid (vapour pressure, viscosity and surface tension). The geyser formation is accompanied by cavitation at the gas–source-liquid interface. When the amplitude of the acoustic vibrations exceeds a certain threshold value, liquid atomisation occurs. Above this threshold, a continuous and regular mist (nebulised spray) is generated. The nebulised spray produced in the first chamber is transported by a carrier gas introduced through a side port. The second zone is a long glass tube which consists of a heater (nichrome wire wound on a ceramic tube), a substrate holder and a chromel–alumel thermocouple fixed to the substrate holder. The nebulised spray is decomposed in the pyrolysis reactor zone on a hot substrate at an appropriate temperature and the pyrolysis reaction product consists of a thin film whose composition, adherence and morphology depend on the experimental conditions.

5.1   Ferroelectric oxide films

Epitaxial thin films of ferroelectric materials such as PZT have received much attention because of their applications in non-volatile memories, sensor and actuator devices, etc. [25]. Using the nebulised spray pyrolysis technique, we have deposited films of $PbZr_xTi_{1-x}O_3$ with $x = 0.5$ on STO(100) substrates at 625 K for 3 hours, which were further annealed for 12 hours in air at the same temperature. The precursors employed were lead acetate, zirconium isopropoxide and titanium isopropoxide. The solution containing the stoichiometric quantities of the precursors in methanol solvent (0.1 mol) was nebulised to obtain the PZT film. The films were first subjected to energy dispersive X-ray analysis (EDAX), which showed a Pb : Zr : Ti ratio close to 1 : 0.5 : 0.5, confirming the film stoichiometry. An X-ray Θ–Θ scan of the PZT film deposited on the STO(100) substrate is shown in Fig. 1.9. The $PbZr_{0.5}Ti_{0.5}O_3$ corresponds to a composition on the tetragonal side of the $PbTiO_3$–$PbZrO_3$ phase diagram. The only phase

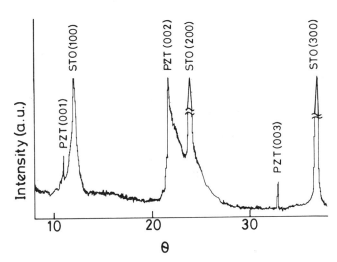

Figure 1.9. X-ray diffraction pattern (Θ–Θ scan) of a $PbZr_{0.5}Ti_{0.5}O_3$ film deposited on $SrTiO_3$(100) substrate by nebulised spray pyrolysis.

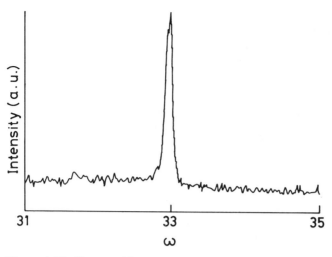

Figure 1.10. X-ray rocking curve (ω scan) of the (003) reflection of a PZT film deposited on STO(100) by nebulised spray pyrolysis.

observed is that of the perovskite type $PbZr_xTi_{1-x}O_3$. The X-ray diffraction (XRD) scan showed that the film on the STO(100) is c-axis oriented since only the reflections appearing in the pattern are those from the (001) plane of PZT along with the (h00) reflections of the STO substrate.

X-ray rocking curve measurements were performed to obtain information on the crystalline perfection of these films and to quantify their planar orientation of these films. Both Θ and ω scans were carried out on the preselected film peaks known from the Θ–Θ XRD pattern. The full width at half maximum (FWHM) of the X-ray rocking curves of the (003) reflection of PZT (Fig. 1.10, ω scan) was c. 0.2°, indicating an excellent alignment in the growth direction among the c-axis oriented grains and implying that the film deposited was indeed epitaxial. This FWHM value compares well with that of epitaxial ferroelectric (PZT, PT) films obtained by processes such as RF magnetron sputtering, MOCVD, etc. This epitaxial deposition of a PZT film on an

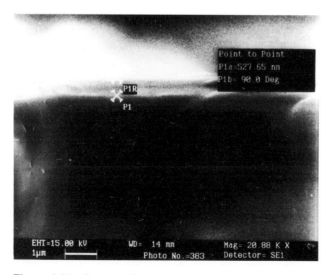

Figure 1.11. Cross-sectional scanning electron micrograph of the PZT film deposited on the STO(100) substrate by nebulised spray pyrolysis. Film thickness is about 530 nm.

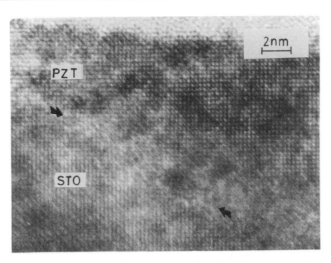

Figure 1.12. HRTEM image of the PZT–STO interface revealing the epitaxial (*c*-oriented) growth of PZT film. The interface is indicated by the arrows.

STO substrate is realised because of reasonably good lattice matching between STO (a = 3.905 Å) and PZT (c = 4.1 Å). A cross-sectional scanning electron micrograph of the PZT film deposited on the STO substrate is shown in Fig. 1.11. The micrograph shows the film to be smooth, uniform and densely packed, having a thickness of 530 nm.

Cross-sectional transmission electron microscopy (TEM) of PZT–STO revealed the film to be a single domain. No 90° domains were observed in the bright field images. This may be due to the fact that PZT films in the present case have been deposited below the Curie temperature T_c. Figure 1.12 shows a high-resolution lattice image of the PZT–STO interface. A direct correspondence between the crystallography in the film and that of the substrate is evident. The (001) planes of both PZT and $SrTiO_3$ are parallel to the interface and the *a*- and *b*-axes, i.e. the ⟨100⟩ and ⟨010⟩ directions of the PZT are aligned with the corresponding directions of STO in the interfacial plane. This clearly demonstrates the epitaxial growth of the PZT on STO. The SAED pattern from the film–substrate (PZT–STO) interface is shown in Fig. 1.13. The interplanar spacings derived from the Bragg spot separations (about 4 Å) correspond to the (100) and (001) reflections. From this analysis, we conclude that the epitaxial growth of $PbZr_{0.5}Ti_{0.5}O_3$ on STO is *c*-axis oriented. This, along with the coincidence of SAED pattern of the PZT film with that of STO(100) substrate, further confirms the epitaxial nature of the PZT film. The epitaxial relationship is (001)PZT//(100)STO.

The epitaxial PZT films deposited by nebulised spray pyrolysis show a hysteresis loop characteristic of ferroelectric films with reasonably good values of coercive field, E_c (*c.* 80 kV) and remnant polarisation, P_r (*c.* 22 C cm^{-2}).

5.2 *Metallic LaNiO$_3$ films*

One of the problems in integrating ferroelectric thin films into devices is the difficulty of growing a single-crystalline thin film with an appropriate electrode material. Perovskite type metallic oxides such as $LaNiO_3$ (LNO) lattice-matched to ferroelectric oxides would be good candidates for this purpose [26]. It is therefore important to study

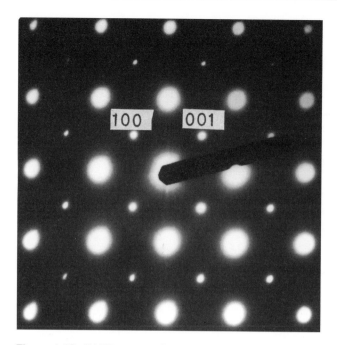

Figure 1.13. SAED pattern from the PZT–STO(100) interface region. The interplanar spacing derived from the spot separations is *c.* 4 Å.

epitaxial thin films of LNO. We have deposited LNO films on STO(100) substrates at 675 K (8 hours) by the nebulised spray pyrolysis of acetylacetonate precursors of lanthanum and nickel.

The deposited films of LNO were stoichiometric, as confirmed by EDAX analysis, which gave the La : Ni ratio as 1 : 1. The X-ray diffraction pattern of LNO deposited on STO (Fig. 1.14) distinctly shows the (h00) reflections of LNO besides the substrate (h00) reflections, indicating that the film deposited is clearly highly oriented or textured. The crystallinity of the LNO film in the growth direction was ascertained by means of the X-ray rocking curve. The ω scan (Fig. 1.15) of the LNO(100) reflection has an FWHM of 0.4°, indicating very good crystallinity and the epitaxial nature of the

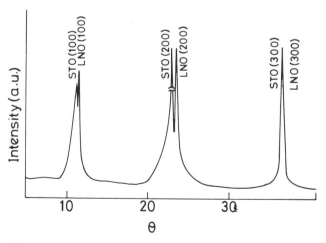

Figure 1.14. X-ray diffraction pattern (Θ–Θ scan) of an LNO film deposited on STO(100) by nebulised spray pyrolysis.

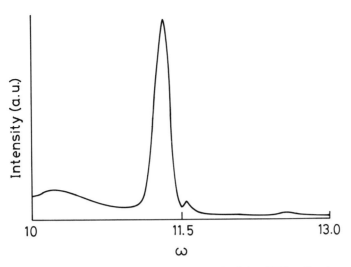

Figure 1.15. X-ray rocking curve (ω scan) of the (100) reflection of the LNO film deposited on STO(100).

film. The STO(100) substrate promotes the epitaxial growth of the LNO film due to closeness of the lattice constants between STO ($a = 3.905$ Å) and that of the LNO ($a = 3.846$ Å, pseudocubic), i.e. a lattice mismatch of only 1.66%.

A cross-sectional scanning electron micrograph of an LNO–STO film revealing the film and the substrate regions is shown in Fig. 1.16. The LNO film is highly dense, without any porosity. The thickness of the LNO film obtained from the micrograph was 19.8 µm.

In Fig. 1.17, we show a cross-sectional low-magnification TEM image (bright field image) of LNO–STO. The interface is clearly visible and is well marked by the presence of the extinction contours arising out of the differential thinning process of the film and the substrate. The film is continuous and is devoid of any grain boundaries or defects. The high-resolution cross-sectional image of the LNO–STO interface is shown in

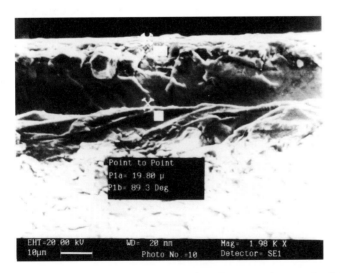

Figure 1.16. Cross-sectional scanning electron micrograph of the LNO film deposited on STO(100) substrate. Film thickness is *c.* 19 µm.

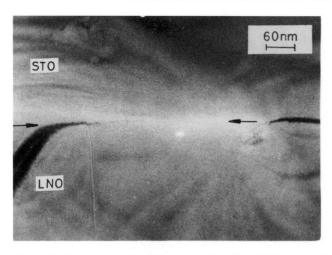

Figure 1.17. Cross-sectional low-magnification TEM bright field image of the LNO film deposited on STO(100) showing the interface. The interface is indicated by the arrows.

Fig. 1.18. The bright spots on either side of the interface are *c.* 2.7 Å apart, showing good lattice matching. The SAED pattern (Fig. 1.19), including diffraction spots of both the LNO film and the STO substrate, shows intense Bragg spots corresponding to the $\sqrt{2}$ type of the perovskite cell. The weak spots corresponding to *c.* 3.9 Å are actually split, thereby showing the slight mismatch of the *a* parameters of the film and the underlying substrate. Thus, the cross-sectional electron microscopic studies clearly show that the films deposited are truly epitaxial and that the epitaxial relationship is (100)LNO//(100)STO.

The epitaxial LNO film showed the expected metallic nature of the film in the four-probe electrical resistivity measurements.

5.3   *Films of LaMnO₃ exhibiting giant magnetoresistance*

Recently, large negative magnetoresistance effects have been observed in doped perovskite $La_{1-x}A_xMnO_3$ (A = Ca, Sr, Ba) epitaxial thin films, which are significantly larger

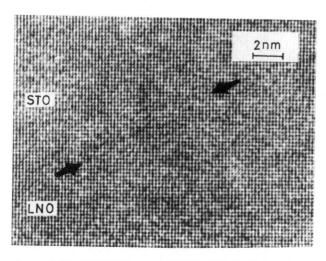

Figure 1.18. HRTEM image of LNO–STO(100) interface showing perfect lattice matching.

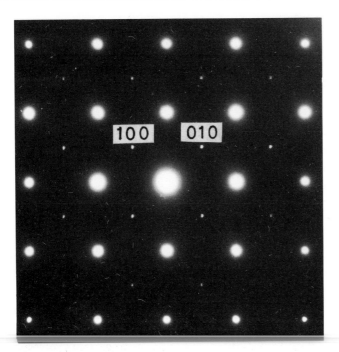

Figure 1.19. SAED pattern from the LNO–STO(100) interface region.

than their bulk counterparts [27–30]. This can be useful for various device applications such as magnetic recording, magnetoresistive sensors and magnetoresistive microphones. In this context, we sought to obtain epitaxial films of lanthanum manganates by nebulised spray pyrolysis. We deposited $LaMnO_3$ (LMO) films on STO(100) substrates at 675 K (5 h) using lanthanum acetylacetonate and manganese acetylacetonate as precursors. The film stoichiometry was first confirmed by EDAX analysis. Figure 1.20 gives the XRD Θ–Θ scan of the LMO–STO film. This pattern reveals the film to be of

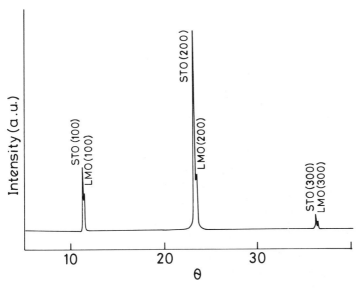

Figure 1.20. X-ray diffraction pattern (Θ–Θ scan) of an LMO film on STO(100) substrate obtained by nebulised spray pyrolysis.

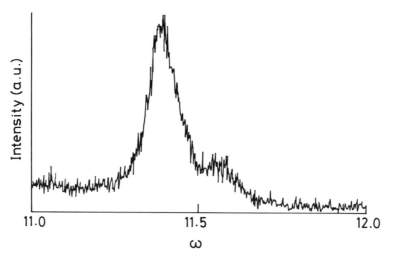

Figure 1.21. X-ray rocking curve (ω scan) of the (100) reflection of the LMO film deposited on STO(100) substrate.

a cubic perovskite phase with $a = 3.88$ Å and to be highly oriented along the (100) direction on the STO(100) substrate.

LMO containing around 30% Mn^{4+} has been shown to have a cubic structure [31,32]. The Mn^{4+} ions arise not because of the oxygen excess but due to the presence of a roughly equal number of vacancies on both the La and the Mn sites. Cation vacancies are present randomly but do not give rise to extended defects. The cubic LMO phase obtained by us can be approximately described as $La_{0.95}Mn_{0.95}O_3$ [33]. We have carried out electrical resistivity measurements on some of these films. The cubic LMO phase prepared by us showed a metal–insulator transition at around 200 K, in agreement with the literature [34].

The X-ray rocking curve (ω scan) of the (100)LMO reflection is shown in Fig. 1.21. The FWHM value of the curve is about 0.15°, indicating a very high degree of alignment/perfection of the film growth plane with respect to the STO(100) substrate, confirming the epitaxial nature of the film. Fig. 1.22 shows an HRTEM image of the

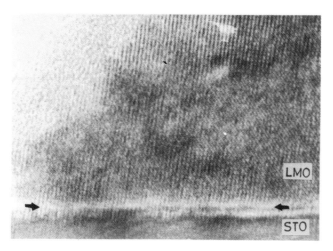

Figure 1.22. HRTEM lattice image of the LMO film deposited on STO(100) substrate showing the epitaxial nature of the film. The interface is indicated by the arrows.

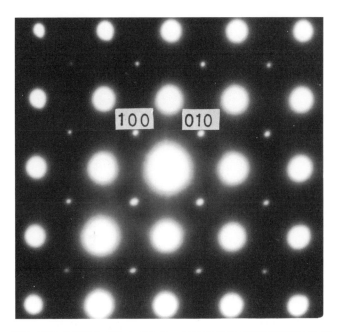

Figure 1.23. SAED pattern from the LMO–STO(100) interface region.

LMO–STO interface. The interfacial boundary region does not show any extra phases. The LMO lattice lines are well resolved and are seen to extend all the way down to the STO surface, closely matching with the STO lattice. The SAED pattern of the LMO–STO interface is given in Fig. 1.23. In this pattern also, the Bragg spots corresponding to c. 3.9 Å are split, showing the slight mismatch in the a parameters. These HRTEM observations establish the epitaxial growth of the LMO film on STO(100) and the epitaxial relationship is (100)LMO//(100)STO.

We have also successfully deposited LMO films on an LAO(100) substrate. The XRD pattern (Θ–Θ scan) of the LMO–LAO film is shown in Fig. 1.24. The epitaxial nature of

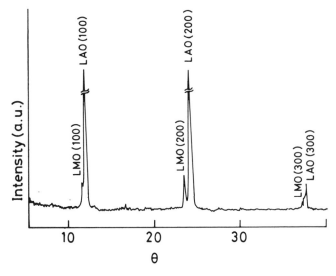

Figure 1.24. X-ray diffraction pattern (Θ–Θ scan) of an LMO film deposited on LAO(100) substrate by nebulised spray pyrolysis.

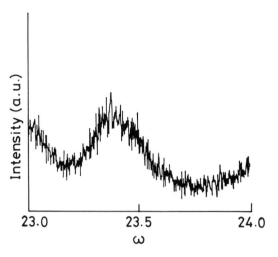

Figure 1.25. X-ray rocking curve (ω scan) of (200) reflection of an LMO film deposited on LAO(100).

the LMO film is clearly evidenced by the presence of only the (h00) pseudocubic reflections of LMO in the Θ–Θ scan and also from the rocking curve of the (200) reflection of LMO (Fig. 1.25), which shows an FWHM of 0.3°.

6   Conclusions

The above discussion of the results on films of various oxide systems deposited by different methods should suffice to demonstrate how one is able to get excellent single-crystalline (epitaxial) films with good interface properties as determined by lattice matching. What is particularly noteworthy is that an inexpensive, chemical method, namely nebulised spray pyrolysis, gives epitaxial films of technologically important oxides. The interface features, as well as the quality of the films obtained by nebulised spray pyrolysis, are as good as those obtained by the more difficult and expensive methods such as PLD and MBE. One can readily verify this observation by comparing the results on ferroelectric films obtained by the different methods presented earlier. Clearly, nebulised spray pyrolysis is likely to emerge as a viable technique for the preparation of epitaxial films of complex metal oxides.

7   References

1 Bai GR, Chang HLM, Foster CM, Shen Z, Lam DJ. *J. Mater. Res.* 1994; **9**: 156.
2 De Veirman AEM, Timmers J, Hakkens FJG, Cillesen JFM, Wolf RM. *Philips J. Res.* 1993; **47**: 185.
3 Greek J, Linker G, Mayer O. *Mater. Sci. Rep.* 1989; **4**: 193.
4 Kentgens APM, Carim AH, Dam B. *J. Cryst. Growth* 1988; **91**: 355.
5 Hseigh YF, Siegal MP, Hull R, Phillips JM. *Appl. Phys. Lett.* 1990; **57**: 2268.
6 Hwang DM, Nazar L, Venkatesan T, Wu XD. *Appl. Phys. Lett.* 1988; **57**: 1834.
7 Lee ST, Chen S, Hung LS, Braunstein G. *Appl. Phys. Lett.* 1989; **55**: 286.
8 Gupta A, Hussey BW, Guloy AM *et al. J. Solid State Chem.* 1994; **108**: 202.
9 Chrisey DB, Hubler GK. *Pulsed Laser Deposition of Thin Films.* New York: Wiley, 1994.

10 Blank DHA, Ijsselteijn RPJ, Out PG, Kuiper HJH, Flokstra J, Rogalla H. *Mater. Sci. Engg.* 1992; **B13**: 67.

11 Singh RK, Narayan J. *Phys. Rev.* 1990; **B41**: 8843.

12 Singh RK, Narayan J. *Mater. Sci. Engg.* 1989; **B3**: 217.

13 Lee AE, Platt CE, Burch JF, Simon RW, Goral JP, Al-Jassim MM. *Appl. Phys. Lett.* 1990; **57**: 2019.

14 Basu SN, Carim AH, Mitchell TE. *J. Mater. Res.* 1991; **6**: 1823.

15 De Veirman AEM, Cillesen JFM, Dekeijser M *et al. Mater. Res. Soc. Symp. Proc.* 1994; **341**: 329.

16 Herman MA, Sitter H (eds). *Molecular Beam Epitaxy: Fundamentals and Current Status.* Berlin: Springer-Verlag, 1989.

17 Joyce BA. *Contemp. Phys.* 1991; **32**: 21.

18 McKee RA, Walker FJ, Specht ED, Alexander KB. *Mater. Res. Soc. Symp. Proc.* 1994; **341**: 309.

19 Wessels BW. *Ann. Rev. Mater. Sci.* 1995; **25**: 525.

20 Kaiser DL, Vaudin MD, Rotter LD *et al. Mater. Res. Soc. Symp. Proc.* 1991; **361**: 355.

21 Foster CM, Csencsits R, Baldo PM *et al. Mater. Res. Soc. Symp. Proc.* 1995; **361**: 307.

22 Langlet M, Joubert JL. In Rao CNR (ed.) *Chemistry of Advanced Materials.* Oxford: Blackwell Scientific, 1993.

23 Raju AR, Aiyer HN, Rao CNR. *Chem. Mater.* 1995; **7**: 225.

24 Raju AR, Rao CNR. *Appl. Phys. Lett.* 1995; **66**: 896.

25 Lines ME, Glass AM (eds). *Principles and Applications of Ferroelectrics and Related Materials.* Oxford: Clarendon Press, 1977.

26 Klein JD, Yen A, Clauson SL. *Mater. Res. Soc. Symp. Proc.* 1994; **341**: 393.

27 Chahara K, Ohno T, Kasai M, Kozono Y. *Appl. Phys. Lett.* 1993; **63**: 1990.

28 von Helmolt R, Wecker J, Samwer K, Haupt L, Barner K. *J. Appl. Phys.* 1994; **76**: 6925.

29 Ju HL, Kwon C, Li Q, Greene RL, Venkatesan T. *Appl. Phys. Lett.* 1994; **65**: 2108.

30 Cheetham AK, Rao CNR. *Science* 1996; **272**: 369.

31 Verelst M, Rangavittal N, Rao CNR, Rousset A. *J. Solid State Chem.* 1993; **104**: 74.

32 Mahesh R, Kannan KR, Rao CNR. *J. Solid State Chem.* 1995; **114**: 294.

33 Hervieu M, Mahesh R, Rangavittal N, Rao CNR. *Euro. J. Solid State Inorg. Chem.* 1995; **32**: 79.

34 Mahendiran R, Tiwary SK, Raychaudhuri AK *et al. Phys. Rev.* 1996; **B53**: 3348.

2 Interfacial Science of Solid Host–Guest Systems

K.D.M. HARRIS

School of Chemistry, University of Birmingham, Edgbaston, Birmingham B15 2TT, UK

1 Introduction

Solid inclusion compounds exhibit a diversity of interesting, unique and important properties that have captivated the attention of scientists for several centuries. The observations of 'boiling stones' (zeolites) by early mineralogists [1], the pioneering scientific researches by Davy [2] and Faraday [3] on clathrate hydrates, the seminal structural studies of Powell and co-workers [4,5] and the modern realisation of the importance of microporous solids in many aspects of applied and industrial science have paved the way for the development of a field of scientific enquiry that is now becoming ripe with the fruits of maturity.

Solid inclusion compounds are defined here to encompass those materials in which *guest* molecules are included, in some way, within the crystal architecture of a *host* solid. These host solids are diverse, in terms of both their chemical nature and their structural character. The 'inclusion spaces' within which the guest molecules are located in these host structures encompass a rich diversity of topological forms (Fig. 2.1), and can be described as zero-dimensional (isolated cages), one-dimensional (tunnels), two-dimensional (interlamellar regions within layered hosts) or three-dimensional (for example, systems of interconnected cages or networks of intersecting tunnels). The chemical constitution of these host solids encompasses both inorganic materials (for example, aluminosilicates, aluminophosphates, metalloaluminophosphates, cyclophosphazenes, metal chalcogenides, metal phosphonates) and organic materials (for example, urea, thiourea, tri-*ortho*-thymotide, deoxycholic acid, cholic acid, perhydrotriphenylene).

In these solid inclusion compounds, the association of host and guest components is strictly a solid-state phenomenon. It is relevant to note that there is also another type of inclusion compound in which the host is a molecule containing an appropriate cavity or binding site that can include guest molecules, atoms or ions; these molecular host–guest complexes can generally exist as associated entities both in solution and in the crystalline state, and thus the inclusion phenomenon is not exclusively a property of the solid state. Examples of these host molecules are crown ethers, cyclodextrins, cryptands and calixarenes. Apart from one example (discussed in Section 3) of a molecular inclusion system of this type, this article focuses entirely on the solid inclusion compounds defined above.

The region of contact between the host and guest components in a solid inclusion compound can be regarded (perhaps loosely, but nevertheless conveniently) as an interface, and many properties of the inclusion compound are dictated by the nature of this interface. This paper aims to highlight some of these properties, and, in doing so, to introduce a variety of materials and some of the fundamental and applied issues that

21

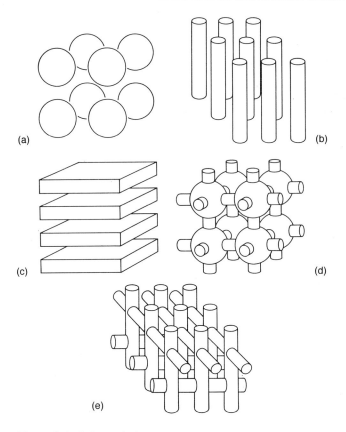

Figure 2.1. Schematic illustration of some typical topologies of 'inclusion space' in the host structures of solid inclusion compounds: (a) isolated cages; (b) linear, parallel tunnels; (c) two-dimensional interlamellar regions within layered hosts; (d) system of interconnected cages; (e) network of intersecting tunnels. The smallest dimension of each of these 'inclusion spaces' is of the order of molecular dimensions.

are currently of interest and importance. The coverage is necessarily selective, but nevertheless it is intended to give a representative glimpse of contemporary issues in this field of research. Among the issues that will be considered are those that devolve upon structural, dynamic, chemical, optical, electronic and magnetic properties of the guest component, with particular interest in understanding the extent to which these properties are influenced, or indeed controlled, by the constraints imposed upon the guest component by the host structure.

Before considering specific properties, it is constructive to comment upon some of the key features of solid inclusion compounds. First, if an inclusion compound is to be formed between a given host structure and a given type of guest molecule, there must, in general, be a sufficient degree of size and shape compatibility between the guest molecule and the 'inclusion space' that is available to it within the host structure. This size and shape specificity is a crucial factor underlying many properties and applications of solid inclusion compounds, and is indeed a general feature of most types of molecular inclusion phenomenon (encompassing, for example, enzyme–substrate binding). In some cases, electronic complementarity between host and guest components (such as the provision of appropriate host–guest binding sites) is an additional prerequisite for inclusion compound formation, although this is not necessarily a major factor in many of the examples highlighted here.

Second, there is an important subdivision between solid inclusion compounds for which the host structure remains stable when the guest component is removed, and those for which the host structure undergoes substantial reorganisation when the guest component is removed. For convenience, this paper will use the term 'hard' host for the former category and the term 'soft' host for the latter category. In the case of the soft hosts, the structural reorganisation generally involves collapse of the low-density 'empty' host structure, with recrystallisation to a more compact structure of higher density. Thus, for the soft hosts, the guest component generally acts as an essential template for the formation of the host structure as well as an essential buttress for maintaining the stability of the host structure; the collapse of the host structure on removal of the guest component is often an irreversible process. The exact structural nature of the soft hosts often varies substantially depending on the structural and chemical attributes of the guest molecules. There is a greater synergy in properties between the host and guest components in the case of the soft hosts, and it is often not a satisfactory approximation to rationalise the properties of these inclusion compounds in terms of the separate behaviour of the host and guest components. Furthermore, even in the case of the hard hosts, it is generally not satisfactory to consider the host structure to be a passive framework, influencing the properties of the guest molecules only through providing rigid boundaries that constrain their spatial distribution. Thus, some degree of local structural relaxation of the host, in response to the presence of the guest molecules within it, is inevitable, even in the case of the hard hosts.

Third, in solid inclusion compounds, the guest molecules often have considerable motional freedom, and these dynamic properties can have a significant bearing on the properties of the inclusion compound. The mechanism and rate of these dynamic processes are dictated to a large extent by characteristics of host–guest interaction and are therefore influenced directly by the nature of the host–guest interface. Although we do not discuss dynamic properties of the guest molecules in detail in this paper, passing reference is nevertheless made to these properties in some of the subsequent sections.

This paper highlights a selection of properties that devolve strongly upon the interfacial aspects of solid inclusion compounds. Emphasis is placed upon issues of contemporary interest, with the aim of identifying areas in which rapid progress is likely, as the field of solid inclusion compounds springboards into the next century.

2   Examples of solid inclusion compounds

Zeolites [6–11] are aluminosilicate materials with three-dimensional open framework structures, constructed from SiO_4 and AlO_4 tetrahedra interconnected by corner-sharing; a multitude of different ways of arranging these building blocks is possible. Each AlO_4 unit is associated with a negative charge, which is balanced by a positive ion; this is often H^+ (leading to applications that exploit the Brønsted acidity of these materials) or another extra-framework cation. Many important properties of zeolites depend on the Si/Al ratio, as this clearly dictates the electrostatic properties of the host framework and the number of extra-framework cations. The opportunity to connect the tetrahedral building blocks in a variety of geometric arrangements leads to a diversity of

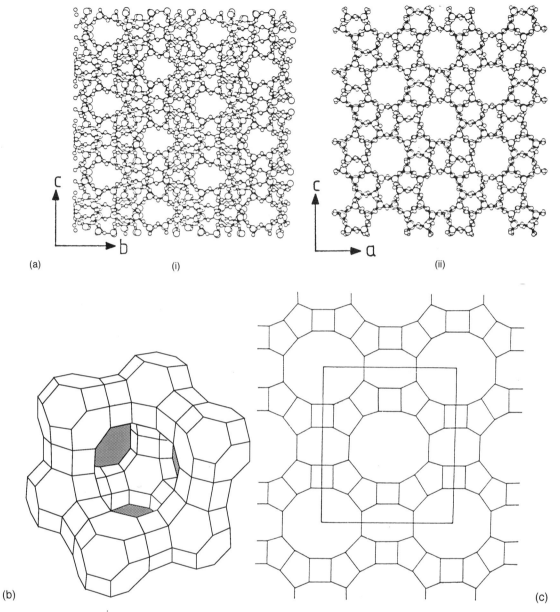

Figure 2.2. Representations of zeolitic host frameworks, illustrating the topological features discussed in the text: (a) ZSM-5, viewed (i) along the [100] direction and (ii) along the [010] direction (straight tunnels); (b) zeolites X and Y; (c) mordenite; (d) zeolite L.

framework structures and topologies, containing 'inclusion spaces' in the form of tunnels, cages and extended networks thereof; these 'inclusion spaces' come in a variety of sizes and dimensions, allowing selective inclusion of specific guest molecules by different zeolite structures. Details of zeolite structures are to be found elsewhere [6,12], although we now highlight briefly the key structural attributes of the 'inclusion spaces' in some of the zeolitic hosts discussed later in this paper (Fig. 2.2): (i) ZSM-5 comprises a set of sinusoidal tunnels (diameter c. 5.1–5.5 Å) intersecting a set of straight tunnels (diameter c. 5.3–5.6 Å); (ii) zeolites X and Y (which are isostructural but differ in Si/Al

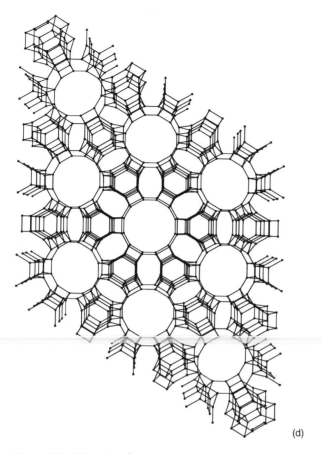

(d)

Figure 2.2. (*Continued*).

ratio) are composed of large supercages (diameter *c.* 13 Å, accessed through windows of diameter *c.* 7.4 Å), smaller sodalite cages (diameter *c.* 5 Å, accessed through windows of diameter *c.* 3 Å), and hexagonal prisms (double 6-rings); (iii) mordenite has one-dimensional tunnels (diameter *c.* 6.5–7.0 Å); (iv) zeolite L has one-dimensional tunnels (diameter *c.* 7.1 Å). Host materials for which the dimensions of the 'inclusion spaces' are of the order of magnitude exemplified by these zeolites are generally referred to as *microporous solids*.

Most zeolites are synthesised with the use of a molecular template, which acts as a precursor guest species. The template can be removed by appropriate calcination procedures, generating the 'empty' host framework which can then adsorb other guest species; clearly these host structures are of the hard type defined above. Ion-exchange processes can also be carried out readily to introduce different extra-framework cationic species.

The novel properties of zeolites have engendered much recent work [7,10] to prepare analogous microporous materials with similar framework structures but different chemical compositions. Thus, the AlO_4 and SiO_4 building blocks can be replaced by other MO_4 units, leading, for example, to families of gallosilicates, aluminophosphates (ALPOs), silicoaluminophosphates (SAPOs) and metalloaluminophosphates (MAPOs; containing metals such as Cr, Mn, Fe, Co and Zn). Clearly this diversity of chemical compositions creates, *inter alia*, the opportunity to fine-tune the charge distribution of the framework, which can be an important determinant controlling many properties of these materials. Furthermore, some of the structures found for these materials have no known analogues among the zeolites.

In recent years, a new family of so-called *mesoporous* solid host materials (typified by MCM-41) has entered the scene [13–15]. These are silicate or aluminosilicate materials with unprecedentedly large tunnel diameters (ranging from *c.* 15 Å to over 100 Å), generated from the use of ordered surfactant micelles as templates. Although these hosts possess well-defined tunnels, the structure of the tunnel walls in the materials prepared so far is actually amorphous. Again, other metals may be incorporated within these frameworks, or grafted on to the walls of the tunnels, and this family of hosts possesses much of the chemical and structural diversity of their smaller-pore zeolitic analogues. The substantially larger 'inclusion spaces' in these materials clearly create exciting new opportunities for exploring a range of novel issues of both fundamental and applied importance.

We now turn our attention to host structures constructed from the crystal packing of organic molecules (Fig. 2.3). In the urea inclusion compounds [16–19], the host structure is an extensively hydrogen-bonded arrangement of urea molecules

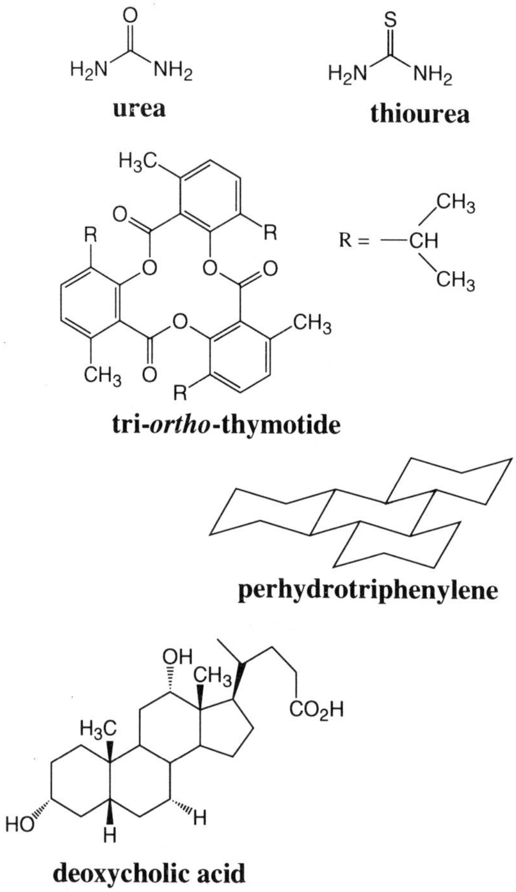

Figure 2.3. Examples of organic molecules that form the host structures of solid inclusion compounds with appropriate guest molecules.

(Fig. 2.4(a)); this structure contains linear, parallel tunnels (diameter *c.* 5.5–5.8 Å (Fig. 2.4(c)) within which the guest molecules are densely packed. As a consequence of the requirement for size and shape compatibility between host and guest components, urea only forms inclusion compounds with guest molecules that are based on a sufficiently long *n*-alkane chain, with a further requirement that the degree of substitution of this chain must be small. The urea tunnel structure is unstable if the guest molecules are removed from the inclusion compound, whereupon the urea recrystallises in its 'native' crystalline phase (which does not contain empty tunnels). The urea host tunnel is a chiral structure, and the degree to which the properties of the guest molecules are affected by this chiral environment has important structural and chemical implications (see Section 6).

Thiourea also forms inclusion compounds [16,18,19] with a tunnel host structure (Fig. 2.4(b)) in the presence of appropriate guest molecules. Whereas the urea tunnel structure has a fairly uniform cross-section, the thiourea tunnel has prominent constrictions (diameter *c.* 5.8 Å) and bulges (diameter *c.* 7.1 Å) at certain positions along the tunnel (Fig. 2.4(c)), and many properties of thiourea inclusion compounds are better understood by regarding the host structure as cage-type rather than tunnel-type. As a consequence of these structural differences, the types of guest molecule that form inclusion compounds with urea and thiourea are generally different. For example, the thiourea host structure can accommodate guest molecules such as cyclohexane and some of its derivatives, ferrocene and other organometallics, and certain compounds containing a benzene ring; such guest molecules do not generally form inclusion compounds with urea.

Examples of other organic host structures discussed in this article are tri-*ortho*-thymotide [20,21] (which forms either a tunnel-type or cage-type host structure, depending on the nature of the guest component), perhydrotriphenylene [22,23] (tunnel host structure) and deoxycholic acid [24,25] (tunnel host structure).

Finally, attention is also drawn to recent progress in the design of new families of host structures, including the use of cyclic peptides to generate nanotubular structures [26,27], the application of concepts of metal coordination chemistry and organometallic chemistry in the construction of open framework structures [28–31] and the assembly of tunnel host structures using porphyrin building blocks [32].

3   Applications involving separation of guests

As a consequence of the requirements for size and shape compatibility between host and guest components, solid inclusion compounds have very clear potential in applications involving separation. Thus, inclusion compound formation may provide an efficient and convenient means to separate two or more potential guest molecules, provided one of these guest molecules forms a more stable inclusion compound. The potential guest molecules may indeed be geometric isomers or stereoisomers. Depending on the relative energetics associated with inclusion of the competing guest molecules (host–guest interaction is generally the main discriminatory factor), a significant degree of selectivity may be achieved. However, in addition to having this energetic driving force for selectivity between different potential guest molecules, it is important that a mechanism exists to allow the selection process to be accomplished.

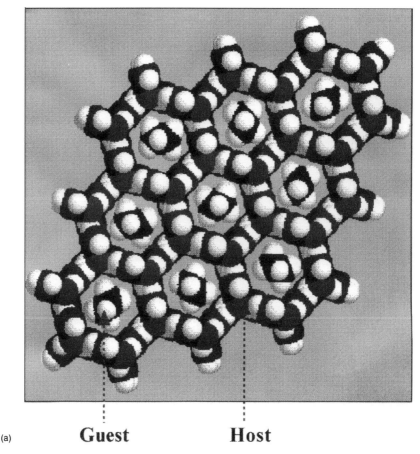

(a)   **Guest   Host**

Figure 2.4. (a) Representation of the hexadecane/urea inclusion compound, showing nine complete tunnels with van der Waals radii, viewed along the tunnel axis. (b, *facing page*) Representation of the cyclohexane/thiourea inclusion compound, showing ten complete tunnels with van der Waals radii, viewed along the tunnel axis. (c, *facing page*) Graphs showing the minimum tunnel diameter (d_{min}) as a function of position (z) along the tunnel for the urea tunnel structure and the thiourea tunnel structure; in both cases, the range of z considered in the graph corresponds to just over one lattice period of the host structure along the tunnel axis. In (a) and (b), the host structure has been determined from X-ray diffraction data recorded at room temperature. The guest molecules have been inserted into the tunnels on the diagram, illustrating orientational disorder of the guest molecules at room temperature (the positions of the guest molecules are not actually determined from X-ray diffraction data recorded at room temperature). Note the size and shape compatibility between the host and guest components.

In general, the strategy for applications involving separation of guests is different for the hard hosts and the soft hosts. For the hard hosts, the 'empty' host structure can be exposed to the mixture of potential guests, leading to selective adsorption of the preferred guest. For the soft hosts, on the other hand, the 'empty' host structure does not exist as such, and it is usually necessary to construct the inclusion compound 'from scratch' (for example, by crystal growth) in the presence of the mixture of potential guests; for certain soft hosts, guest exchange reactions may provide a viable alternative approach.

There are numerous examples of the successful application of inclusion phenomena in applications involving separation, typified by the separation of linear and branched

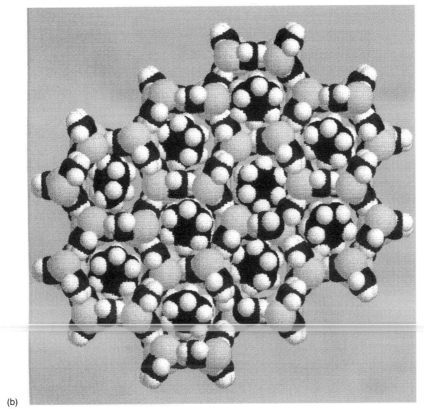

(b)

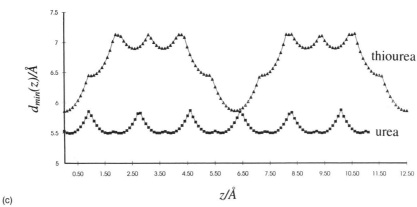

(c)

Figure 2.4. (*Continued*).

alkanes by selective inclusion in an appropriate host structure (for example, urea or ZSM-5). Related applications of urea inclusion compounds find everyday use in the laboratories of chemists working with fatty acids.

A striking example [33] of the use of inclusion in molecular separation concerns the selective inclusion of conformational isomers of α-ionone and β-ionone by different host materials (Fig. 2.5). In part, interest in these guests is motivated by the fact that β-ionone is related to the retinal chromophore, the binding of which within rhodopsin is conformationally dependent. The inclusion compound formed between β-ionone and the host **1** isolates the s-*trans* conformation of β-ionone, whereas the inclusion compound formed between β-ionone and the host **2** isolates the s-*cis* conformation of

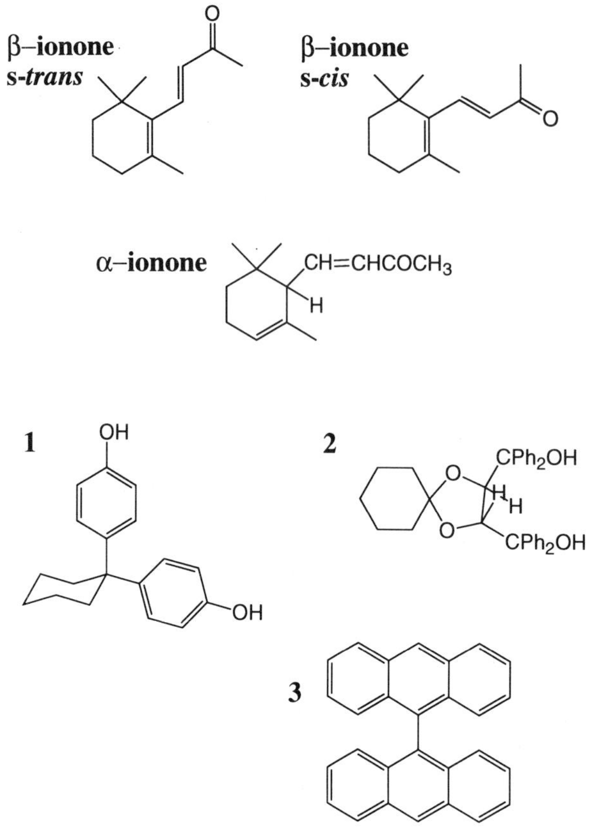

Figure 2.5. Inclusion compounds of the host materials **1**, **2** and **3** exhibit high selectivity, respectively, for the following guest species: the s-*trans* conformation of β-ionone, the s-*cis* conformation of β-ionone and the axial quasi-chair conformation of α-ionone.

β-ionone. The preferential inclusion of these conformational isomers of β-ionone within the host materials **1** and **2** can be rationalised on the basis of host–guest interactions. Additionally, it has been shown that guest molecules of α-ionone within the host **3** exist exclusively in the axial quasi-chair conformation; again, this conformation of the guest molecule has been selected from among several different possibilities, including equatorial quasi-chair, axial quasi-boat and equatorial quasi-boat conformations.

Another illustrative example [34] concerns the purification of C_{60} and C_{70} by selective inclusion within calixarenes (these are macrocyclic host molecules, with phenolic OH groups directed towards the centre of the macrocycle). The host–guest complex formed between C_{60} and p-Bu^t-calix[8]arene exists both in solution and crystalline phases, and crystallisation of this complex serves as a means of separating C_{60} from other fullerenes that do not readily form complexes with this host. Thus, addition of p-Bu^t-calix[8]arene to a toluene solution containing a crude fullerene mixture leads to the production of crystalline inclusion complexes; the guest component can be recovered from these inclusion complexes by addition of chloroform. As an illustration, this procedure led to the recovery of a mixture of C_{60} (89%) and C_{70} (11%) from the original fullerene mixture used in the reported experiments. On repeating the procedure of adding p-Bu^t-calix[8]arene followed by crystallisation, the C_{60} content was increased to 96% and then to greater than 99.5%. Clearly C_{60} has a greater affinity than C_{70} for the p-Bu^t-calix[8]arene host, as the cavity size within the host molecule is optimum for the efficient inclusion of C_{60}. On the other hand, the host molecule p-Bu^t-calix[6]arene forms an inclusion compound selectively with C_{70}, and treatment of the residues from the experiments using p-Bu^t-calix[8]arene to isolate C_{60} serves as a convenient means of selective purification of C_{70} from these residues.

4 Selection and isolation of guest molecules in specific conformations

Building upon the theme of structural selectivity between host and guest components, it is important to note that inclusion within a solid host structure can serve to select an uncharacteristic conformation of the guest species. This can be important in allowing spectroscopic characterisation of conformations that may not be significantly populated in dispersed phases or in the 'native' crystalline state of the molecule. In addition (as discussed in Section 9), the attainment of uncharacteristic conformations could open up reaction pathways for constrained guest molecules that may be improbable for the same molecules in their normal conformational state.

A striking illustration of the constraints that can be imposed on the conformational properties of guest molecules by a host structure is provided by monohalocyclohexane ($C_6H_{11}X$; X = Cl, Br, I) guest molecules in the thiourea tunnel structure. For monohalo-cyclohexanes in the liquid and vapour phases, the dynamic equilibrium between the equatorial and axial conformations favours the equatorial conformer, and in the solid state (at sufficiently low temperature or high pressure) these molecules exist almost entirely as the equatorial conformer. On the other hand, when included as guest molecules within the thiourea tunnel structure, $C_6H_{11}Cl$, $C_6H_{11}Br$ and $C_6H_{11}I$ exist predominantly in the axial conformation; these results have been established from infrared [35,36], Raman [37] and high-resolution solid-state ^{13}C nuclear magnetic resonance (NMR) [38–40] techniques. The mole fraction of the equatorial conformer is $c.$ 0.05–0.15 from ^{13}C NMR results [40], which also demonstrate that a 'chair–chair' ring inversion process occurs for these guest molecules. Bromine K-edge extended X-ray absorption fine-structure spectroscopy (EXAFS), which provides a direct mea-surement of the intramolecular Br$\cdots$C(3) distance ($c.$ 3.27 Å), confirms that the axial conformation of bromocyclohexane predominates within the thiourea tunnel structure [41] (Fig. 2.6(a)). Interestingly, for the guest molecules $C_6H_{11}X$ = CH$_3$, NH$_2$, OH included within the thiourea tunnel structure, the equatorial conformer predominates (mole fraction $c.$ 0.82–0.97) [40]. The conformational properties for these guests resemble those for the same molecules in solution, and contrast markedly with the behaviour of the monohalocyclohexane guests in the thiourea tunnel structure. There is also a marked contrast between the conformational behaviour of the monohalocyclo-hexanes in thiourea and in various zeolitic hosts, within which the equatorial con-former predominates [42].

Certain disubstituted cyclohexanes also exist in uncharacteristic conformational states within the thiourea tunnel structure [43]. For the $trans$-1-bromo-2-chlorocyclo-hexane/thiourea inclusion compound, the intramolecular Br$\cdots$Cl and Br$\cdots$C(3) dis-tances of $c.$ 4.50 Å and 3.27 Å determined from bromine K-edge EXAFS spectra [41]

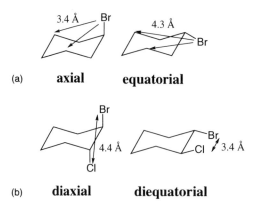

(a) axial   equatorial

(b) diaxial   diequatorial

Figure 2.6. (a) Idealised Br···C(3) distances in the axial and equatorial conformations of bromocyclohexane. (b) Idealised Br···Cl distances in the diaxial and diequatorial conformations of *trans*-1-bromo-2-chlorocyclohexane (note that measurement of the Br···C(3) distance (as in (a)) also allows the diaxial and diequatorial conformations of *trans*-1-bromo-2-chlorocyclohexane to be distinguished).

demonstrate clearly the preference for the diaxial conformation of the guest molecule (Fig. 2.6(b)); in contrast, the diequatorial conformation is preferred in dispersed phases.

The fact, demonstrated above, that guest molecules trapped within solid host structures may be constrained to exhibit unconventional conformational properties, can be exploited as a means of spectroscopic characterisation of these conformations. The recent demonstration [44] that the guest molecules in the 1,6-dibromohexane/urea inclusion compound exist exclusively with the bromine end-groups in a *gauche* conformation has allowed the definitive characterisation [45] of the vibrational properties of this type of end-group conformation.

5 Generation of orientationally well-ordered systems

Inclusion within one-dimensional tunnel structures may be exploited as a means of generating an orientationally well-ordered ensemble of molecules, which may be difficult to achieve in other phases. An example based on this fact concerns measurement of the orientation of the electronic transition dipole moment for conjugated polyenes; this property is important in relation to the use of these molecules in nonlinear optoelectronics and other applications (including their use as probes of biophysical systems). Simple theoretical approaches to predict this property have generated differing results (some suggesting that the transition dipole moment is essentially parallel to the long axis of the molecule, others suggesting an angle of 30° with respect to this axis), and experimental verification of these predictions has been hindered by the difficulty of preparing perfectly oriented samples of the polyenes. This problem has been addressed [46,47] by constraining these molecules as guests within the urea tunnel structure, thus ensuring that the molecular axes of all guests in a given single crystal are parallel and orientationally well defined with respect to the external morphology of the crystal. Specifically, octadeca-9,11,13,15-tetraenoic acid was considered as a dilute guest within the hexadecane/urea inclusion compound (dilution ensures that absorbance is low and that exciton effects are eliminated). Polarised

octadeca-9,11,13,15-tetraenoic acid

fluorescence excitation spectra of a single crystal of this material have shown that the transition dipole does not lie strictly along the molecular axis, but at an angle of $c.\ 20\pm1°$ with respect to this axis. On taking into account the effects of the surrounding medium for the guest in the urea inclusion compound, this value is modified to $c.\ 15\pm1°$ for the isolated molecule. This result has important implications with regard to the applications of these chromophoric materials.

It is clear that there are important prospects for exploiting uni-directional tunnel host structures in applications of this type, in which a highly anisotropic (uni-directional) orientation of guest molecules is required for the measurement of electronic or other properties. However, many solid inclusion compounds tend to exhibit orientational disorder of the guest molecules with respect to reorientation about the tunnel axis, and this feature may cause potential problems in the interpretation of results from such investigations. Nevertheless, if the nature of this orientational disorder can be determined from other experiments, and appropriate corrections made, the use of solid inclusion compounds in such applications is promising.

6 Chirality of solid inclusion compounds

In many solid inclusion compounds (particularly for organic hosts), the host crystal structure is chiral. Although in some cases the building blocks of the host structure are themselves chiral (and therefore the construction of a chiral host crystal structure is obligatory), in many cases chirality is introduced by spontaneous assembly of achiral building blocks into a chiral packing arrangement in the host crystal structure. As an illustration of the latter case, the host structure in the conventional urea inclusion compounds comprises a spiral hydrogen-bonded arrangement of urea molecules; the urea molecules themselves are achiral. The symmetry of the host structure in any given single crystal is either $P6_122$ (the inclusion compound contains only right-handed spirals of urea molecules) or $P6_522$ (the inclusion compound contains only left-handed spirals of urea molecules); this chiral character of the urea tunnel structure is generated spontaneously during crystal growth of the inclusion compound.

Clearly chiral host structures can exert an important influence on the structural and chemical properties of chiral guest molecules. As the R-guest/(+)-host and S-guest/(+)-host inclusion compounds have a diastereoisomeric relationship, they will generally differ in energy, and a given crystal of a chiral host ((+)-host, for example) should therefore have a preference for incorporating one enantiomer of a chiral guest species. Significant enantiomeric excesses (approaching 100% in some cases [48]) of chiral guest molecules within single crystals of chiral host solids have indeed been demonstrated [49].

In addition to experimental investigations of this phenomenon, computational investigations provide detailed insights into the characteristics of host–guest interaction

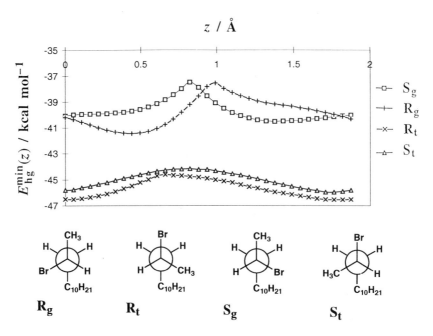

Figure 2.7. Host–guest interaction energy ($E_{hg}^{min}(z)$) for 2-bromotridecane guest molecules as a function of position (z) along the tunnel of the $P6_122$ urea host structure (for the range $0 \leqslant z < c_h/6$). The R_t, R_g, S_t and S_g types of 2-bromoalkane molecule are defined in the Newman projections (R and S represent the R and S enantiomers; subscript g represents the Br *gauche*/CH_3 *trans* conformation; subscript t represents the Br *trans*/CH_3 *gauche* conformation).

that underlie such chiral recognition. A recent computational study [50] of 2-bromo-alkane/urea inclusion compounds has demonstrated (Fig. 2.7) a clear preference for the R-enantiomer of 2-bromoalkane guest molecules within the $P6_122$ urea tunnel struc-ture, with the proportion of R-2-bromoalkane guest molecules at 300 K predicted to be $c.$ 0.75 for 2-bromotridecane/urea and $c.$ 0.82 for 2-bromotetradecane/urea. Interest-ingly (see Fig. 2.7), for the lowest energy conformation (Br *trans*/CH_3 *gauche*) of the 2-bromotridecane guest molecule within the urea tunnel, the same enantiomer (R) of the guest is preferred at all positions along the tunnel (this is, of course, not obligatory, and indeed for other conformations (for example, Br *gauche*/CH_3 *trans*) of the 2-bromotridecane guest molecule, the R enantiomer is preferred at some positions and the S enantiomer is preferred at others (Fig. 2.7). In assessing the enantiomeric excesses for incommensurate inclusion compounds (see Section 7), such as 2-bromoalkane/urea inclusion compounds, it is important to consider the characteristics of host–guest interaction as a function of the position of the guest molecule along the host tunnel.

A classic example [51] demonstrating the way in which a chiral host structure can dictate the chirality of included guests, concerns the inclusion compound of methyl methanesulfinate (MMS: Fig. 2.8) within the tri-*ortho*-thymotide (TOT) host structure. In solution, the TOT molecule adopts a chiral propeller-shaped conformation, with rapid interconversion between the P and M forms of this propeller. When TOT is crystallised in the presence of appropriate guests, it forms either a cage-containing or a tunnel-containing host structure, depending on the size and shape of the guest molecule. In both cases, a given single crystal contains only one enantiomer of TOT and the host

Figure 2.8. Definition of the two enantiomers of methyl methanesulfinate. Interconversion between these two enantiomers occurs at sufficiently high temperature within the tri-*ortho*-thymotide host structure.

structure is chiral. When crystallised from a solution containing racemic MMS and TOT, those single crystals containing (+)-TOT preferentially incorporate (+)-MMS, in a cage structure, with an enantiomeric excess of *c.* 14%. A chiral sample of MMS in chloroform solution racemises at 388 K, but no appreciable racemisation occurs in the MMS/TOT inclusion compound when it is heated to this temperature. Structural constraints imposed by the host clearly inhibit racemisation of the guest molecules. When the inclusion compound is heated to 398 K, however, the host cage becomes sufficiently flexible to allow interconversion between the two enantiomers of MMS and, moreover, to impose no significant preference for one enantiomer of MMS. When the inclusion compound at 398 K is subsequently cooled, the enantiomeric excess remains zero. Thus, on cooling the inclusion compound, the point at which the host becomes sufficiently rigid to prevent interconversion between the two enantiomers of the guest is presumably passed before host–guest chiral discrimination becomes significant. The racemic guest mixture produced at high temperature consequently becomes 'frozen in' upon cooling.

Host–guest chiral recognition in inclusion compound formation clearly leads to many potential applications, including the use of inclusion compound formation as a basis for resolving racemic samples into their two separate enantiomers. An illustrative example [52] concerns separation of the enantiomers of amphetamine (β-phenyl-isopropyl amine), in nearly quantitative yield, by exploiting selective inclusion compound formation between the *R*-enantiomer of amphetamine and the chiral host (*R,R*)-4,5-bis(hydroxydiphenylmethyl)-2,2-dimethyl-1,3-dioxolane.

Most chiral host structures are of the soft type, and the fact (see Section 3) that it is usually necessary to crystallise the host–guest inclusion compound (rather than simply to adsorb guest molecules within the 'empty' host structure) lends inefficiency to the inclusion approach for chiral resolution in these cases. There are clear advantages to the use of hard host materials in applications involving separation of enantiomers, allowing the host to be used in a preformed state, and to be recycled readily. However, the current paucity of chiral hosts of the hard type (most zeolites and related microporous materials have achiral structures, although polymorph A of zeolite beta is an important exception [10]) suggests considerable scope for future development. In this regard, the use of chiral template molecules [53] in synthesis of microporous and mesoporous solids of the hard type, and the grafting of chiral substituents [54] within the architectures of such materials, are attractive prospects.

The chiral nature of host structures can also be exploited in asymmetric synthesis, by carrying out reactions involving achiral guest molecules within a chiral host structure. Such reactions, generating a chiral centre in the product guest molecule, could, for example, involve reaction between adjacent guest molecules or reaction between guest molecules and reagents introduced within the host structure from the gas phase. The basis of these applications is the role of the chiral host structure in directing a preference, through the specificity of the host–guest interactions, for one particular chiral transition state and/or reaction product. Examples of such reactions are discussed in Section 9.

7 Commensurate versus incommensurate structural properties in one-dimensional solid inclusion compounds

An important question that is intimately linked with the energetic properties of the host–guest interface concerns the degree of structural registry between the host and guest substructures in solid inclusion compounds. Here we focus only on the simplest case of one-dimensional tunnel host structures in which both host and guest substructures have well-defined periodicities (denoted c_h and c_g, respectively) along the tunnel (Fig. 2.9). In such cases, the relationship between c_h and c_g is of interest. Conventionally, the ratio c_g/c_h is used as a basis for dividing these systems into two categories — commensurate and incommensurate — with the classical definition terming the system commensurate if c_g/c_h is rational and incommensurate if c_g/c_h is irrational. However, since c_g and c_h can never be measured with absolute precision, it is more practical to define an inclusion compound as commensurate if and only if c_g/c_h is sufficiently close to a rational number with a low denominator. However, this structural definition of incommensurate and commensurate systems is still far from satisfactory, as it does not quantify the terms 'sufficiently close to' and 'low denominator'. To understand the commensurate versus incommensurate nature of such one-dimensional inclusion compounds at a more fundamental level, new directions have led to the development [55] of a commensurate/incommensurate classification that reflects a division in the energetic 'behaviour' of the inclusion compounds within each category. Specifically, the magnitude of fluctuations in the average host–guest interaction energy per guest molecule, as the guest substructure is moved along the tunnel (keeping the guest periodicity c_g fixed), is considered. If these fluctuations are sufficiently small (i.e. within $\pm\varepsilon$, where ε is some physically meaningful energy term) the inclusion compound is taken to exhibit incommensurate behaviour, whereas if these fluctuations are sufficiently large (i.e. larger than $\pm\varepsilon$) the inclusion compound is taken to exhibit commensurate behaviour. In the commensurate case, a significant energetic 'lock-in' between the host and guest substructures will occur for a specific position of the guest sub-

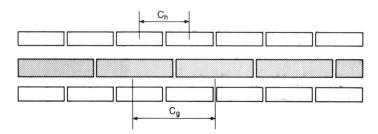

Figure 2.9. Schematic illustration of a solid one-dimensional inclusion compound, with the tunnel axis horizontal. The periodic repeat distance of the host structure is denoted c_h, and the periodic repeat distance of the guest molecules is denoted c_g.

structure relative to the host substructure, whereas for the incommensurate case, the energy of the inclusion compound is essentially independent of the position of the guest substructure relative to the host substructure.

A methodology has been developed [56] for applying these new concepts to predict structural properties of one-dimensional inclusion compounds from a knowledge of potential energy functions for the inclusion compound (with known host structure and fixed c_h). Fundamental to this approach is the definition of an appropriate energy expression — the 'characteristic energy' of the inclusion compound — that directly indicates the relative energetic favourability of inclusion compounds with different guest periodicities. The characterisitic energy $\hat{E}(\alpha,n)$ is defined as:

$$\hat{E}(\alpha,\, n) = \frac{1}{\alpha}\left(\inf_{\lambda}\left(\frac{1}{n}\sum_{k=0}^{n-1} E_h(k\alpha + \lambda) \right) + \hat{E}_{\text{guest}}(\alpha) + \hat{E}_{\text{intra}} \right) \qquad (1)$$

where n is the number of guest molecules within the host tunnel; α is the ratio c_g/c_h; the first guest molecule is located at position $t = \lambda$ along the tunnel; $E_h(t)$ represents the energy of an individual guest molecule, due to host–guest interaction, when the guest molecule is at position t along the tunnel; $\hat{E}_{\text{guest}}(\alpha)$ is the guest–guest interaction energy, per guest molecule, when the periodicity of the guest structure is α; and $\hat{E}_{\text{intra}}$ is the intramolecular potential energy of the guest molecule. The optimum guest structure for the inclusion compound corresponds to minimum characteristic energy, and the methodology allows the following structural properties to be established from computed potential energy functions $E_h(t)$, $\hat{E}_{\text{guest}}(\alpha)$ and $\hat{E}_{\text{intra}}$ for the inclusion compound of interest: (i) the optimum guest periodicity (c_g); (ii) whether this value of c_g corresponds to commensurate or incommensurate behaviour; (iii) the optimum conformation of the guest molecules within the host structure. Importantly, the methodology can handle tunnels of finite length, allowing the properties of 'real' one-dimensional inclusion compounds to be predicted directly.

The method has been applied successfully to predict structural properties of alkane/ urea inclusion compounds [57], giving predictions in excellent agreement with experimental results, and leading to new insights concerning the energetic properties of these inclusion compounds. *Inter alia,* the results demonstrate that, in the optimum structure of these incommensurate inclusion compounds, the interaction between neighbouring guest molecules in the same tunnel is repulsive.

The approach has also been applied [58] to determine, from first principles, the preferred conformation of chlorocyclohexane guest molecules within the thiourea tunnel structure. For *axial*-chlorocyclohexane/thiourea, the optimum guest periodicity corresponds to $\alpha = \frac{1}{2}$, representing commensurate behaviour and corresponding to a lower characteristic energy than any guest periodicity for *equatorial*-chlorocyclohexane/ thiourea. This predicted preference for the axial conformation is in direct agreement with the experimental results discussed in Section 4. In essence, *axial*-chlorocyclo-hexane can be packed more efficiently (smaller α) than *equatorial*-chlorocyclohexane within the constrained environment of the thiourea tunnel, and this contributes (via the factor $1/\alpha$ in the equation (1)) to the more favourable characteristic energy for the axial conformation. The optimum guest period ($c_g = \alpha c_h = c_h/2 = 6.24$ Å) predicted for *axial*-chlorocyclohexane/thiourea is in good agreement with information inferred from X-ray diffraction data [59].

We now consider the question of obtaining direct experimental evidence of incommensurateness in solid inclusion compounds, recalling that direct measurement of c_g and c_h from diffraction data is generally not a satisfactory approach. An alternative approach is based on recognising that conventional crystals (including commensurate inclusion compounds) have three translation invariances, whereas an incommensurate one-dimensional inclusion compound has four translation invariances; the extra translation invariance corresponds to the shift of the guest substructure relative to the host substructure along the incommensurate direction (as discussed above, the energy of an incommensurate inclusion compound is, in principle, independent of the shift of the guest substructure relative to the host substructure along this direction). There is an acoustic phonon corresponding to each translation invariance in a crystal, and therefore an incommensurate one-dimensional inclusion compound should have four acoustic phonons and a commensurate inclusion compound should have three acoustic phonons. The additional acoustic mode in the incommensurate system is called the 'sliding mode', and observation of the sliding mode can be taken as direct experimental evidence for the incommensurate behaviour of the inclusion compound. With this motivation, Brillouin scattering investigations [60] of the heptadecane/urea inclusion compound have provided evidence for a fourth acoustic mode, assigned as the sliding mode, thus substantiating the incommensurate nature of this inclusion compound. It is interesting to reflect that the new (energetic) definition of commensurate versus incommensurate behaviour discussed earlier in this section is in fact directly akin to the concept of a sliding mode for an incommensurate material.

8   Phase transitions

In many solid inclusion compounds, the cooperative behaviour of host and guest components is manifested by the occurrence of solid-state phase transitions, usually below ambient temperature. As may be expected, such phenomena tend to be more prevalent for the soft host materials for which the coexistence of both host and guest components is essential for maintaining the integrity of the host structure, and for which the behaviour of the guest molecules can exert a significant influence on the behaviour of the host structure. For these inclusion compounds, it is common for the guest molecules to be dynamic at ambient temperature, and for the average host structure (as determined by diffraction-based techniques) to have a high symmetry that reflects the time-averaged distribution of guest molecules within it. At sufficiently low temperature, the dynamics of the guest molecules diminishes, and the guest molecules adopt a well-defined orientation with respect to the host; concomitantly, the host structure distorts to a lower symmetry that reflects the static distribution of the guest molecules (which may or may not be disordered). Clearly this description represents a gross over-simplification of the true situation in any particular case, but it nevertheless embodies the key elements of the low-temperature phase transitions in many solid inclusion compounds. Central to these ideas is the intimate interplay of both host and guest components in dictating the overall behaviour of the inclusion compound.

These concepts are well established for the urea inclusion compounds, many of which undergo phase transitions below ambient temperature. The phase transitions in the alkane/urea inclusion compounds have received particular attention, and are known to be associated with alterations in the structural characteristics of the host [61,62] as well as changes in the motional behaviour of the guest molecules [63,64]. There have been various attempts [65–67] to rationalise the phase transition in the alkane/urea inclusion compounds. The most recent of these approaches [67] embodies certain crucial features of the experimental behaviour, and draws an analogy between the phase transition in alkane/urea inclusion compounds and the order–disorder phase transitions in alkali cyanide crystals. Specifically, it has been proposed that, in the alkane/urea inclusion compounds, coupling between transverse acoustic phonons of the host structure and the orientational order of the guest molecules provides an indirect mechanism for orientational ordering of the guest molecules in the low-temperature phase (it being suggested that the inter-tunnel guest–guest interaction is too weak to provide a direct mechanism for orientational ordering of the guest molecules at the experimentally observed transition temperatures). While there is currently no available experimental evidence to assess the importance of the proposed translation–rotation coupling mechanism for the alkane/urea inclusion compounds (single-crystal neutron spectroscopy on a triple-axis spectrometer should significantly enlighten this issue), it has nevertheless been shown that such a mechanism can explain successfully some known features of the phase transition in these systems (Fig. 2.10). Nevertheless, several aspects of the phase transitions in alkane/urea inclusion compounds remain to be understood, and the development of a fundamental understanding of the mechanism of these phase transitions is still one of the major challenges in this field.

Another example [68] which illustrates the close cooperativity in the behaviour of host and guest components in solid inclusion compounds concerns the study of stress-induced domain reorientation in the 2,10-undecanedione/urea inclusion compound. In contrast to most other urea inclusion compounds discussed in this paper, 2,10-undecanedione/urea is a commensurate system ($2c_g = 3c_h$) with a well-defined stoichiometric guest/host ratio (1 : 9) and a well-defined positioning of the guest molecules with respect to the host. In this structure (Fig. 2.11), one out of every three

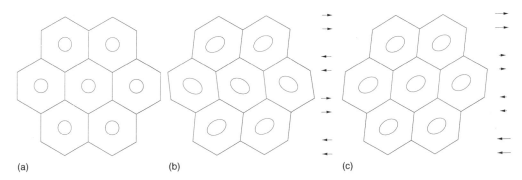

(a)　　　　　　　　　(b)　　　　　　　　　(c)

Figure 2.10. Schematic illustrations of the structure of an alkane/urea inclusion compound, viewed along the tunnel axis, for: (a) the high-temperature phase (hexagonal; guest molecules orientationally disordered); (b) the low-temperature phase (orthorhombic; herringbone orientational order of guest molecules). Another hypothetical structure (orthorhombic; parallel orientational order of guest molecules) for the low-temperature phase is shown in (c). The rationalisation of the phase transition discussed by Lynden-Bell [67] explains the production of structure (b) rather than structure (c) in the low-temperature phase. Diagram modified from [67].

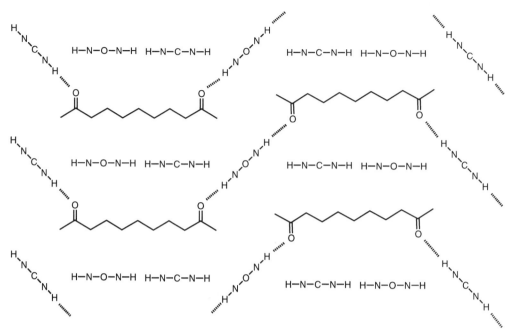

Figure 2.11. Schematic illustration of part of the structure of the 2,10-undecanedione/urea inclusion compound, showing the network of host–guest hydrogen bonding. The urea tunnels containing the 2,10-undecanedione guest molecules are horizontal. Urea molecules are represented as H—N—C—N—H (C → O vector directed into plane of page) and H—N—O—N—H (C → O vector directed out from plane of page).

urea molecules is oriented such that its two *syn* N—H groups form N—H$\cdots$O hydrogen bonds with C=O groups of guest molecules in different tunnels (in contrast, in the conventional (incommensurate) urea inclusion compounds there is no hydrogen bonding between urea molecules and guest molecules). All C=O groups of the guest molecules are involved in host–guest hydrogen bonding in this way. There is thus an extended hydrogen-bonded network connecting the host and guest components throughout the structure, and providing a direct mechanism for guest molecules in different tunnels to communicate with each other. Not surprisingly, the structure is distorted from the hexagonal structure of the conventional urea inclusion compounds, and has orthorhombic symmetry. Crystals of this inclusion compound consist of domains; within a given domain, the planes of all guest molecules are oriented in the same direction, with different domains related by rotation of the guest molecules by a multiple of *c.* 60° about the tunnel axis. These domains are birefringent, allowing the different domain orientations to be observed directly by optical microscopy (striking optical micrographs illustrating the well-defined sectoring of crystals of this inclusion compound are shown in a paper by Brown and Hollingsworth [68]). A remarkable property of these inclusion compounds is that compressive stresses applied to the crystal can bring about reorientation of the domains. Clearly the reorientation of the guest molecules associated with this domain reorientation requires breakage and formation of hydrogen bonds (guest–host and host–host), and the requirement to reconstruct the complete guest–host hydrogen-bonded network underlies the large-scale cooperativity of this process. Interestingly, by incorporating 2-undecanone guest impurities (*c.* 20%), the domain reorientation is spontaneously reversible, with the daughter phase reverting to the mother phase when the applied stress is removed. The 2-undecanone impurities clearly disrupt the guest–host hydrogen-bonding scheme, influencing the energetic properties of the mother and daughter phases and the interface between them; this provides the basis for a comprehensive rationalisation of the experimental observations.

9   Reactions of guest molecules in solid inclusion compounds

The reactivity of guest molecules in solid inclusion compounds can often differ substantially from the reactivity of the same molecules in solution, or in other solid phases (such as their 'native' crystalline phase or as guest molecules in other host solids). These differences devolve, to a large extent, on the structural and geometric constraints imposed on the guest molecules by the host environment, and specific factors include the following: (i) the guest molecule may be constrained to adopt an uncharacteristic conformation within the inclusion compound; (ii) the guest molecules are generally less mobile than in the liquid phase, but generally more mobile than in their 'native' crystalline phase; (iii) the specific intermolecular guest–guest 'contacts' in the inclusion compound may differ from those in the 'native' crystalline phase, and may represent low-probability trajectories of approach of two free molecules in dispersed phases; (iv) the relative energies of transition states with different geometries (representing competing reactions) within a host solid may be very different from the relative energies of the transition states for the corresponding reactions in other phases.

A classic example [69], which illustrates many of the contrasts that can arise, concerns the photolysis of *N,N*-dialkyl pyruvamide guest molecules within the tunnels of the deoxycholic acid host structure (Fig. 2.12). The main reaction products are

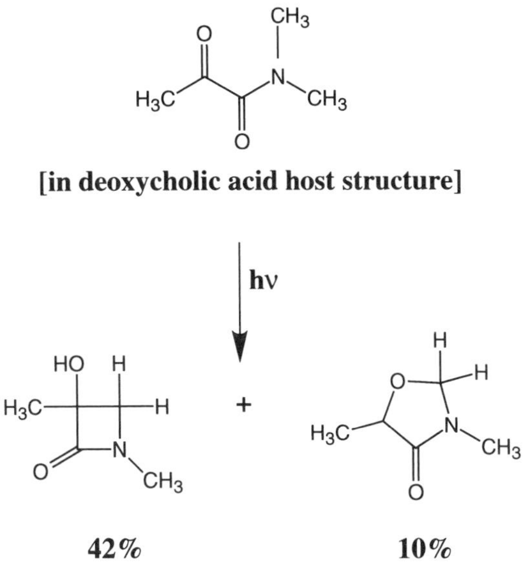

Figure 2.12. Photolysis of *N,N*-dialkyl pyruvamide guest molecules within the tunnels of the deoxycholic acid host structure.

β-lactams, which are not formed in the corresponding solution state reactions. Furthermore, the β-lactams produced in these reactions are optically active, and enantiomeric excesses of up to 15% have been obtained. The preferential formation of one enantiomer is a consequence of the chirality of the host tunnel, illustrating the role of the host structure in mediating the course of reactions of the guest molecules.

Polymerisation of monomer guest molecules arranged along the one-dimensional tunnels in crystalline host solids has been explored extensively [70,71], particularly for perhydrotriphenylene and other organic hosts such as urea, thiourea and deoxycholic acid. As may be expected from the highly constrained nature of these inclusion compounds, such 'inclusion polymerisation' generates polymers with a high degree of regularity. The resultant polymers can be isolated readily by dissolving the host component in an appropriate solvent. A classic example [72] is the γ-ray-induced polymerisation of *trans*-1,3-pentadiene within perhydrotriphenylene, which yields an isotactic product. Optically active polymers have also been produced by inclusion polymerisation [73]. Zeolitic host materials have also been exploited for inclusion polymerisation, exemplified by the production of *trans*-polyacetylene within mordenite [74]; other examples (including the use of the mesoporous host MCM-41) are discussed in Section 12.

A novel approach for effecting asymmetric induction within zeolitic host solids has been developed recently [75]. This work focused on the photochemical conversion (Norrish–Yang type II reaction) of *cis*-4-*tert*-butylcyclohexyl ketones (Fig. 2.13) to the corresponding cyclobutanols within Na–Y zeolite, noting that this reaction converts an achiral reactant to a chiral product. To bring about asymmetric induction in this reaction, the Na–Y zeolite was first treated with chiral amines or alcohols (for example, (–)-ephedrine, (–)-menthol, (–)-borneol, (+)-bornylamine, L-proline *tert*-butylester), which become incorporated in the supercages of the zeolite Y framework. The reactant (ketone) molecules were then adsorbed into the supercages. Typically, the number of guest molecules per zeolite supercage was *c.* 0.3–2.6 for the chiral inductors and *c.* 0.14 for the ketones. Following photolysis of these inclusion compounds, enantiomeric excesses as high as 30% were found for the cyclobutanol photoproducts. Of the different chiral inductors considered, ephedrine generally led to the highest enantiomeric excesses. Important control experiments vindicate the role of the chiral environment created within the zeolite host structures; thus, the use of (+)-ephedrine rather than (–)-ephedrine led to a preference for the opposite enantiomers of the chiral photoproducts,

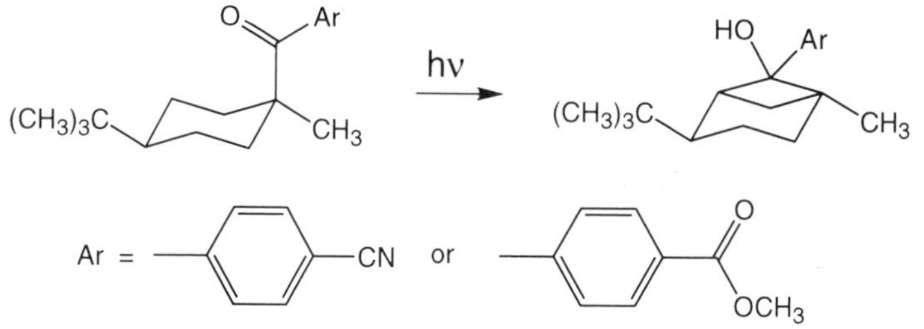

Figure 2.13. Photochemical conversion of *cis*-4-*tert*-butylcyclohexyl ketones to the corresponding cyclobutanols.

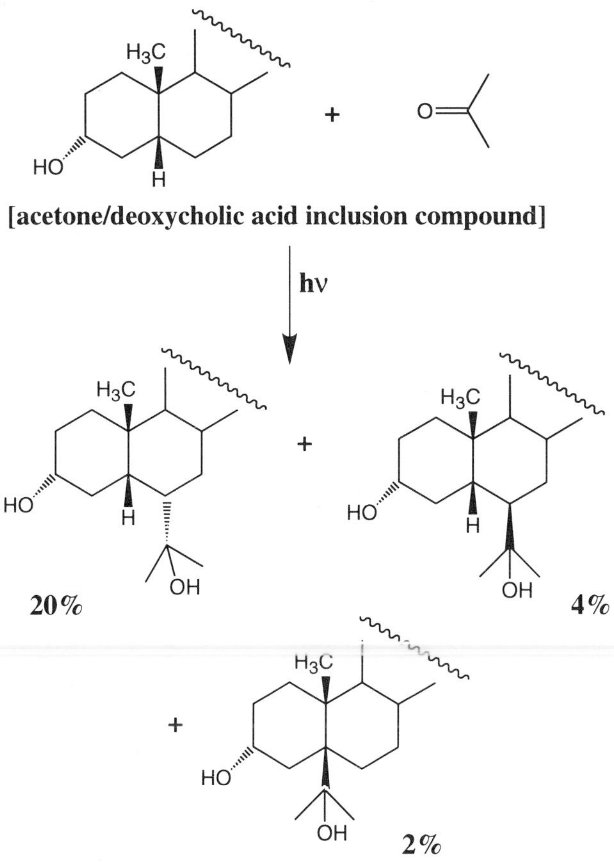

[acetone/deoxycholic acid inclusion compound]

hν

20%

+

4%

+

2%

Figure 2.14. Reaction between acetone and deoxycholic acid, carried out by photoexcitation of acetone guest molecules included within the deoxycholic acid host structure.

and photolysis of the ketones in zeolite Na–Y in the absence of a chiral inductor led to racemic photoproducts.

We now consider chemical reactions between host and guest components in solid inclusion compounds. For these processes, well-defined reaction pathways, differing from those of the same molecules in other phases, may arise due to the establishment of specific host–guest interactions within the inclusion compound. The functionalisation of deoxycholic acid, via photoexcitation of guest molecules (for example, ketones) included within the deoxycholic acid host structure [76], provides a striking example of the high degree of regiospecificity and stereospecificity that can be associated with such reactions (Fig. 2.14). Although many highly specific reactions could, in principle, be carried out using this approach, some reactions between host and guest components may be associated with degradation of the host framework (as a consequence of changing the chemical identity of the host molecules) leading to a loss of control and specificity as the reaction proceeds. Furthermore, in incommensurate inclusion compounds, different guest molecules experience a wide distribution of environments with respect to the host structure; for this reason, the prospects for carrying out well-defined host–guest reactions exhibiting a high degree of specificity are more promising for commensurate inclusion compounds.

10 Heterogeneous catalysis

A major area of research that exploits many of the unique facets of solid host–guest systems is heterogeneous catalysis. Although we make only passing reference to catalysis in this paper (readers are directed to another monograph in this series [77]), it is nevertheless pertinent to recall some of the key concepts in this field. One of the most fruitful aspects of this area is the extent to which the structural characteristics and electronic properties of the host structure (both of which are fundamental to its catalytic action) can be controlled, and indeed fine-tuned, by premeditated design. The geometric constraints imposed by the host structure can have a significant bearing on the course of reactions of the guest species, and can therefore strongly influence the product distributions. This shape selectivity may manifest itself with regard to selectivity of the reactant guest molecules (for example, only one specific geometric isomer or conformation may be able to access the catalytic active site) and/or the transition state for the reaction of the guest molecules and/or the product molecules (for example, only one geometric isomer of the product molecule may be able to exit from the host solid following the reaction). Examples of shape-selective catalysis by zeolites are numerous, a classic example being the preferential generation of the *para* isomer of xylene in the disproportionation of toluene in H-ZSM-5 zeolite. Importantly, analogies between the concepts underlying the action of microporous solid catalysts and biological catalysts have been recognised [78].

An interesting example which illustrates the important interplay between host and guest components in catalytic contexts concerns the cyclotrimerisation of acetylene to form benzene, using Ni–Y zeolite as catalyst [79–82]. When the Ni–Y zeolite is prepared (for example, by ion exchange from Na–Y zeolite) it is inactive for the cyclotrimerisation of acetylene, but becomes active on prolonged exposure to nitrogen gas saturated in acetylene. Activation of the catalyst is associated with substantial rearrangements of the Ni^{2+} cations within the host structure (Fig. 2.15), involving migration of these cations from the S_1 site in the hexagonal prism (which is 'inaccessible' to guest molecules) to the S_2 site in the supercage (which is 'exposed' to the guest molecules). Computational studies [81,82] have shown that the energetic requirements for migration of the Ni^{2+} cation between the S_1 and S_2 sites can be overcome by the acetylene–Ni^{2+} interaction that is established when Ni^{2+} is in the S_2 site. Furthermore, a plausible mechanism for the cyclotrimerisation reaction, involving successively one, two and three acetylene molecules interacting with the Ni^{2+} cation and ultimately departing as a benzene molecule, has been proposed on the basis of local density functional calculations. The close synergy in the behaviour of the host and guest components in activating and effecting this reaction is noteworthy, and underlines the fact that a catalyst may itself become structurally and electronically perturbed by its interaction with the reactant species.

The new mesoporous solid hosts (typified by MCM-41) have much to offer in the future development of catalytic applications, not least because of the prospects for grafting catalytically active centres (for example, based on organometallic complexes) on to the tunnel walls. As an example of the potential of this approach [83], attachment of a titanocene-derived catalyst precursor on to the tunnel walls of MCM-41 generates a catalyst that is active for the epoxidation of cyclohexene and other cyclic alkenes.

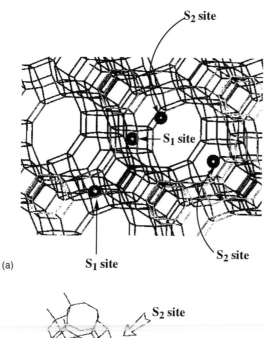

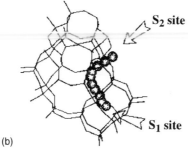

(a)

(b)

Figure 2.15. (a) Specification of S_1 (in hexagonal prism) and S_2 (in supercage) sites within the structure of zeolite Y. (b) Computational prediction of the trajectory for migration of Ni^{2+} ions from the S_1 site to the S_2 site of Ni–Y zeolite. Modified from [82].

Specifically, the procedure (Fig. 2.16; characterised structurally at each stage using Ti X-ray absorption spectroscopy and Fourier-transform infrared spectroscopy) involves diffusion of $(C_5H_5)_2TiCl_2$ into the MCM-41 host structure, anchoring to the tunnel walls (following activation of the surface silanols using triethylamine) and calcination to remove the cyclopentadienyl moieties. The resultant catalyst is superior in catalytic performance to materials with Ti incorporated into the MCM-41 framework [84,85] (in these materials, the catalytic centres (Ti) are buried within the inner walls of the host structure, limiting the access of reactant molecules to these sites). A number of factors contribute to the improved performance of the Ti-grafted catalyst, including the fact that all Ti sites are surface species and therefore directly accessible to the reactant molecules, and the fact that the $(C_5H_5)_2TiCl_2$ precursor can diffuse uniformly through the MCM-41 host structure, resulting in highly dispersed catalytic sites which provide maximum access to reactant molecules. This demonstration of the success and potential of the grafting approach augurs well for the development of a wide range of potent catalysts based on mesoporous host frameworks. Furthermore, the ability to graft organometallic complexes on to the walls of these hosts creates attractive prospects to design and develop materials that blend together the advantages inherent in homogeneous catalysis and heterogeneous catalysis.

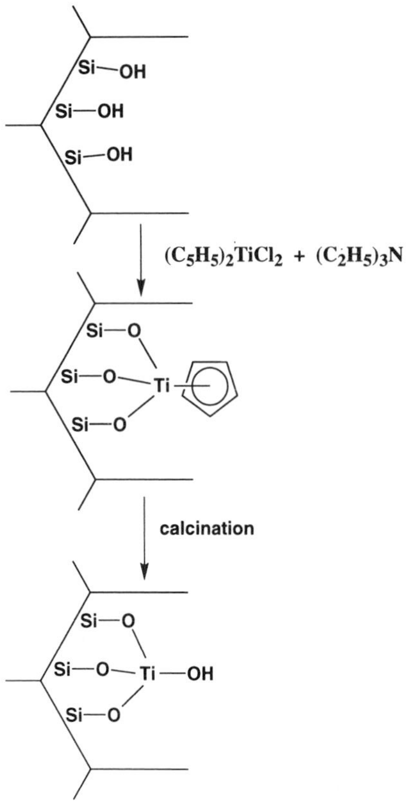

Figure 2.16. Schematic illustration of the procedure for grafting catalytically active Ti centres on to the framework of the mesoporous host material MCM-41.

11   Applications in nonlinear optics

The potential to exploit the phenomenon of inclusion in the field of nonlinear optics has received considerable attention in recent years. We focus here on second harmonic generation (SHG), which involves doubling of the frequency of light as it passes through a material. Materials that exhibit SHG are important in many device applications (including extending the frequency range of lasers) and have an important role in the field of optoelectronics. For a material to exhibit SHG, the component molecules must have a high value of second-order molecular hyperpolarisability (β), and in addition the molecules must aggregate in a non-centrosymmetric arrangement. Molecules with large β often have large degrees of intramolecular charge transfer and usually possess a large dipole moment in their ground state; such molecules, however, often have a tendency to crystallise in centrosymmetric structures. There is therefore substantial impetus to find ways to induce these molecules into non-centrosymmetric environments, and an attractive prospect is to include them as guest molecules within appropriate host materials. Thus, parallel alignment of guest molecules (with high values of β) within tunnel host structures has been particularly exploited, and a wide variety of hosts (including thiourea [86], tri-*ortho*-thymotide [86], perhydrotriphenylene [87], zeolites [88] and layered hosts [89]) have been used successfully in this regard.

As an illustration [86,90], thiourea inclusion compounds containing appropriate organometallic guests (for example $(\eta^6\text{-}C_6H_6)Cr(CO)_3$) have been shown to exhibit pronounced SHG. These organometallics possess large values of β, but their 'native' crystalline phases are centrosymmetric and are therefore inactive for SHG. For the thiourea inclusion compounds, it was shown that the SHG arises predominantly from the organometallic guests rather than the thiourea molecules (which also have significant β). The structures of the inclusion compounds are non-centrosymmetric, in accord with the idea that dipole organisation of the guest molecules should be favoured both within and between tunnels. It is interesting that the host structure in many other thiourea inclusion compounds (for example, chlorocyclohexane/thiourea) is centrosymmetric, and the results here illustrate that the structure of a given host material can differ, often substantially, depending on the identity of the guest molecules within it. Such observations are particularly prevalent for hosts of the soft type.

Although inclusion compound formation is now well established as a general strategy for the generation of materials for applications in nonlinear optics (particularly for cases in which the 'native' crystalline phase of the active (guest) molecule is centrosymmetric) it is pertinent to recognise some of the important considerations in designing inclusion compounds for such applications. First, the host structures used in these applications generally have little or no activity with respect to the nonlinear optical response, and inclusion compound formation therefore reduces the density of the active component of the material. Second, while appropriate functional groups at the ends of the guest molecule can promote head-to-tail interactions between adjacent guest molecules, giving rise to a non-centrosymmetric parallel alignment of guest molecules within each individual tunnel, the relative orientations of guest molecules in different tunnels are also crucially important; construction of an inclusion compound that is active for SHG clearly requires that the SHG responses from the guest molecules in different tunnels do not cancel each other out. Third, it is important to note that the absence of a centre of symmetry is not the only crystallographic criterion dictating the usefulness of a material for SHG. For example, optimum phase-matched SHG is achieved when the angle between the molecular charge transfer axis and the polar crystal axis is 54.7°. Thus, while parallel alignment of guest molecules in tunnel host structures is conducive to the nonlinear optical response, it does not necessarily ensure that the performance of the material for SHG is optimised.

12   Electronic properties

A considerable amount of research has been devoted to the study of conducting and semiconducting guest components incorporated within solid host structures [91,92], particularly for zeolitic and related hosts. One advantage here is to exploit the structural and geometric features of the host framework to control the size and shape of functional guest components assembled within it. Typically, these guest components have a very uniform size and shape distribution. Such 'nanoparticles' of controlled dimensions provide an ideal basis for studying quantum size effects on optical, electrical and magnetic properties (which may differ substantially from the corresponding properties of the bulk material and may depend critically on the actual size of the particles).

Classical work in this area has included the study [93] of cadmium sulfide particles synthesised within zeolites A, X and Y. In its bulk phase, cadmium sulfide is a semiconductor with a narrow bandgap. Within the zeolitic frameworks, X-ray diffraction and EXAFS spectroscopy have shown that well-defined discrete Cd_4S_4 clusters (Fig. 2.17(a)) exist within the sodalite cage due to a favourable interaction between cadmium atoms of the cluster and oxygen atoms of the zeolite framework. At low loading, these clusters are isolated from each other, and have an optical absorption at c. 290 nm. At higher loading (Fig. 2.17(b)), the clusters begin to interconnect, with a $Cd\cdots Cd$ distance of c. 6 Å between clusters in adjacent sodalite cages. With this interaction between adjacent clusters, a description as a 'supercluster' is appropriate; the behaviour of these superclusters is intermediate between that of discrete Cd_4S_4 clusters and that of bulk cadmium sulfide. For such superclusters, the optical absorption is at c. 350 nm; above a certain threshold concentration, this optical absorption is independent of loading, and percolation behaviour is strongly implicated. A subsequent computational study [94] of cadmium sulfide particles inside zeolite Y has furnished results in excellent agreement with the experimental observations. In addition to the fact that the size and geometry of the individual Cd_4S_4 clusters are controlled by the specific geometry of the sodalite cages, the architecture of the superclusters depends on the spatial arrangement of these sodalite cages; in both of these aspects, the structure, and ultimately the electronic properties, of the semiconducting guest components are controlled critically by the structural character of the host framework.

Another important field, relating to guest components with interesting electronic properties, concerns the construction and confinement of so-called 'molecular wires' within tunnel host structures. For polymeric chains constructed within narrow tunnel structures, adjacent chains can be regarded as electronically isolated from one another

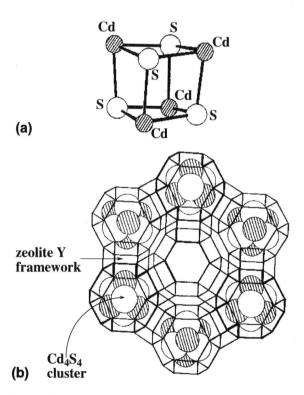

Figure 2.17. (a) Structure of the Cd_4S_4 cluster generated within the sodalite cages of zeolite Y. (b) Schematic illustration of the cadmium sulfide 'supercluster' generated within the zeolite Y framework at high loading. Modified from [92].

by the intervening host framework, creating new possibilities to characterise the electronic properties of conducting polymers in the absence of inter-chain interactions. Examples include the study of polypyrrole, polyaniline and polythiophene within the zeolite Y and mordenite host structures [95–97].

Recent work has focused on the larger tunnels of mesoporous hosts, within which groups of polymeric chains (rather than individual chains) may be expected. For example, polyaniline wires have been generated within the host material MCM-41 [98,99] by first adsorbing aniline within the host to saturation and immersing the resulting inclusion compound in an aqueous solution containing the oxidant $(NH_4)_2S_2O_8$. A combination of techniques has confirmed that polyaniline is indeed generated inside the host material, Raman spectroscopy suggests that the included polymer is in the conducting (protonated) state of polyaniline, and sorption data imply that a monolayer of polyaniline covers the walls of the tunnels in the MCM-41 host structure (suggesting, from geometric considerations, that each tunnel (diameter $c.$ 30 Å) should contain a maximum of about 20 polyaniline chains). To investigate charge transport properties of the polyaniline chains in MCM-41, the contactless microwave absorption technique was used. The microwave conductivity of the polyaniline/ MCM-41 system (0.0014 S cm^{-1} at 2.6 GHz (after correction for the volume fraction of polyaniline in the inclusion compound)) is significant (it is only a factor of about 4 smaller than that of bulk polyaniline (0.0057 S cm^{-1} at 2.6 GHz)), demonstrating that conjugated polymers included within the MCM-41 host can support mobile charge carriers. It is interesting to compare this with the observed lack of significant a.c. conductivity for polypyrrole in zeolite Y and mordenite [100], for which the charge carriers are believed to be trapped because of electrostatic interactions with the tunnel walls and the lack of contact between polymer chains (which are isolated in these hosts). Clearly, the wider tunnels in MCM-41 allow significant interaction between different polymer chains. Interestingly, the d.c. conductivity of the polyaniline/MCM-41 inclusion compound is only $c.$ 10^{-8} S cm^{-1} (comparable to that of the empty MCM-41 host) whereas the d.c. conductivity of the polyaniline extracted from this inclusion compound is $c.$ 10^{-2} S cm^{-1}. The low d.c. conductivity of the inclusion compound originates from the insulating nature of the tunnel walls of the host, and indicates that there are no percolating conduction paths on the external crystal surfaces.

Subsequent research [101] has generated conducting carbon wires within MCM-41; the carbon wires were produced by pyrolysis of polyacrylonitrile prepared *in situ* within the host material. The microwave conductivity of the carbon filaments generated within the host is higher, by a factor of about 10, than that for bulk carbonised polyacrylonitrile. Clearly, in all these examples, the host structure plays a key role in controlling the construction of the 'molecular wires' and mediating their interesting electrical properties.

A related field of interest concerns alkali metals 'dissolved' in zeolites. As an illustration, the formation of Na_4^{3+} clusters within sodalite cages of zeolite Na–Y is well established (Fig. 2.18), and relevant issues for these materials include the question of whether metallic behaviour can be achieved at sufficiently high cluster loading. The current state of the art in this field has been reviewed recently [102].

13 Magnetic properties

In recent years, there has been considerable interest [103] in molecular solids that

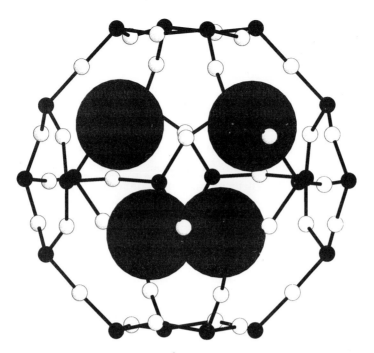

Figure 2.18. Structure of a Na_4^{3+} cluster within the sodalite cage of zeolite Y.

exhibit magnetic properties. With regard to solid inclusion compounds, the use of organic radicals as guest molecules clearly creates opportunities for the establishment of different types of intermolecular spin interactions, controlled by the specific guest–guest and guest–host contacts in the inclusion compound. A diversity of magnetic properties may be expected, depending on the structural characteristics of the inclusion compound.

These ideas have been illustrated [104] by comparison of two inclusion compounds containing the stable organic radical 4-amino-2,2,6,6-tetramethyl piperidine 1-oxyl (ATEMPO) as the guest component. The two host compounds are shown in Fig. 2.19 (the sulfur atoms in these molecules are clearly of potential importance with regard to their interactions with the spin system of the guest molecules). Host **4** forms an inclusion compound **4**/(ATEMPO)$_2$, which behaves as a paramagnet down to the lowest temperature (2 K) investigated, implying that there is little intermolecular interaction involving the spin systems. This is consistent with the known crystal structure of this inclusion compound, in which there are no close contacts between the N—O bonds of different guest molecules or any short intermolecular contacts between host and guest molecules. In contrast, host **5** forms an inclusion compound **5**/(ATEMPO)$_2$ which is paramagnetic down to *c.* 4 K and exhibits antiferromagnetic behaviour below this temperature. It is proposed that, as a consequence of structural differences between the **4**/(ATEMPO)$_2$ and **5**/(ATEMPO)$_2$ inclusion compounds, there are magnetic interactions only in the latter case. However, in the absence of a crystal structure determination of **5**/(ATEMPO)$_2$, no definitive rationalisation of the observed antiferromagnetism can be made at this stage. Nevertheless, this work alludes to the encouraging prospects for achieving a measure of control over the magnetic properties of radical guest molecules constrained within solid host structures.

4

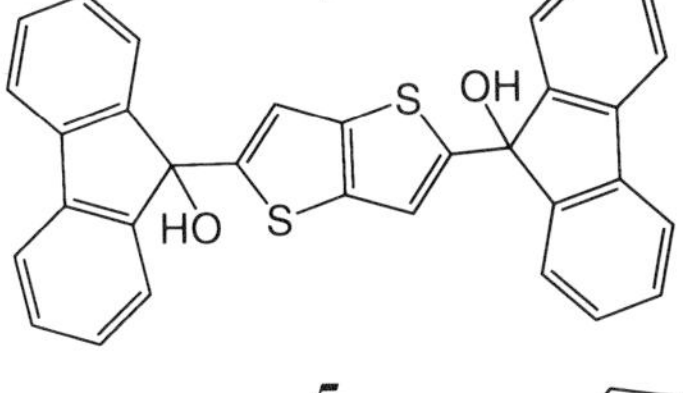

5

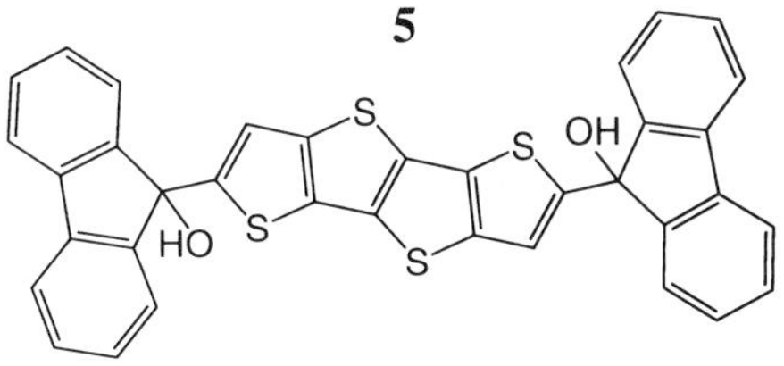

ATEMPO

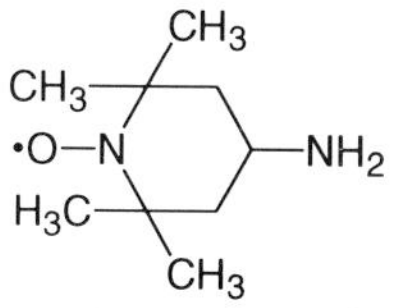

Figure 2.19. The inclusion compound formed between the stable organic radical ATEMPO and the host material **4** is paramagnetic, whereas the inclusion compound formed between ATEMPO and the host material **5** exhibits antiferromagnetic behaviour at sufficiently low temperature.

14 Inclusion phenomena as the basis of molecular sensors

There is clear potential, as yet largely under-exploited, for using solid inclusion phenomena as the basis of sensor devices. Again, the characteristic selectivity of the inclusion process is the key feature underlying the potential success of such applications. Essentially, the requirement is to have a host material that will selectively incorporate, as the guest component, the molecule that is to be detected in the external environment (for example, in the atmosphere). If the formation of the host–guest system is to be used in a quantitative manner, then, first, the extent of uptake of the guest molecules by the host material should be related via a well-defined (and known) relationship to the concentration of these molecules in the external environment, and second, the extent of uptake of the guest molecules by the host material should be directly measurable by a well-defined change in some property of the inclusion compound (electrical and optical properties may be particularly convenient).

An example illustrating these concepts concerns inclusion of the cationic dye thionin within zeolitic hosts [105]. The aggregation state and local environment of this guest depends on the characteristics of the zeolite. For example, in tunnel-containing zeolites

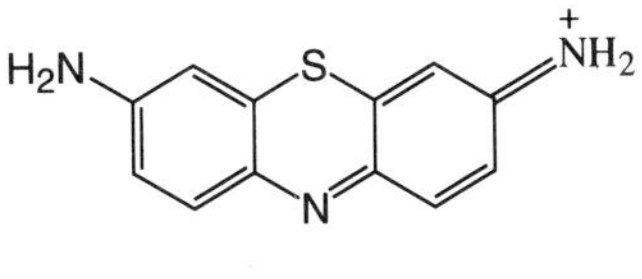

thionin

(for example, zeolite L), the guest is present as isolated molecules, whereas in cage-containing zeolites (for example, zeolites X and Y) the guest molecules are aggregated (probably as dimers) even at low concentration. When the inclusion compound of thionin in zeolite Y is dried, there is a significant colour change (pink to blue) and the material becomes highly fluorescent. This new state remains stable, provided the material is not exposed to any moisture. Diffuse reflectance spectra suggest that the thionin is present as monomers when the material is dehydrated and is aggregated as dimers when the material is hydrated. Clearly this inclusion compound has potential for application as a sensor for water. Water was found to be unique (among a wide range of other solvents considered) in bringing about these changes in the aggregation state of the guest molecules and the associated colour changes; this is probably related to (though not explained solely by) the high dielectric constant of water. High selectivity in the potential application of this material as a sensor for water is therefore implicated.

15   Future perspectives

It is clear that the field of solid-state inclusion chemistry encompasses a broad spectrum of concepts and phenomena, ranging from chemical reactivity to molecular electronics, from structural design to catalysis, and from chirality to magnetism. Underlying virtually all fundamental features and applied aspects of solid inclusion compounds is the crucial interplay between the host and guest components, each influencing the properties of the other, and combining together in a synergistic manner to determine the properties of the material as a whole.

Although many of the examples highlighted in this paper represent significant strides forward both in our fundamental understanding of these materials and in our perception of their potential applications, they are perhaps only hints as to the exciting developments and discoveries that lie ahead. With the increasing power to design materials with specifically targeted structural properties, with an evolving understanding of the fundamental concepts underlying inclusion phenomena in solids, and with an acquired foresight to design such materials with specific properties and applications in mind, we forecast with confidence that there will be further exciting developments in this fertile area of science as we enter the 21st century.

16   Acknowledgements

I am grateful to Lily Yeo for help in preparation of this paper and to Joe Hriljac for useful discussions.

17   References

1 Cronstedt AF. *Kongl. Vetenskaps Acad. Handl. Stockholm* 1756; **17**: 120.
2 Davy H. *Phil. Trans. Roy. Soc.* 1811; **101**: 30.
3 Faraday M. *Quart. J. Sci.* 1823; **15**: 71.
4 Palin DE, Powell HM. *J. Chem. Soc.* 1947; 208.
5 Powell HM. *J. Chem. Soc.* 1948; 61.

6 Smith JV. *Chem. Rev.* 1988; **88**: 149.

7 Thomas JM. *Angew. Chem., Int. Ed. Engl.* 1988; **27**: 1673.

8 Thomas JM, Vaughan DEW. *J. Phys. Chem. Solids* 1989; **50**: 449.

9 Szostak R. *Molecular Sieves: Principles of Synthesis and Identification*, Van Nostrand Reinhold, New York, 1989.

10 Davis ME, Lobo RF. *Chem. Mater.* 1992; **4**: 756.

11 Suib SL. *Chem. Rev.* 1993; **93**: 803.

12 Meier WM, Olson DH. *Atlas of Zeolite Structure Types*, International Zeolite Association, Butterworth-Heinemann, 3rd revised edition, 1992.

13 Kresge CT, Leonowicz ME, Roth WJ, Vartuli JC, Beck JS. *Nature* 1992; **359**: 710.

14 Beck JS, Vartuli JC, Roth WJ *et al. J. Am. Chem. Soc.* 1992; **114**: 10834.

15 Behrens P, Stucky GD. *Angew. Chem., Int. Ed. Engl.* 1993; **32**: 696.

16 Takemoto K, Sonoda N. In Atwood JL, Davies JED, MacNicol DD (eds) *Inclusion Compounds,* Vol. 2. New York: Academic Press, 1984, p. 47.

17 Harris KDM. *J. Solid State Chem.* 1993; **106**: 83.

18 Harris KDM. *J. Mol. Struct.* 1996; **374**: 241.

19 Hollingsworth MD, Harris KDM. In MacNicol DD, Toda F, Bishop R (eds) *Comprehensive Supramolecular Chemistry,* Vol. 6. Oxford: Pergamon Press, 1996, p. 177.

20 Gerdil R. *Topics in Current Chem.* 1987; **140**: 71.

21 Gerdil R. In MacNicol DD, Toda F, Bishop R (eds) *Comprehensive Supramolecular Chemistry,* Vol. 6. Oxford: Pergamon Press, 1996, p. 239.

22 Farina M. In Atwood JL, Davies JED, MacNicol DD (eds) *Inclusion Compounds* Vol. 2. New York: Academic Press, 1984, p. 69.

23 Farina M, Di Silvestro G, Sozzani P. In MacNicol DD, Toda F, Bishop R (eds) *Comprehensive Supramolecular Chemistry*, Vol. 6. Oxford: Pergamon Press, 1996, p. 371.

24 Giglio E. In Atwood JL, Davies JED, MacNicol DD (eds) *Inclusion Compounds,* Vol. 2. New York: Academic Press, 1984, p. 207.

25 Miyata M, Sada K. In MacNicol DD, Toda F, Bishop R (eds) *Comprehensive Supramolecular Chemistry,* Vol. 6. Oxford: Pergamon Press, 1996, p. 147.

26 Ghadiri MR, Granja JR, Milligan RA, McRee DE, Khazanovich N. *Nature* 1993; **366**: 324.

27 Ghadiri MR. *Adv. Mater.* 1995; **7**: 675.

28 Gardner GB, Venkataraman D, Moore JS, Lee S. *Nature* 1995; **374**: 792.

29 Venkataraman D, Gardner GB, Lee S, Moore JS. *J. Am. Chem. Soc.* 1995; **117**: 11600.

30 Brimah AK, Siebel E, Fischer RD, Davies NA, Apperley DC, Harris RK. *J. Organomet. Chem.* 1994; **475**: 85.

31 Schwarz P, Siebel E, Fischer RD, Apperley DC, Davies NA, Harris RK. *Angew. Chem., Int. Ed. Engl.* 1995; **34**: 1197.

32 Abrahams BF, Hoskins BF, Michail DM, Robson R. *Nature* 1994; **369**: 727.

33 Toda F, Tanaka K, Fujiwara T. *Angew. Chem., Int. Ed. Engl.* 1990; **29**: 662.

34 Atwood JL, Koutsantonis GA, Raston CL. *Nature* 1994; **368**: 229.

35 Nishikawa M. *Chem. Pharm. Bull.* 1963; **11**: 977.

36 Fukushima K. *J. Mol. Struct.* 1976; **34**: 67.

37 Allen A, Fawcett V, Long DA. *J. Raman Spectrosc.* 1976; **4**: 285.

38 McKinnon MS, Wasylishen RE, *Chem. Phys. Lett.* 1986; **130**: 565.

39 Müller K. *J. Phys. Chem.* 1992; **96**: 5733.

40 Aliev AE, Harris KDM. *J. Am. Chem. Soc.* 1993; **115**: 6369.

41 Shannon IJ, Jones MJ, Harris KDM, Siddiqui MRH, Joyner RW. *J. Chem. Soc., Faraday Trans.* 1995; **91**: 1497.

42 Aliev AE, Harris KDM, Mordi RC. *J. Chem. Soc., Faraday Trans.* 1994; **90**: 1323.

43 Gustavsen JE, Klæboe P, Kvila H. *Acta Chem. Scand., Sect. A* 1978; **32**: 25.

44 Hollingsworth MD, Harris KDM, Chaney JD *et al.* Manuscript in preparation.

45 Elizabe L, Smart SP, El Baghdadi A, Guillaume F, Harris KDM. *J. Chem. Soc., Faraday Trans.* 1996; **92**: 267.

46 Shang QY, Dou X, Hudson BS. *Nature* 1991; **352**: 703.

47 Lee SK, Shang QY, Hudson BS. *Mol. Cryst. Liq. Cryst. Sci. Technol., Sect. A* 1992; **211**: 147.

48 Weber E, Wimmer C. *Chirality* 1993; **5**: 315.

49 Arad-Yellin R, Green BS, Knossow M, Tsoucaris G. In Atwood JL, Davies JED, MacNicol DD (eds) *Inclusion Compounds,* Vol. 3. New York: Academic Press, 1984, p. 263.

50 Yeo L, Harris KDM. *Tetrahedron Asymmetry* 1977; **7**: 1891.

51 Arad-Yellin R, Green BS, Knossow M, Tsoucaris G. *J. Am. Chem. Soc.* 1983; **105**: 4561.

52 Ács M, Mravik A, Fogassy E, Böcskei Z. *Chirality* 1994; **6**: 314.

53 Newsam JM, Treacy MMJ, Koetsier WT, De Gruyter CB. *Proc. Royal Soc. A* 1988; **420**: 375.

54 Thomas JM. *Faraday Discuss.* 1995; **100**: C9.

55 Rennie AJO, Harris KDM. *Proc. Royal Soc. A* 1990; **430**: 615.

56 Rennie AJO, Harris KDM. *J. Chem. Phys.* 1992; **96**: 7117.

57 Shannon IJ, Harris KDM, Rennie AJO, Webster MB. *J. Chem. Soc., Faraday Trans.* 1993; **89**: 2023.

58 Schofield PA, Harris KDM, Shannon IJ, Rennie AJO. *J. Chem. Soc., Chem. Commun.* 1993; 1293.

59 Harris KDM, Thomas JM. *J. Chem. Soc., Faraday Trans.* 1990; **86**: 1095.

60 Schmicker D, van Smaalen S, de Boer JL, Haas C, Harris KDM. *Phys. Rev. Lett.* 1995; **74**: 734.

61 Chatani Y, Taki Y, Tadokoro H. *Acta Crystallogr.* 1977; **B33**: 309.

62 Harris KDM, Gameson I, Thomas JM. *J. Chem. Soc., Faraday Trans.* 1990; **86**: 3135.

63 Harris KDM, Jonsen P. *Chem. Phys. Lett.* 1989; **154**: 593.

64 Guillaume F, Sourisseau C, Dianoux AJ. *J. Chim. Phys. (Paris)* 1991; **88**: 1721.

65 Parsonage NG, Pemberton RC. *Trans. Faraday Soc.* 1967; **63**: 311.

66 Fukao K. *J. Chem. Phys.* 1990; **92**: 6867.

67 Lynden-Bell RM. *Mol. Phys.* 1993; **79**: 313.

68 Brown ME, Hollingsworth MD. *Nature* 1995; **376**: 323.

69 Aoyama H, Miyazaki K, Sakemoto M, Omote Y. *J. Chem. Soc., Chem. Commun.* 1983; 333.

70 Farina M. In Atwood JL, Davies JED, MacNicol DD (eds) *Inclusion Compounds,* Vol. 3. New York: Academic Press, 1984, p. 297.

71 Miyata M. In Reinhoudt DN (ed.) *Comprehensive Supramolecular Chemistry,* Vol. 10. Oxford: Pergamon Press, 1996, p. 557.

72 Farina M, Natta G, Allegra G, Loffelholz M. *J. Polym. Sci., Part C* 1967; **16**: 2517.

73 Farina M, Audisio G, Natta G. *J. Am. Chem. Soc.* 1967; **89**: 5071.

74 Lewis AR, Millar GJ, Cooney RP, Bowmaker CA. *Chem. Mater.* 1993; **5**: 1509.

75 Leibovitch M, Olovsson G, Sundarababu G, Ramamurthy V, Scheffer JR, Trotter J. *J. Am. Chem. Soc.* 1996; **118**: 1219.

76 Addadi L, Ariel S, Lahav M, Leiserowitz L, Popovitz-Biro R, Tang CP. In Roberts MW, Thomas JM (eds) *Chemical Physics of Solids and Their Surfaces,* Vol. 8. London: Royal Society of Chemistry (Specialist Periodical Reports), 1980; 202.

77 Thomas JM, Zamaraev KI (eds). *Perspectives in Catalysis.* Oxford: Blackwell Science (IUPAC Chemistry for the 21st Century monographs), 1992.

78 Thomas JM. *Angew. Chem., Int. Ed. Engl.* 1994; **33**: 913.

79 Maddox PJ, Stachurski J, Thomas JM. *Catal. Lett.* 1988; **1**: 191.

80 Dooryhee E, Greaves GN, Steel AT *et al. Faraday Discuss. Chem. Soc.* 1990; **89**: 119.

81 George AR, Sanderson JS, Catlow CRA. *J. Computer-Aided Mater. Design* 1993; **1**: 169.

82 George AR, Catlow CRA, Thomas JM. *J. Chem. Soc., Faraday Trans.* 1995; **91**: 3975.

83 Maschmeyer T, Rey F, Sankar G, Thomas JM. *Nature* 1995; **378**: 159.

84 Tanev PT, Chibwe M, Pinnavaia TJ. *Nature* 1994; **368**: 321.

85 Corma A, Navarro MT, Perez-Pariente J. *J. Chem. Soc., Chem. Commun.* 1994; 147.

86 Tam W, Eaton DF, Calabrese JC, Williams ID, Wang Y, Anderson AG. *Chem. Mater.* 1989; **1**: 128.

87 Hulliger J, König O, Hoss R. *Adv. Mater.* 1995; **7**: 719.

88 Cox SD, Gier TE, Stucky GD, Bierlein JD. *J. Am. Chem. Soc.* 1988; **110**: 2986.

89 Cooper S, Dutta PK. *J. Phys. Chem.* 1990; **94**: 114.

90 Anderson AG, Calabrese JC, Tam W, Williams ID. *Chem. Phys. Lett.* 1987; **134**: 392.

91 Stucky GD, MacDougall JE. *Science* 1990; **247**: 669.

92 Herron N. In Atwood JL, Davies JED MacNicol DD (eds) *Inclusion Compounds*, Vol. 5. Oxford: Oxford University Press, 1991; 90.

93 Herron N, Wang Y, Eddy MM *et al. J. Am. Chem. Soc.* 1989; **111**: 530.

94 Jentys A, Grimes RW, Gale JD, Catlow CRA. *J. Phys. Chem.* 1993; **97**: 13535.

95 Enzel P, Bein T. *J. Phys. Chem.* 1989; **93**: 6270.

96 Enzel P, Bein T. *J. Chem. Soc., Chem. Commun.* 1989; 1326.

97 Bein T, Enzel P. *Angew. Chem., Int. Ed. Engl.* 1989; **28**: 1692.

98 Wu C-G, Bein T. *Chem. Mater.* 1994; **6**: 1109.

99 Wu C-G, Bein T. *Science* 1994; **264**: 1757.

100 Zuppiroli L, Beuneu F, Mory J, Enzel P, Bein T. *Synth. Met.* 1993; **55–57**: 5081.

101 Wu C-G, Bein T. *Science* 1994; **266**: 1013.

102 Edwards PP, Anderson PA, Thomas JM. *Acc. Chem. Res.* 1996; **29**: 23.

103 Miller JS, Epstein AJ, Reiff WM. *Chem. Rev.* 1988; **88**: 201.

104 Mazaki Y, Awaga K, Kobayashi K. *J. Chem. Soc., Chem. Commun.* 1992; 1661.

105 Ramamurthy V, Sanderson DR, Eaton DF. *J. Am. Chem. Soc.* 1993; **115**: 10438.

3 Interfacial Chemistry on Metal-Oxide Single Crystals Relevant to Oxide Catalysis

H. ONISHI and Y. IWASAWA

Department of Chemistry, Graduate School of Science, University of Tokyo, Hongo, Bunkyo-ku, Tokyo 113, Japan

1 Introduction

Metal oxides have found applications in a variety of technologies where interfacial chemistry is critical to success, including catalysis, gas sensors, photoelectrolysis, electronic ceramics, semiconductor devices, pigments and cosmetics. Understanding and controlling oxide surfaces are the key issues for the development of industrial oxide catalysts and related advanced materials [1]. Oxide surfaces are, however, in general heterogeneous and complicated, and hence have been little studied, so that the attempt to put them on a scientific basis using traditional approaches has not proceeded very far. The inherent compositional and structural inhomogeneity of oxide surfaces makes the problem of identifying the essential issues for their catalytic performance extremely difficult. In order to reduce the level of complexity, a common approach is to study model catalysts such as single-crystal oxide surfaces and epitaxial oxide flat surfaces. Characterisation of catalytic phenomena at model oxide surfaces includes: (i) characterisation of specific model surfaces to transfer the knowledge so obtained to catalytic systems or even to create a new type of catalyst; and (ii) characterisation of catalysis for the understanding of reaction mechanisms at the atomic or molecular scale, optimising catalytic operations and extracting essential factors for the genesis of the best catalysis [1].

While studies of surface structures have focused on metals and semiconductors since the 1960s, the application of surface science techniques to metal oxides has partly blossomed only within the last decade. Many oxide surfaces undergo thermal fracture, reconstruction, and particularly faceting, which phenomena arise mainly from the need for charge balancing and minimisation of surface polarity and energy. The properties of oxide catalysts pre-treated at different temperatures and ambient conditions may be relevant to such restructuring. Excellent comprehensive compilations of works in this field may be found in recent books [2–5] and reviews [6–12]. Most of these studies have mainly focused on static characterisation of clean surfaces, adsorbed molecules, deposited metal particles or films, and non-catalytic reactions on those model surfaces. On the other hand, it is now known that the adsorption of molecules from the gas phase can seriously influence the reactivity of adsorbed species at oxide surfaces [1,3]. *In-situ* observation of surface atoms and adsorbed molecules during catalytic reactions has vital importance in the understanding of catalysis on metal-oxide surfaces.

One of the goals of interfacial chemistry on oxide catalysts is to identify which adsorbed molecule reacts at which surface site. This paper sets out to discuss recent work towards the realisation of this goal. Monitoring the structure of surfaces and adsorbates on a molecular scale during reactions *in situ* under conditions of high reaction temperatures and pressures has become possible on model oxide surfaces. We believe this approach makes possible the prediction of future trends for the next century.

2   *In-situ* characterisation of catalytic phenomena

Recently, the characterisation of oxide surfaces at the atomic and molecular level has received a considerable boost from the development of a variety of sophisticated techniques, including X-ray absorption fine-structure spectroscopy (XAFS) [14–17], high-resolution electron microscopy (HREM) [18,19], ion scattering spectroscopy (ISS) [20], static secondary ion mass spectroscopy (SSIMS) [21,22], solid-state magic angle spinning nuclear magnetic resonance (NMR) [23–25], transmission Fourier-transform infrared (FT-IR) spectroscopy [26,27], reflection–absorption infrared spectroscopy (RAIRS) [27–31], and Raman spectroscopy [32,33]. Besides those techniques, scanning probe microscopy (SPM), like scanning tunnelling microscopy (STM) [34] and scanning force microscopy (SFM) [35], has been very rapidly developing and realising *in-situ* visualisation of catalytic phenomena in recent years. It is thus helpful to survey the applications of SPM on oxide surfaces.

2.1   *Scanning probe microscopy*

Scanning probe microscopy has particularly great potential for *in-situ* chemical studies. While our present knowledge of the atomic structure of catalyst surfaces is largely limited to those structures which are stable in ultrahigh vacuum (UHV) before and after reaction, SPM may provide an insight into both adsorbate and catalyst surface structure during the reaction. The following issues may be most relevant to characterisation of catalysts and catalysis: (i) identification of structural characteristics of the variety of non-equivalent surface sites and observation of site specificity to reactivity; (ii) study of how the reaction at one particular surface site affects the local activity of neighbouring sites; (iii) structural transformation and chemical modification of the surface caused by adsorbates and chemical reactions; (iv) detailed information about the mechanism of surface chemical reaction; and (v) surface diffusion and mobility of reaction intermediates.

A serious limitation of the SPM technique so far is its lack of chemical sensitivity. Generally, SPM is not specific for the elemental species in multi-component systems. The surface area which one may look at by SPM is typically quite small. This 'tunnel vision' might sometimes be dangerous when drawing conclusions on the whole surface from information gained at only a few locations of small lateral dimensions. The problem of the representativeness of the SPM results obtained is, at least partly, solved by considerably increasing the total scan range of SPM instruments.

2.2   *Scanning tunnelling microscopy*

The successful combination of vacuum tunnelling with a piezoelectric drive system to form a scanning tunnelling microscope was first demonstrated in 1981 [36]. A conducting sample and a sharp metal tip, which acts as a local probe, are brought to within a distance of 1 nm or less, resulting in a significant overlap of the wave functions. The exponential dependence of the tunnelling current on the tip-to-sample spacing is the key to the high spatial resolution which can be achieved with STM. By using a piezoelectric drive system for the tip and a feedback loop, a map of the surface topography is obtained. Under favourable conditions, a vertical resolution of 1 pm and a lateral resolution of about 10 pm can be reached. Thus STM provides real-space topography of surfaces of conducting materials down to the atomic scale. In contrast to other electron microscopes and surface analytical techniques using electrons, STM can be operated in air, liquids, vacuum and reaction conditions. Recently, it has been shown that *in-situ* STM studies at high pressures and temperatures are indeed feasible [37,38]. *Ex-situ* STM images showing direct evidence for surface restructuring of Co(0001) at the atomic level caused by actual reaction conditions have also been obtained by combining a high-pressure reaction cell with a UHV system equipped with STM [39].

Following the invention of STM, the technique has been used to observe the surface structures of several transition metal oxides such as $TiO_2(110)$ [40–45], $TiO_2(100)$ [46], $V_2O_5(010)$ [47], $V_2O_5(001)$ [48], $Rb_{0.3}WO_3(0001)$ [49,50], $WO_3(001)$ [51], $Fe_3O_4(001)$ [52], $Mo_{18}O_{52}(100)$ [53] and $SrTiO_3(100)$ [54] in real space with atomic resolution. Figure 3.1 shows representative results on the sputter-annealed $TiO_2(110)(1 \times 1)$ surface [43,55]. Stacked (110) terraces and regularly aligned Ti^{4+} ions on a terrace are imaged in Fig. 3.1(a) and 3.1(b), respectively. Electron tunnelling to the unoccupied surface states localised on the exposed Ti ions has been claimed for imaging [55].

2.3   *Scanning force microscopy*

An important class of oxides (Al_2O_3, SiO_2, MgO, zeolites, etc.) lies outside the scope of tunnelling microscopy due to non-conductance. Scanning force microscopy can be applied to insulators as well as conductors; it does not use electron tunnelling to probe local properties of sample surfaces, but rather the tip–surface force interaction [35]. Nanometre-scale topography is routinely visualised by SFM in a contact force mode. The probing tip is in 'contact' with the sample surface, and is scanned over the surface to keep constant the short-range repulsive force between the tip and the surface. The loading on the tip is normally 1–10 nN. Experimentally, atomic-scale periodicities have been resolved by contact SFM. Under the assumption of a monoatomic tip, the relatively large loading of 10 nN theoretically results in the destruction of the tip apex [56]. It is hence assumed that the effective tip is composed of several atoms, to explain the atomically resolution of contact SFM. Thus the atomic-scale image obtained by contact SFM is usually interpreted as the periodicity of the ordered surface.

On the other hand, non-contact force microscopy provides an opportunity to observe atomic-scale singularities, such as isolated adatoms and admolecules. This mode uses longer-range forces (van der Waals force, electrostatic force, etc.) for the regulation of tip–surface separation, by increasing the tip–surface separation to 10–100 nm. Since the magnitude of the long-range forces at the relatively large distances may be considerably smaller than that of the short-range repulsion in contact SFM, the method of interaction force detection in non-contact SFM is different from contact SFM [35]. Atomically resolved imaging of single atom vacancies has been demonstrated on Si(111) [57,58], Si(100) [59] and InP(110) [60]. By extrapolating the rapid development

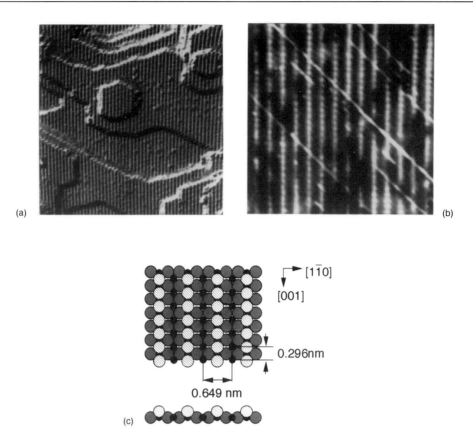

Figure 3.1 STM topography of the sputter-annealed $TiO_2(110)(1 \times 1)$ surface: (a) a variable-current image (35×35 nm) of stacked (110) terraces; (b) a constant-current image (9×9 nm) of individual Ti^{4+} ions on a terrace; (c) top and side views of the structure proposed for the (1×1) surface. Hatched and filled symbols represent oxygen and titanium ions, respectively.

in STM technology, we may expect non-contact SFM to become a powerful technique for determining the atomic-scale structure of insulating oxide surfaces.

3 Acid–base and redox properties of model catalysts

Acid–base reactivity is an important property of oxide catalysts, and its control is of interest in surface chemistry as well as being of importance in industrial applications [61,62]. The exposed cations and anions on oxide surfaces have long been described as acid–base pairs [63,64]. The polar planes of ZnO showed dissociative adsorption and subsequent decomposition of methanol and formic acid related to their surface acid–base properties [65–67]. Further examples related to the topic of acid–base properties have been accumulated to date [2].

In contrast to the extensive studies of heterogeneous acidic oxides, less effort has been devoted to the study of heterogeneous basic oxides [1]. The first study of heterogeneous basic catalysts, in which sodium metal dispersed on alumina acted as an effective catalyst for double-bond migration of alkenes, was reported by Pines *et al.* [68]. A number of materials have since been reported to act as heterogeneous basic catalysts: alkaline–earth oxides, alkaline metal oxides, rare earth oxides, ZrO_2, ZnO and TiO_2, alkaline ion-exchanged zeolites, alkaline metal ions on oxides, hydrotalcite, chrysotile, sepiolite, KF supported on alumina, etc. [69,70]. A superbasic catalyst (γ-Al$_2$O$_3$– NaOH–Na) is prepared by addition of NaOH to alumina, followed by further addition of Na. The resulting catalyst, which possesses basic sites stronger than H$^-$ = 37 and a distorted β-NaAlO$_2$ phase at the catalyst surface as characterised by solid NMR and X-ray photoelectron spectroscopy (XPS), is industrially employed, with nearly 100% yield in the commercial plant for synthesis of 5-ethylidene-2-norbornene (additive to ethene-propene copolymer rubber) from 5-vinyl-2-norbornene that is obtained from dicyclopentadiene [71]. Considering the tendency of Na to donate electrons, it seems natural that Na dispersed on alumina acts as a heterogeneous basic catalyst. It is, however, found that the basic property of oxide surfaces is not naturally proportional to the quantity of Na deposited on the surface [72,73].

Using atom-resolved STM, it has been possible to see the mechanism producing structural sensitivity in the reaction of CO$_2$, a reaction probe at basic sites, with Na-deposited TiO$_2$(110) surfaces [72]. The amount of CO$_2$ adsorption to form carbonates varies with Na coverage, exhibiting an S-shaped dependence. An STM image of the 0.1 ML Na-deposited TiO$_2$(110) surface depicts a randomly dispersed geometry due to repulsive forces between the Na atoms (Fig. 3.2(a)). The surface is converted to a nearly

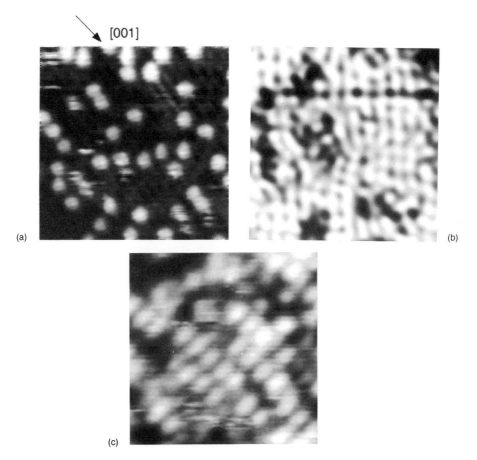

Figure 3.2 STM topographs (14 × 14 nm) of Na-deposited TiO$_2$(110) surfaces: (a) randomly dispersed Na adatoms; (b) c(4 × 2)–Na overlayer; (c) carbonate chains ordered in a p(3 × 2) order observed following exposure of the c(4 × 2) surface to 10^2 L CO$_2$ at room temperature.

complete c(4 × 2) overlayer (Fig. 3.2(b)) via a p(4 × 2) order locally formed on the surface by increasing Na coverage. The Na atoms in the c(4 × 2) surface were found to be reactive to CO_2, as shown in the STM topography (Fig. 3.2(c)) in which the c(4 × 2) order disappeared and the chains for uniformly oriented CO_3^{2-} along the [001] direction were locally ordered in a p(3 × 2) periodicity. In contrast, CO_2 did not absorb on surfaces with randomly dispersed Na atoms. The genesis of strongly basic sites is thus suggested not to be linearly correlated with Na quantity, but to be correlated with the ordered structure or at least suitable ensembles of Na atoms.

Redox property is another key in reactions on oxide surfaces. Redox properties may be scaled by using the oxygen affinity, which is the negative of the logarithm of the oxygen pressure in equilibrium at 1000 K. The oxygen affinities of metals are plotted against the groups in the periodic table (Fig. 3.3). Chemical and catalytic trends in oxide surfaces where metal atoms are isolated by oxide ligands resemble those observed in homogeneous metal oxo-complexes [12].

Recently, it has been demonstrated that coordination vacancies on the surface metal cations are relevant to the unique redox reactivity of oxide surfaces [12,74]. Oxidation of formaldehyde and methyl formate to adsorbed formate intermediates on ZnO(0001) [75] and reductive C–C coupling of aliphatic and aromatic aldehydes and cyclic ketones on TiO_2(001) surfaces reduced by Ar^+ bombardment are observed in temperature-programmed reaction (TPR) [76–78]. The thermally reduced TiO_2(110) surface, which is a less heavily damaged surface than that obtained by bombardment and contains Ti cations in only the 3+ and 4+ states, also shows activity for the reductive coupling of formaldehyde to ethene [79]. Interestingly, the catalytic cyclotrimerisation of alkynes on bombarded TiO_2(001) is also traced in UHV conditions, where cation coordination and oxidation state appear to be closely linked to activity and selectivity [80]. The non-polar Cu_2O(111) surface shows a maximum selectivity for complete oxidative dehydrogenation of methanol to CO, while the polar Cu^+-terminated (100) surface shows a maximum selectivity for partial dehydrogenation to HCHO [81]. Methanol on ZrO_2(100) decomposes near 630 K to produce CO and CH_4, whereas on the (110) surface the primary methoxide decomposition pathway is oxidation to produce HCHO.

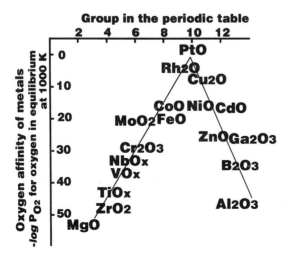

Figure 3.3 Oxygen affinity of metals in the periodic table.

These differences in reactivity can be related to the local atomic structure of each surface [82] in addition to the oxygen affinity of the corresponding metal [1].

4   Catalysis by oxide single crystals

Most of the experimental studies on metal-oxide single-crystal surfaces have focused mainly on the characterisation of adsorbates in static states and their non-catalytic reactions under vacuum by means of temperature-programmed desorption (TPD)/TPR and photoelectron spectroscopic techniques. Considerable effort has been invested in applying UHV methods to basic catalytic problems on metal single-crystal surfaces, by combining in a single apparatus the ability to measure kinetics at elevated pressures on single-crystal catalysts with the ability to carry out surface analytical measurements. This approach has allowed the direct comparison of reaction rates measured on metal single-crystal surfaces with those measured on more realistic catalysts [83,84]. The relationships between the structure, composition, and electronic properties of oxide surfaces and catalytic activity/selectivity averaged over the surface of a sample wafer can be addressed by applying this approach to catalytic reactions over oxide single crystals.

Kung and co-workers have pioneered the examination of catalytic performance on ZnO single-crystal surfaces, where reaction sites were assigned into defects generated by the reduction of the surfaces [85,86]. Very recently cyclotrimerisation of alkynes on Ar^+-bombarded $TiO_2(001)$ surfaces has been found to proceed catalytically, where more than 20 turnovers could be obtained without deactivation of the single-crystal catalyst [87].

A new aspect of acid–base catalysis was found in the steady-state decomposition reaction of formic acid on $TiO_2(110)$, where a switchover of the reaction paths from unimolecular dehydration to bimolecular dehydrogenation took place when a formic acid molecule participated in the decomposition step of a surface formate (below)

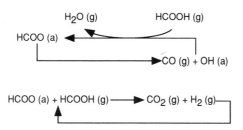

[88,89]. This contrasts with the traditional acid–base concept. It has been thought that acid–base character is an intrinsic property of oxide substrates. The selectivity in the catalytic decomposition reaction of formic acid has been used to scale the acid–base property: dehydration occurs over acidic oxide and dehydrogenation proceeds over basic oxide, though this classification is oversimplified (see below). The catalytic dehydration reaction is suggested to involve the unimolecular decomposition of formates as a rate-determining step. In other words, the formate–surface interaction activates the unimolecular decomposition of formate (HCOO(a)) to yield preferentially CO(g) and OH(a). An acidic proton of an HCOOH molecule, which encounters the surface in a steady state, reacts with the resultant OH(a) to form H_2O. The two reaction

steps form a catalytic dehydration cycle with an activation energy of 120 kJ mol^{-1}. At lower temperatures a formic acid molecule selectively reacts with the formate before the unimolecular decomposition, to open a new basic-catalysis path with a much lower activation energy of 15 kJ mol^{-1} by a bimolecular reaction mechanism. In the latter case, the intermolecular interaction between H$^{\delta-}$ of HCOO(a) and acidic H$^{\delta+}$ of HCOOH(g) may become more important than the intrinsic acidic character of the TiO$_2$ surface. It is to be noted that acidic molecules induce basic catalysis [1]. Although TiO$_2$ powder is a selective catalyst for dehydration [90], the TiO$_2$(110) single-crystal surface favours dehydrogenation at low temperatures, which might be an unexpected feature. It is possible on oxide single crystals to find new catalytic reaction paths by uniformly controlling the coverage and bond arrangement of intermediate and reactant, but this situation rarely applies to powder catalysts [1].

In fact the latter feature is achieved on the TiO$_2$(110) surface, where the alternative alignment of the exposed Ti row and the bridging O-ridge row results in the characteristic anisotropy on this surface, as illustrated in Fig. 3.1(c). Formate ions on TiO$_2$(110) form a monolayer ordered in a (2×1) periodicity, when saturating the surface. The (2×1)–formate surface was observed by STM and individual ions were resolved as shown in Fig. 3.4(a). The anisotropic STM image (elongated in the $[1\bar{1}0]$ direction) of the formates has been related to the tunnelling to the lowest unoccupied molecular orbital (LUMO) of the formate in a bridge conformation as illustrated in Fig. 3.4(b), where the C–H bond of a formate is normal to the surface, while the O–C–O plane is parallel to the [001] axis (Fig. 3.4(c)) [91].

The transport of adsorbed species often affects strongly the rate of reaction on oxide catalysts. Time-resolved STM observation has been used to see the migration of formate ions [92]. The formates in the (2×1) monolayer were removed without damage to the substrate, by rastering the STM tip biased at appropriate voltages to achieve a patchwork of uncovered substrate (Fig. 3.5(a)) [93]. The residual formates diffused to the square patch, driven by the mutual repulsion in the monolayer. Sequential STM images revealed the molecular-scale kinetics of formate migration; the transport of the formate ion along the one-dimensional row of Ti cations was an order of magnitude faster than that across the Ti row. The void shrank with time and disappeared, as Fig. 3.5 shows. The boundaries between the void and the migrating monolayer remained straight and clear. The boundaries parallel to the $[1\bar{1}0]$ direction advanced along the [001] direction at a rate of 0.15 nm min^{-1}, whereas the boundaries parallel to the [001] axis shifted at a much smaller rate, 0.02 nm^{-1} or less, along the $[1\bar{1}0]$ direction. The anisotropic transport suggests that the formate ions slide on Ti rows with their O–C–O plane parallel to the row. It might be possible to control the surface reaction in orientation-controlled conditions, through the highly regulated transport of absorbed species on anisotropic oxide surfaces [1,94].

5 Surface/gas-phase reaction

A surface/gas-phase reaction on TiO$_2$(110) was first seen *in situ* by STM at high temperature [95]. We cannot assume a priori that the catalyst surface has the same composition and structure as the bulk. The composition of the surface phases of oxide catalysts is frequently affected by the presence of gaseous ambient at high reaction

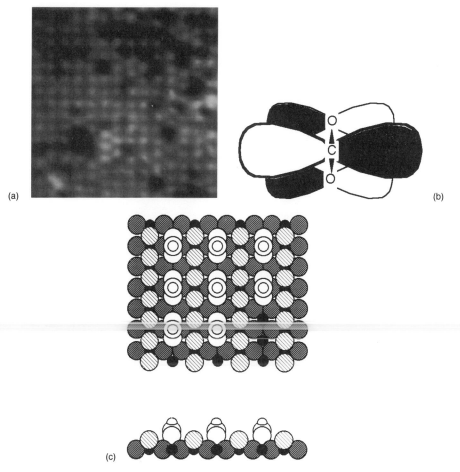

Figure 3.4 STM image of the formate-covered $TiO_2(110)$ surface: (a) a constant-current topography of a 9×9 nm square; (b) an illustration of the lowest unoccupied molecular orbital $(2b_1)$ of a free formate ion; (c) top and side views of the bridging formate ions proposed for the surface in (a). Open, hatched and filled symbols represent formate, oxygen and titanium ions, respectively.

temperatures. Real-time STM imaging dramatically demonstrates the metamorphosis of a $TiO_2(110)$ surface by the reaction with an oxygen ambient.

Figure 3.6 presents some of the sequential images of a nearly square region (96×93 nm) of the $TiO_2(110)$ surface maintained at 800 K. In UHV (Fig. 3.6(a)), added-row structures comprising Ti_2O_3 (1.2 nm wide and 0.2 nm high) and small dots (1.2 nm in diameter and 0.3 nm high) are observed. When an O_2 atmosphere (1×10^{-5} Pa) was introduced into the STM chamber, many hills were randomly nucleated over the terraces (Fig. 3.6(b,c)). Then they were transformed into new terraces and added-row structures (Fig. 3.6(d)). The added rows fluctuated along their [001]-axis, while the terrace boundaries progressed and regressed scan by scan. A set of small area scans (Fig. 3.6(e,f)) depicts the fluctuation of terraces and added rows.

A reoxidation scheme is proposed to interpret the dynamic behaviour of the surface: partially reduced Ti^{n+} ions ($n \leqslant 3$), which had been accumulated at interstitial positions in the vacuum-annealed crystal, were oxidised at the surface to form the hills, added rows of Ti_2O_3, and new terraces [95]. This scheme claims that the low-valency Ti

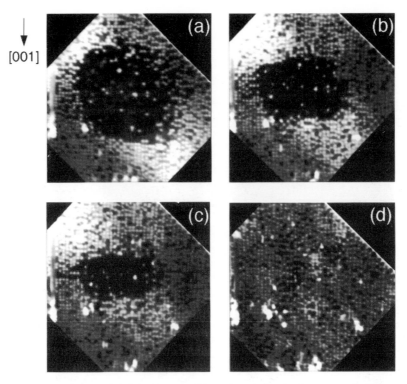

Figure 3.5 Sequential STM images (29 × 28 nm) for the formate migration on $TiO_2(110)$. Images (a)–(d) were recorded at 10, 26, 35 and 63 min after the rastering of a 14 × 14 nm square at the centre.

ions are oxidised by the O_2 ambient as follows:

interstitial ions in bulk $TiO_2 \rightarrow$ hill-like $TiO_x \rightarrow$ added Ti_2O_3 row $\rightarrow TiO_2(110)(1 \times 1)$ terrace.

The reoxidation reaction continues until the Ti^{n+} ions from the bulk accessible to the surface have been exhausted. The random nucleation of hills suggests that the interstitial Ti^{n+} ions to be oxidised migrated vertically from the bulk to the surface, by hopping from one O-octahedron to another. The vertical transport of Ti ions contrasts with the surface migration of neutral Cu atoms on the Cu surface, where mobile metal atoms migrate across the terraces even at room temperature [96].

Ti^{3+} ions of electron donor character are formed by high-temperature activation of TiO_2 powder [97,98]. The *cis/trans* ratio in the isomerisation reaction of 1-butene shows a maximum around 750 K, which coincides with the rise in the electron spin resonance (ESR) signal intensity of Ti^{3+} species. It is thus suggested that the 1-butene isomerisation occurs through a π-allyl carbanion on the Ti^{3+} sites. The dynamic features observed on $TiO_2(110)$ may be conceived to visualise the surface activation process of oxides by a reactive ambient at elevated temperatures.

6 Dynamic visualisation of reacting intermediate

Reaction intermediates (adsorbed molecules and atoms) are often so mobile on metal surfaces that they are invisible by STM at reaction temperatures. Overcoming this

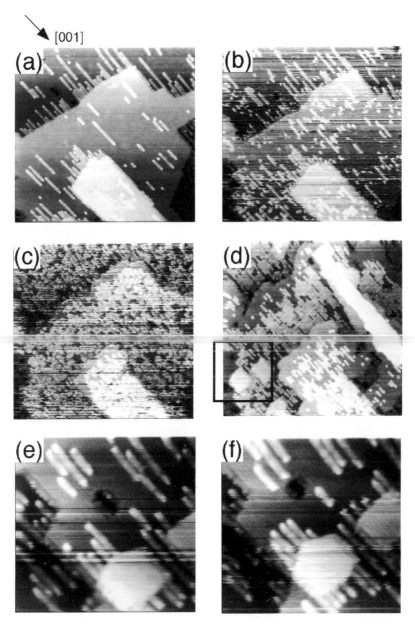

[001]

Figure 3.6 *In-situ* STM images sequentially recorded on a TiO$_2$(110) surface maintained at 800 K. (a)–(d) Large-area (96 × 93 nm) scans performed at t = 0, 550, 630, and 740 s, respectively. An O$_2$ atmosphere of 1 × 10^{-5} Pa was dosed at t = 260 s. (e), (f): Small-area (31 × 30 nm) scans observed at t = 2220 and 2370 s, respectively. A solid square over (d) depicts the area imaged in (e) and (f).

difficulty of high mobility on metals, Land *et al.* were the first to observe the conversion of ethylene to ethylidyne on Pt(111) [99]. One-dimensional reactivity of metal–oxygen chains with H$_2$S, NH$_3$, CH$_3$OH, and CO was reported on Ni(110) [100,101], Cu(110) [102–105], and Rh(110) [106]. These papers have dealt with ordered reactant domains (two-dimensional ethylene domain and one-dimensional metal–oxygen chains) consumed in the reactions. It was, however, difficult to identify which intermediate reacted at which site in the receding domains. On the other hand, ionic adsorbates on oxide surfaces are visible by STM at room temperature due to restricted mobility, as demonstrated in the previous sections. Thus, we believe that there is an opportunity for visualising site-specific reactions, including the position and orientation of intermediates on metal oxides, by means of *in-situ* STM imaging under reactive atmospheres.

The ability of STM to monitor the reaction of intermediates liberated at oxide surfaces has been demonstrated very recently. The reaction of an acetate-covered $TiO_2(110)$ surface was successfully monitored by STM operated in a temperature-jump mode [107]. Acetate ions were individually resolved at high temperatures, where they started decomposing. Reaction of the acetate-covered surface was monitored by a temperature-jump method: the sample wafer was heated step-wise, while STM topography was continuously determined. The topography of the same area of the surface could be sequentially recorded following a temperature jump by careful tracking against thermal drift. In this operation, surface reaction on the acetate-covered surface activated by the temperature increments was monitored in a quasi-isothermal condition.

The reaction of acetate ions was thereby monitored at 540 K. Figure 3.7 shows four frames (frames 1, 3, 8 and 14) selected from a sequence of 15 images recorded on the same area. Acetate ions were observed as small bright spots at different positions frame by frame due to surface migration. String-like Ti_2O_3 rows are spectators resident on the surface. The acetates have been suggested to be consumed in the unimolecular decomposition reaction to ketene [108]:

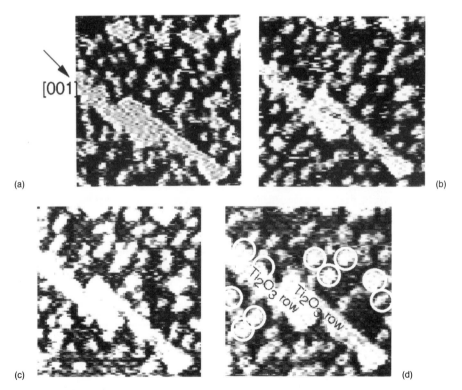

(a)   (b)

(c)   (d)

Figure 3.7 Sequential constant-current STM images (9 × 9 nm) of the acetate-covered $TiO_2(110)$ surface heated at 540 K. (a)–(d) are frames 1, 3, 8, and 14 selected from a sequence of 15 images. Frames 1–15 were sequentially recorded with a scan rate of 16.6 s per frame. The immobile particles are marked by solid circles in (d).

$$CH_3COO^-(a) \rightarrow CH_2{=}C{=}O(g) + OH^-(a) \tag{1}$$

The number of acetates decreased with the reaction time at 540 K in line with the first-order rate law. The reaction rate deduced by acetate counting in the images agreed with macroscopic kinetics observed in TPD spectra [108]. Particles larger than the acetate appeared on the surface instead. Ten particles were distinguished and marked in Figure 3.7(d). The particles, probably carbonaceous residuals yielded in a side reaction, did not migrate on the surface until they disappeared at higher temperatures. The migrating acetates and immobile particles were easily distinguished by fast viewing of the sequential frames. This suggests the ability of STM to distinguish surface species by dynamically observing migration without examining tunnelling spectroscopic data.

7   Characterisation of the anisotropic structure of active sites

X-ray absorption fine-structure spectroscopy is a most powerful technique in the characterisation of structures and electronic states of any sort of catalytic material, though its analysis and interpretation should not be overestimated [15,16]. XAFS can provide quantitative information on bond length and coordination number around atoms of a specific element, not only in a static state but also in a dynamic state (*in-situ* conditions) that is not given by other techniques used in the characterisation of catalysts and related surfaces [14–17]. Metal-oxide phases in a highly dispersed form or thin layers on inorganic oxide supports are not naturally symmetric/isotropic. Dispersion, growth mode and morphology of the catalytically active sites in supported metal-oxide catalysts are affected by many factors: such as the chemical state and surface energy of the oxide phase, the texture and morphology of the support, and the interaction between metal oxide and support. The Mo-dimer–SiO$_2$ catalyst is a typical sample which shows an anisotropic change of 0.04 nm in Mo–Mo separation and 0.01 nm in Mo–surface separation during an ethanol oxidation cycle [109]. In the case of powder samples such as typical heterogeneous catalysts, however, the structures determined by XAFS are averaged structures in every direction of the sample.

When single crystals or flat surfaces are employed as supports, characterisation of reaction sites on them can be achieved separately in three different bond directions parallel (two in-plane directions) and normal (one direction) to the surface by polarised X-ray stemming from synchrotron radiation. For K-, and L$_{II}$-, L$_{III}$-edge extended X-ray absorption fine-structure (EXAFS) spectra of oriented samples, effective coordination numbers N^* are given by $N^* = 3\Sigma\cos^2\theta_j$ and $N^* = 0.7N + 0.9\Sigma\cos^2\theta_j$, respectively, where θ_j is the angle between the electric field vector of the incident X-ray and the bond direction (see [110]). Thus, when the electric field polarisation of the incident X-ray is normal to the surface (p-polarisation), the X-ray absorber's neighbours which lie along lines parallel to the surface do not contribute to EXAFS signals. These bonds are preferentially observable when s-polarised X-rays parallel to the surface are used. In reverse, the bonds normal to the surface are favourably detectable by p-polarised EXAFS.

For single-crystal surfaces, we can obtain XAFS spectra from the surface phase in total reflection conditions because the penetration depth of X-rays is small, less than

3 nm. Furthermore, the XAFS signals with a high signal-to-background ratio can be read by a fluorescence XAFS technique. By changing the orientation of the sample to polarised synchrotron radiation and using the total reflection fluorescence method we can measure polarisation-dependent total reflection fluorescence XAFS (PTRF–XAFS), to determine the anisotropic or asymmetric structures on surfaces [17,94]; e.g. Cu atoms on α-quartz(0001) [111], CoO_x on α-Al_2O_3(0001) [112], Pt clusters on α-Al_2O_3(0001) [113], VO_x on ZrO_2(100) [114], and MoO_x on TiO_2(110) [115]. A schematic diagram of the PTRF–XAFS for *in-situ* characterisations is shown in Fig. 3.8 [116].

Small $[Co_3O_4]_n$ particles on α-Al_2O_3, which were prepared by chemical vapour deposition (CVD) of $Co_2(CO)_8$ on the α-Al_2O_3(0001), followed by calcination at 873 K, showed extremely high activity for CO oxidation at 273 K as compared to a more usual impregnation Co_3O_4 catalyst [112]. The PTRF–EXAFS analysis reveals the asymmetric growth of seven Co-oxide layers in the $[Co_3O_4]_n$ spinel structure with the (001) plane parallel to the α-Al_2O_3(001) surface, which is relevant to the high catalytic performance.

Surface V=O sites are known to play a key role in defining the catalytic performance for many reactions, such as methane oxidation on V_2O_5–SiO_2 [117,118], ammoxidation of toluene to benzonitrile on V_2O_5–Al_2O_3 [119], NO–NH_3–O_2 reaction on V_2O_5–TiO_2 [120], and partial oxidation of alcohols [121]. The oriented V=O bonds of vanadium oxides supported on ZrO_2(100) were studied by *in-situ* PTRF X-ray absorption near-edge structure (XANES). The pre-edge peak in the XANES spectra reflects the local structure around the V=O bond. Its intensity appears to be strong when vanadium atoms are located in a tetrahedral symmetry. The pre-edge peak is assigned to the transition from a V(1s) binding orbital to a V=O(3d + 2p) composite antibonding orbital, which is spread to the bond direction. As a consequence, the pre-edge peak intensity of the oriented vanadium oxide shows a maximum when the electric field vector of the incident X-ray is parallel to the V=O bond direction and zero when the vector is perpendicular to the V=O bond direction [122]. The PTRF–XANES results on the VO_x–ZrO_2(100) model catalyst showed that the V=O bonds were inclined to the ZrO_2(100) surface by about 45° [114].

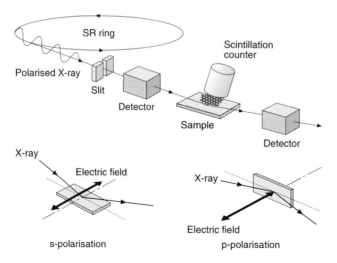

Figure 3.8 Schematic diagram of polarisation-dependent total-reflection fluorescence XAFS.

It has been demonstrated that both Brønsted acid sites and V=O groups play an important role in the selective catalytic reduction of NO by ammonia over V_2O_5–TiO_2 catalysts. NH_3 adsorption leads to a downward frequency shift of the V=O bond [123,124], reflecting a weakening of the V=O bond, probably by the transfer of a hydrogen atom from the adsorbed NH_3 on Brønsted acid sites to the surface vanadyl groups [125]. However, despite extensive studies, the reaction mechanism and the nature of the active sites are still much debated. One of the most important problems to be solved is where and when ammonia dissociates on the vanadium oxides. Upon NH_3 adsorption at 423 K, the pre-edge peak intensities of both s- and p-polarisation spectra decreased in a similar way and became zero. The decrease of the pre-edge peak intensity is due to the structural change from V=O to V−OH. The similar decrease in the peak intensity of both polarisations reflects a nearly uniform orientation of the V=O bonds on $ZrO_2(100)$. All of the tilted V=O bonds on $ZrO_2(100)$ were active for the reaction with NH_3 at 423 K [114]. The PTRF–XANES results identify the V=O bonds with 45° orientation to be capable of reacting with ammonia.

The PTRF–XAFS technique is advantageous for the study of the asymmetric and anisotropic structure and orientation of metal and metal-oxide sites supported on single crystal surfaces as models for supported catalysts *in situ* under the reaction conditions. However, this technique takes a relatively long time to obtain good signal-to-noise ratio spectra. The problem will be solved with the advent of a new-generation high-intensity synchrotron source and/or an increase in the detection efficiency. Another disadvantage may be the need for large-area flat samples to conduct measurements under total reflection conditions. This sort of problem is now less important because many kinds of large metal-oxide single crystals as substrates such as SiO_2, Al_2O_3, TiO_2, ZrO_2, are commercially available, and metal-oxide thin films are capable of being grown epitaxially on, for example, Si wafers.

8   Future prospects

Metal-oxide surface science is a rapidly expanding frontier area, while the development of techniques and procedures for producing ordered, well-defined oxide surfaces with controlled structure, composition, oxidation state, distribution and morphology remains an important challenge to researchers in this field. While unifying concepts on well-defined surfaces may on occasion produce oversimplifications or omissions, they clearly provide a more useful framework for better understanding of catalysis and the design of catalysts. The trend that is particularly relevant to future progress in catalysis is the observation of catalytic performance on the surface in real space and in real time under reaction conditions, as outlined in the previous sections.

With the further development of *in-situ* STM methods, future experiments should enable morphological changes to catalyst surfaces to be followed in real time as the reaction proceeds, in order to establish an accurate relationship between surface structure and catalytic performance. A breakthrough is needed for real-time imaging for faster reactions.

X-ray photoemission electron microscopy (XPEEM) can provide information on the composition and chemical state of surface species, and their spatial and temporal resolution during cross-diffusion of multi-elements at oxide surfaces, growth of mixed-oxide layers, selective adsorption and catalytic reactions. This new technique is applicable to complex and inhomogenous mixed-oxide catalysts of vital practical importance, which have been considered to be ill suited to fundamental studies because their surfaces are ill defined and poorly characterised [1].

The most popular method for probing solid surfaces, which has gained wide recognition for its ability to determine the bond-specific nature of surfaces and adsorbates, is vibrational spectroscopy. Reflection–absorption infrared spectroscopy [27–31] is particularly suitable for those experiments that require high resolution, strict polarisation rules, or high pressures. For semiconductor and insulator substrates which are infrared-transparent, however, this method has not yet been well developed, mainly because the sensitivity was much lower than that for the metal surfaces where the infrared electromagnetic field is enhanced due to free electrons. Recently, to overcome the disadvantage, buried metal-layer substrates have been proposed [126–130]. The metal layer is made sufficiently thick to appear semi-infinite in the infrared, and the thickness of the semiconductor or insulator overlayer is less than the infrared wavelength in the region of interest. In this case, the substrate is expected to show a metallic-like nature towards the infrared beam. Thus, buried metal enhanced RAIRS is also applicable to insulator substrates, which are hardly used in ordinary RAIRS measurements [130].

Despite the success in modelling catalysts with single crystals and well-defined surfaces, there is a clear need to develop models with higher levels of complexity, to address the catalytically important issues specifically related to mixed-oxide surfaces. The characterisation and design of oxide surfaces have not proven to be easy tasks, but recent progress in identification of the key issues in catalytic phenomena on single-crystal oxide surfaces by *in-situ* characterisation techniques on an atomic and molecular scale will bring us to a vintage century of interfacial chemistry of metal-oxides.

9   References

1 Iwasawa Y. In Hightower JE, Delgass WN, Iglesia E, Bell AT (eds) *11th International Congress on Catalysis — 40th Anniversary.* Amsterdam: Elsevier, 1996; 21.
2 Henrich VE, Cox PA. *The Surface Science of Metal Oxides.* Cambridge: Cambridge University Press, 1994.
3 Kung HH. *Transition Metal Oxides: Surface Chemistry and Catalysis.* Amsterdam: Elsevier, 1990.
4 Nowotny J, Dufour L-C (eds). *Surface and Near-Surface Chemistry of Oxide Materials.* Amsterdam: Elsevier, 1988.
5 Umbach E, Freund H-J (eds). *Adsorption on Ordered Surfaces of Ionic Solids and Thin Films.* Berlin: Springer-Verlag, 1993.
6 Goodman DW. *Chem. Rev.* 1995; **95**: 523.
7 Goodman DW. *Surf. Rev. Lett.* 1995; **2**: 9.
8 Lad RJ. *Surf. Rev. Lett.* 1995; **2**: 109.
9 Diebold U, Pan J-M, Madey TE. *Surf. Sci.* 1995; **331–333**: 845.
10 Linsebigler AL, Lu G, Yates JT Jr. *Chem. Rev.* 1995; **95**: 735.
11 Campbell CT. *J. Chem. Soc. Faraday Trans.* 1996; **92**: 1435.
12 Barteau MA. *Chem. Rev.* 1996; **96**: 1413.
13 Iwasawa Y. In Joyner RW, van Santen RA (eds) *Elementary Reaction Steps in Heterogeneous Catalysis.* NATO ASI Ser. C 1993; **398**: 287.
14 Teo BK. *EXAFS: Basic Principles and Data Analysis.* Berlin: Springer-Verlag, 1986.

15 Koningsberger DC, Prins R. *X-Ray Absorption: Principles, Applications, Techniques of EXAFS, SEXAFS and XANES.* New York: Wiley, 1988.

16 Stöhr J. *NEXAFS Spectroscopy.* Berlin: Springer-Verlag, 1992.

17 Iwasawa Y (ed.) *X-ray Absorption Fine Structure for Catalysts and Surfaces.* Singapore: World Scientific, 1996.

18 Gai-Boyes PL. *Catal. Rev.* 1992; **34**: 1.

19 Gai-Boyes PL, Kourtakis K. *Science* 1995; **267**: 661.

20 Brongersma HH, Groenen PAC, Jacobs J-P. In Nowtony J (ed.) *Science of Ceramic Interfaces II.* Amsterdam: Elsevier, 1994; 113.

21 Borg HJ, Niemantsverdriet JW. In Spivery JJ, Agarwarl SK (eds) *Specialist Periodical Report.* Cambridge: Royal Society of Chemistry, 1994; 1.

22 Henderson MA. *J. Phys. Chem.* 1995; **99**: 15253.

23 DeCanio EC, Edwards JC, Bruno JW. *J. Catal.* 1994; **148**: 76.

24 Jacobsen CJH, Topsøe N-Y, Topsøe H, Kellberg L, Jakobsen HJ. *J. Catal.* 1995; **154**: 65.

25 Mastikhim VM, Terskikh VV, Lapina OB, Filimonova SV, Seidl M, Knözinger H. *J. Catal.* 1995; **156**: 1.

26 Boehm H-P, Knözinger H. In Anderson JR, Boudart M (eds) *Catalysis: Science and Technology.* Berlin: Springer-Verlag, 1983; 39.

27 Yates JT Jr, Madey TE (eds). *Vibrational Spectroscopy of Molecules on Surfaces.* New York: Plenum, 1987.

28 Hoffman FM. *Surf. Sci. Rep.* 1983; **3**: 107.

29 Hollins P, Pritchard J. *Progr. Surf. Sci.* 1985; **19**: 275.

30 Bradshaw AM, Schweizer E. In Hester RE (ed.) *Advances in Spectroscopy: Spectroscopy of Surfaces.* New York: Wiley, 1988.

31 Chabal YJ. *Surf. Sci. Rep.* 1988; **8**: 211.

32 Lunsford JH, Yang X, Haller K, Laane J, Mestl G, Knözinger H. *J. Phys. Chem.* 1993; **97**: 13810.

33 Kapteijn FK, van Langeveld AD, Moulijn JA *et al. J. Catal.* 1994; **150**: 94.

34 Güntherodt H-J, Wiesendanger R (eds). *Scanning Tunneling Microscopy I.* Berlin: Springer-Verlag, 1994.

35 Wiesendanger R. *Scanning Probe Microscopy and Spectroscopy.* Cambridge: Cambridge University Press, 1994.

36 Binnig G, Rohrer H, Gerber C, Weibel E. *Physica (Utrecht)* 1981; **107B + C**: 1335.

37 McIntyre BJ, Salmeron MB, Somorjai GA. *Catal. Lett.* 1992; **14**: 263.

38 McIntyre BJ, Salmeron MB, Somorjai GA. *Rev. Sci. Instrum.* 1993; **64**: 687.

39 Wilson J, Groot C de. *J. Phys. Chem.* 1995; **99**: 7860.

40 Rohrer GS, Henrich VE, Bonnell DA. *Science* 1990; **250**: 1239.

41 Novak D, Garfunkel E, Gustafsson T. *Phys. Rev. B* 1994; **50**: 5000.

42 Sander M, Engel T. *Surf. Sci. Lett.* 1994; **302**: L263.

43 Onishi H, Iwasawa Y. *Surf. Sci.* 1994; **313**: L783.

44 Murray PW, Condon NG, Thornton G. *Phys. Rev. B* 1995; **51**: 10989.

45 Fischer S, Munz AW, Schierbaum K-D, Göpel W. *Surf. Sci.* 1995; **337**: 17.

46 Murray PW, Leibsle FM, Muryn CA, Fischer HJ, Flipse CFJ, Thornton G. *Surf. Sci.* 1994; **321**: 217.

47 Oshio T, Sakai Y, Moriya T, Ehara S. *Scanning Microscopy* 1993; 7: 33.

48 Smith RL, Lu W, Rohrer GS. *Surf. Sci.* 1995; **322**: 293.

49 Garfunkel E, Rudd G, Novak D *et al. Science* 1989; **246**: 99.

50 Lu W, Nevins N, Norton ML, Rohrer GS. *Surf. Sci.* 1993; **291**: 395.

51 Jones FH, Rawlings K, Foord JS *et al. Surf. Sci.* 1996; **359**: 107.

52 Tarrach G, Bürgler D, Schaub T, Wiesendanger R, Güntherodt H-J. *Surf. Sci.* 1993; **285**: 1.

53 Rohrer GS, Lu W, Smith RL, Hutchinson A. *Surf. Sci.* 1993; **292**: 261.

54 Matsumoto T, Tanaka H, Kawai T, Kawai S. *Surf. Sci. Lett.* 1992; **278**: L153.

55 Onishi H, Fukui K, Iwasawa Y. *Bull. Chem. Soc. Jpn* 1995; **68**: 2447.

56 Abraham FF, Batra IP. *Surf. Sci.* 1989; **209**: L125.

57 Giessibl FJ. *Science* 1995; **267**: 68.

58 Kitamura S, Iwatsuki M. *Jpn J. Appl. Phys.* 1995; **34**: L145.

59 Kitamura S, Iwatsuki M. *Jpn J. Appl. Phys.* 1996; **35**: L668.

60 Ueyama H, Ohta M, Sugawara Y, Morita S. *Jpn J. Appl. Phys.* 1995; **34**: L1086.

61 Tanabe K. In Imelik B, Naccache C, Couduier G, Taarit YB, Vedrine JC (eds) *Solid Acids and Bases.* Amsterdam: Elsevier, 1970; 1.

62 Benei HA, Winquist BHC. *Adv. Catal.* 1978; **27**: 97.

63 Burwell RL Jr, Haller GL, Taylor KC, Read JF. *Adv. Catal.* 1969; **29**: 1.

64 Kokes RJ. *Intra-Sci. Chem. Rep.* 1972; **6**: 77.

65 Akhter S, Cheng WH, Lui K, Kung HH. *J. Catal.* 1983; **85**: 341.

66 Vohs JM, Barteau MA. *Surf. Sci.* 1986; **176**: 91.

67 Vohs JM, Barteau MA. *Surf. Sci.* 1988; **201**: 481.

68 Pines H, Veseley JA, Ipatieff VN. *J. Am. Chem. Soc.* 1955; **77**: 6314.

69 Hattori H. *Chem. Rev.* 1995; **95**: 537.

70 Tanabe K. In Imelik B, Naccache C, Couduier G, Taarit YB, Vedrine JC (eds). *Catalysis by Acids and Bases.* Amsterdam: Elsevier, 1984; 1.

71 Suzukamo G, Fukao M, Hibi T, Chikaishi K. In Tanabe K, Hattori H (eds). *Acid–Base Catalysis.* Tokyo: Kodansha-VCH, 1989; 405.

72 Onishi H, Iwasawa Y. *Catal. Lett.* 1996; **38**: 89.

73 Onishi H, Aruga T, Egawa C, Iwasawa Y. *J. Chem. Soc. Faraday Trans. I* 1989; **85**: 2597.

74 Barteau MA. *J. Vac. Sci. Technol. A* 1993; **11**: 2162.

75 Vohs JM, Barteau MA. *Surf. Sci.* 1988; **197**: 109.

76 Idriss H, Pierce KG, Barteau MA. *J. Am. Chem. Soc.* 1994; **116**: 3063.

77 Idriss H, Barteau MA. *Langmuir* 1994; **10**: 3693.

78 Pierce KG, Barteau MA. *J. Org. Chem.* 1995; **60**: 2405.

79 Lu G, Linsebigler A, Yates JT Jr. *J. Phys. Chem.* 1994; **98**: 11733.

80 Pierce KG, Barteau MA. *Surf. Sci.* 1995; **323**: L473.

81 Cox F, Schulz KH. *J. Vac. Sci. Technol. A* 1990; **8**: 2599.

82 Dilara PA, Vohs JM. *Surf. Sci.* 1994; **321**: 8.

83 Somorjai GA. *Introduction to Surface Chemistry and Catalysis.* New York: Wiley, 1994.

84 Rodriguez JA, Goodman DW. *Surf. Sci. Rep.* 1991; **14**: 1.

85 Vest MA, Lui KC, Kung HH. *J. Catal.* 1989; **120**: 231.

86 Berlowitz P, Kung HH. *J. Am. Chem. Soc.* 1986; **108**: 3532.

87 Pierce KG, Lusvardi VS, Barteau MA. In Hightower JE, Delgass WN, Iglesia E, Bell AT (eds) *11th International Congress on Catalysis — 40th Anniversary.* Amsterdam: Elsevier, 1996; 297.

88 Onishi H, Aruga T, Iwasawa Y. *J. Am. Chem. Soc.* 1993; **115**: 10460.

89 Onishi H, Aruga T, Iwasawa Y. *J. Catal.* 1994; **146**: 557.

90 Trillo JM, Munuera G, Criado JM. *Catal. Rev.* 1972; **7**: 51.

91 Onishi H, Iwasawa Y. *Chem. Phys. Lett.* 1994; **226**: 111.

92 Onishi H, Iwasawa Y. *Langmuir* 1994; **10**: 4414.

93 Onishi H, Iwasawa Y. *Jpn J. Appl. Phys.* 1994; **33**: L1338.

94 Iwasawa Y. *Catalysis Surveys Japan* (in press).

95 Onishi H, Iwasawa Y. *Phys. Rev. Lett.* 1996; **76**: 791.

96 Wintterlin J, Schuster R, Coulman DJ, Ertl G, Behm RJ. *J. Vac. Sci. Technol. B* 1991; **9**: 902.

97 Che M, Naccache C, Imelik B. *J. Catal.* 1972; **24**: 328.

98 Hattori H, Itoh M, Tanabe K. *J. Catal.* 1975; **38**: 172.

99 Land TA, Michely T, Behm RJ, Heminger JC, Comsa G. *J. Chem. Phys.* 1992; **97**: 6774.

100 Ruan L, Besenbacher F, Stensgaard I, Lægsgaard E. *Phys. Rev. Lett.* 1992; **69**: 3523.

101 Ruan L, Stensgaard I, Lægsgaard E, Besenbacher F. *Surf. Sci.* 1994; **312**: 31.

102 Leibsle FM, Francis SM, Davis R, Xiand N, Haq S, Bowker M. *Phys. Rev. Lett.* 1994; **72**: 2569.

103 Leibsle FM, Francis SM, Haq S, Bowker M. *Surf. Sci.* 1994; **318**: 46.
104 Crew WW, Madix RJ. *Surf. Sci.* 1994; **319**: L34.
105 Bowker M, Leibsle F. *Catal. Lett.* 1996; **38**: 123.
106 Leibsle FM, Murray PW, Francis SM, Thornton G, Bowker M. *Nature* 1993; **363**: 706.
107 Onishi H, Yamaguchi Y, Fukui K, Iwasawa Y. *J. Phys. Chem.* 1996; **100**: 9582.
108 Kim KS, Barteau MA. *J. Catal.* 1990; **125**: 353.
109 Iwasawa Y. *Adv. Catal.* 1987; **35**: 187.
110 Stern EA. *Phys. Rev. B* 1974; **10**: 3027.
111 Shirai M, Asakura K, Iwasawa Y. *Chem. Lett.* 1992; 1037.
112 Shirai M, Inoue T, Onishi H, Asakura K, Iwasawa Y. *J. Catal.* 1994; **145**: 159.
113 Asakura K, Shirai M, Iwasawa Y. *Catal. Lett.* 1993; **20**: 117.
114 Shirai M, Asakura K, Iwasawa Y. *Catal. Lett.* 1994; **26**: 229.
115 Chun W-J, Asakura K, Iwasawa Y. *Appl. Surf. Sci.* 1996; **100/101**: 143.
116 Shirai M, Nomura M, Asakura K, Iwasawa Y. *Rev. Sci Instrum.* 1995; **66**: 5493.
117 Irusta S, Lombardo E, Miro E. *Catal. Lett.* 1994; **29**: 339.
118 Irusta S, Cornaglia LM, Miro EE, Lombardo EA. *J. Catal.* 1995; **156**: 167.
119 Niwa M, Ando H, Murakami Y. *J. Catal.* 1977; **49**: 92.
120 Miyamoto A, Kobayashi K, Inomata M, Murakami Y. *J. Phys. Chem.* 1982; **86**: 2945.
121 Deo G, Wachs IE. *J. Catal.* 1994; **146**: 335.
122 Stizza S, Mancini G, Benfatto M, Natori CR, Garcia J, Bianconi A. *Phys Rev B* 1989; **40**: 12229.
123 Went GT, Oyama ST, Bell AT. *J. Phys. Chem.* 1990; **94**: 4240.
124 Busca G. *Langmuir* 1986; **2**: 577.
125 Topsøe N-Y. *Science* 1994; **265**: 1217.
126 Finke SJ, Schrader GL. *Spectrochim. Acta A* 1990; **46**: 91.
127 Ehrley W, Butz R, Mantl S. *Surf. Sci.* 1991; **248**: 193.
128 Bermudez VM. *J. Vac. Sci. Technol. A* 1992; **10**: 152.
129 Sato S, Minoura S, Urisu T, Takasu Y. *Appl. Surf. Sci.* 1995; **90**: 29.
130 Fukui K, Miyauchi H, Iwasawa Y. Submitted.

4 Oxygen States at Metal Surfaces

A.F. CARLEY and P.R. DAVIES

Department of Chemistry, University of Cardiff, PO Box 912, Cardiff CF1 3TB, UK

1 Introduction

The scheme below summarises the events occurring during oxygen adsorption and dissociation at metal surfaces, a subject of central importance in selective oxidation catalysis:

$$O\ (g) \rightleftharpoons O_2\ (s) \rightleftharpoons O_2^{\delta-}(s) \quad \Rightarrow \quad O^{\delta-}(s) \quad \Rightarrow \quad O^{2-}(a)$$
$$\Downarrow \qquad\qquad \Downarrow$$
$$O_2^{\delta-}(a) \qquad O^{\delta-}(a)$$

(s) here denotes a short-lived transient and (a) a stable or metastable species with a sufficiently long lifetime for it to be characterised spectroscopically. For a discussion of short-lived transient oxygen species, both molecular and atomic, and their role in surface reactions, the reader is directed elsewhere [1,2]. The present paper will focus instead on the long-lived species and the combination of spectroscopic and reactivity measurements that is beginning to unravel the complexities of oxygen chemisorption at metal surfaces.

Since the chemisorption of oxygen on metals is probably the most thoroughly studied subject in surface science, we are obliged to limit our detailed discussion to a few selected metal–oxygen systems which we feel best illustrate aspects of this important topic. We discuss the structures of the chemisorbed oxygen adlayers at metal surfaces only briefly. Studies in this area involve idealised adlayers at low-index faces and give little direct information about the chemistry of the absorbed oxygen; they do, however, provide a framework from which the behaviour of chemisorbed oxygen under different conditions can be understood. Furthermore, the advent of scanning tunnelling microscopy (STM) has provided a unique insight into the development of chemisorbed adlayers [3–5], a subject of great relevance to the understanding of competitive reactions involving oxygen at metal surfaces. More detailed discussions of the structures of oxygen adlayers at metal surfaces can be found elsewhere [3,4,6–8].

Although designated as $O^{2-}(a)$, the final oxygen state in the above scheme (which corresponds to the oxygen species that dominates at most metal surfaces as the oxygen coverage approaches one monolayer) is unlikely to attain the full $2-$ charge implied. The $O^{2-}(a)$ label for this state has very wide usage, however, and it does indicate the higher charge expected for this species relative to its precursors. We shall therefore use this notation throughout our discussion. We discuss the issue of charge and reactivity in more detail in Section 6.

2 Techniques for characterisation of the nature of oxygen states

Early work-function measurements indicated the complex nature of oxygen adlayers at metal surfaces [9,10], providing information about the surface dipole and evidence for the instability of chemisorbed oxygen leading to incorporation and surface reconstruction. However, further progress awaited the development and application of X-ray photoelectron spectroscopy (XPS) as a quantitative surface-sensitive technique during the 1970s [2,11,12].

Photoelectron spectroscopy is probably the most powerful single technique for studying surface species. XPS, through measurements of O(1s) binding energies, can probe the charge on the oxygen species, provided due cognisance is given to the possible influence of final-state relaxation effects. Furthermore, through appropriate data manipulation procedures which are now well established [13], metal core-level shifts can be used to study the defect nature of oxygen overlayers. For example, at nickel surfaces [14] Ni^{3+} and associated O^- species have been observed, and Ti^{4+}, Ti^{3+} and Ti^{2+} states have been identified during the growth of thin oxide layers on titanium surfaces [15]. A method for deriving the surface concentrations of adsorbed species from measured spectra was developed [12,16] which significantly enhances the usefulness of XPS in surface chemical studies. Ultraviolet photoelectron spectroscopy (UPS) provides complementary information on the molecular orbital structure of adsorbates, and can be used to distinguish between dissociated and molecular states. Combining O(2p) valence-level and O(1s) core-level binding energies has been claimed to be a more reliable identifier of oxygen species than core-level binding energies alone [17].

Vibrational spectroscopy — in the form of high-resolution electron energy-loss spectroscopy (HREELS) — can probe the bonding of oxygen adatoms to metal surfaces. It is possible to distinguish between different oxygen states, and to identify adsorption sites through comparisons with calculated spectra. HREELS and, less commonly, reflection–absorption infrared spectroscopy (RAIRS), have been used to distinguish between dissociated and molecular oxygen species through the presence (or not) of a ν(O–O) loss feature. Moreover, the frequency of this loss depends on the strength of the O–O bond, which in turn depends on the charge on the molecular oxygen species. A correlation between the frequency of the ν(O–O) loss feature and molecular charge and bond order has been reported [18], but theoretical calculations for oxygen adsorption on Ag(110) suggest that this approach may be too simplistic [19].

Temperature-programmed desorption spectroscopy (TPD) has been applied to the investigation of oxygen states at metal surfaces, but there are inherent difficulties in distinguishing between the desorption of molecular oxygen from a molecularly adsorbed state and recombination of adsorbed atomic oxygen followed by molecular desorption.

Structural information about the adlayer can be obtained from low-energy electron diffraction (LEED), and this can be very useful in following, for example, surface reconstruction induced by oxygen chemisorption. However, LEED provides no chemical discrimination between different oxygen species, and is best used in conjunction with other surface spectroscopies. Furthermore, it has been shown in optical simulation studies that up to 40% or more of an ordered array of scatterers may be randomly removed and an acceptable LEED pattern still be obtained [20]. This casts doubt on a commonly used approach for estimating surface concentrations, where observed diffraction patterns are interpreted assuming full occupancy of sites in real space.

STM is now the most powerful technique for studying structural changes at metal surfaces and for obtaining information on active sites in, for example, chemisorptive replacement reactions. The inability of STM to distinguish between chemically distinct oxygen species is a drawback, but attempts to apply scanning tunnelling spectroscopy to adsorbate studies promise to lead to improvements in this area.

One of the most fruitful approaches for characterising oxygen states at metal surfaces is to exploit their different chemical reactivities, by means of probe molecules. Examples of such systems will be presented throughout this paper, but one case study will be mentioned now for illustrative purposes. The oxidation of a caesium film at 80 K leads to the formation of three adspecies, $O_2^{\delta-}(a)$, $O^{\delta-}(a)$ and $O^{2-}(a)$, the distribution of which depends on exposure to dioxygen (Fig. 4.1) [21,22]. The oxygen species were distinguished not only by their O(1s) binding energies, but also by their reactivities at 80 K to CO and CO_2: the $O^{\delta-}(a)$ species reacted with CO to give $CO_2^{\delta-}(a)$ and with CO_2 to form $CO_3(a)$, whereas the peroxo-like species reacted with CO to give $CO_3(a)$. Product identification was made from observations of O(1s) and C(1s) spectral changes, and also from the calculated C : O stoichiometry of the product species. This system illustrates the rich chemistry associated with oxygen species at alkali metal and alkali metal doped surfaces, which for reasons of space we are unable to include in this paper. A detailed review of structural aspects of these systems has been published recently [23].

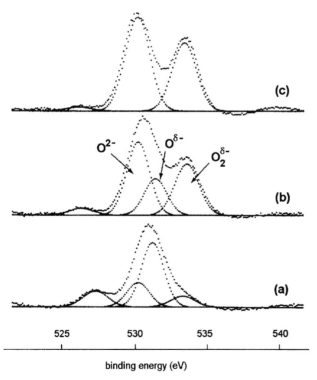

Figure 4.1. O(1s) spectra observed after the exposure of a caesium multilayer to dioxygen at 80 K; oxygen exposures are: (a) 2 L; (b) 10 L; (c) 20 L. The assignments [4] are based on the measured binding energy for each species and the reactivity of the species towards CO(g) and CO_2(g).

3   Background

In this section we briefly review the pioneering studies which challenged the accepted view of chemisorbed oxygen in the 1960s and 1970s. In Section 4 we describe some selected metal–oxygen systems in greater detail, and then finally we discuss the range of surface oxygen species revealed by recent studies.

3.1   *Early studies of the reactivity of the $O^{2-}(a)$ species*

Chemisorbed oxygen was conventionally thought to be a poison for catalytic reactions at metal surfaces, and indeed this perception encouraged the development of methods for preparing and studying clean metal surfaces [2]. The first direct evidence for a surface reaction actually activated by chemisorbed oxygen species was reported by Roberts *et al.* in 1975 [24], and was one of the first studies where XPS revealed the mechanism of a surface reaction. In a study of the interaction of evaporated Pb films with hydrogen sulfide, it was shown that although the clean Pb film was unreactive towards H_2S at 295 K, the surface pre-exposed to oxygen reacted steadily with H_2S to form a sulfide adlayer, the oxygen desorbing as H_2O:

$$O^{2-}(a) + H_2S(g) \Rightarrow S^{2-}(a) + H_2O(g)$$

Spectroscopic evidence for the formation of H_2O was obtained by studying the reaction in the temperature range 80–295 K; for $T < 180$ K the water produced via the above scheme was physically absorbed at the surface. Since the Pb–S bond is weaker than the Pb–O bond, it follows that it is the thermodynamic stability of the water which is the driving force for the reaction. Indeed, the standard free energy change, ΔG^O, for the reaction

$$PbO(s) + H_2S(g) \Rightarrow PbS(s) + H_2O(g)$$

where (s) denotes a surface species, is -99 kJ mol^{-1} at 290 K. Subsequent LEED and XPS studies on well-defined Pb(100) and Pb(110) single-crystal surfaces [25] confirmed this observation but also showed that the sulfide layer thus produced had an unusual structure. On Pb(100) the structure of the adlayer was the same as that formed on the clean surface after exposure to H_2S at 420 K, whereas on Pb(110) the structure was different. Thus, a unique, low-energy reaction pathway to a novel sulfide adlayer was shown to exist (a surface example of the solid-state *chimie douce* principle), accessible only through interaction with chemisorbed oxygen. This study also highlighted the advantages provided by the emerging surface spectroscopies, especially XPS, in obtaining quantitative data relevant to reaction mechanisms, the model deriving from an earlier volumetric investigation [26] being shown to be incorrect. The significant benefits of studying reactions at low and variable temperatures, in order to isolate individual steps of reaction mechanisms, were also demonstrated for the first time.

3.2   *Chemisorptive replacement reactions*

An XPS study of the adsorption of hydrogen halides at a Pb(110) surface [27–29] showed that 'chemisorptive replacement' reactions, such as that between $O^{2-}(a)$ and

H_2S, were likely to be a general phenomenon, both chloride and bromide adlayers being formed by exposure of a Pb(110)–O surface to HCl and HBr, respectively. The stoichiometry of the replacement was shown to be 2 : 1 (Cl(a) adsorbed : O(a) removed). Hydrogen abstraction occurs, with the surface oxygen acting as a strong Brønsted base [30], through the formation of a strong hydrogen bond between the surface oxygen and the adsorbate. This leads to a weakening of the intramolecular bond between the hydrogen and, in this case, the halide atom.

Other molecules have since been used to probe the nature of oxygen states, including water. For the Cu(111)–O surface, strong hydrogen bonding results in a degree of surface hydroxylation at 150 K, dehydroxylation only occurring at 170 K [31]. In contrast, water adsorbed at the clean Cu(111) surface at 80 K desorbs on warming to 150 K. Sexton and Fisher [32] reported similar observations for Pt(111)–O. It is worth noting that for the Cu(111)–O–H_2O system, the distinct chemical identities of the two O(1s) components were established using the probe molecule approach [33]. After exposing the adlayer to CO(g), the intensity of the O(1s) feature due to O^{2-}(a) was selectively reduced, whereas the OH species was unaffected. In 1980 Au and Roberts [33] showed that ammonia could also be activated at copper surfaces by the presence of chemisorbed oxygen, with an observed reaction stoichiometry of approximately 1 : 1. Hydrogen abstraction from ethylene by a Cu(111)–O surface at 373 K was also reported [34,35], with clear implications for the role of specific oxygen species in the selective oxidation of hydrocarbons.

3.3 *Novel reaction pathways with chemisorptive replacement reactions*

Two examples (the reactions of H_2S and HCl with Pb(110)) have already been mentioned, in which the nature of the adsorbed adlayer produced via the chemisorptive replacement route differs from that produced by direct reaction with the clean metal. In the case of HCl, the chloride adlayer formed by hydrogen abstraction from HCl differed from that produced by direct chlorination of the clean metal using Cl_2 in two ways: the Pb(4f) chemical shift for the Pb atoms bonded to the Cl(a) was significantly lower, and, furthermore, if the surface oxygen concentration decreased below a critical value (*c.* 10% of a monolayer), the adlayer generated via chemisorptive replacement was unstable in the presence of HCl(g). It was suggested that the chloride desorbed as a volatile $PbCl_2(HCl)_2$ species, the observed rapid removal of the species being an example of what Madix has termed [36,37] 'surface kinetic explosions'. In contrast, the surface directly chlorinated using Cl_2 was stable. The role of surface oxygen in controlling reaction pathways was evidenced by the observation that direct chlorination of the pre-oxidised Pb(110) surface with Cl_2 resulted in a chloride adlayer indistinguishable, both spectroscopically and chemically, from that generated via HCl adsorption; in other words, the chloride layer was now unstable. Reconstruction of the Pb(110) surface after oxygen chemisorption clearly plays a part in determining the structure and reactivity of subsequently formed adlayers. In this system adsorbed oxygen is thus seen to have both a structural and electronic role in determining the subsequent surface chemistry.

A low-energy pathway for surface nitride formation was observed in the dehydrogenation of ammonia at a Pt(111)–O surface. On the basis of the known correlation between surface oxygen coordination and reactivity (Section 6), it was suggested [38] that the high reactivity of the Pt(111)–O surface was consistent with a surface oxygen population consisting predominantly of isolated oxygen atoms or small clusters, rather than oxide-like islands. This model has been confirmed in a recent STM study [39].

We discuss in Section 4.1.4 examples where it is the nature of the *oxygen* adlayer that is influenced by the coadsorbate.

3.4   *The effect of oxygen coverage*

The effect of oxygen pre-coverage on chemisorptive replacement reactions was explored by Madix and Wachs in a study of methanol oxidation at copper surfaces [40]. Clean Cu(110) is unreactive towards methanol but a partially covered surface reacts at 300 K to form adsorbed methoxy species

$$O(a) + 2 CH_3OH(g) \rightarrow 2 CH_3O(a) + H_2O(g)$$

the extent of reaction being highest for $\theta_O = 0.5$ (see Section 4 for the definition of Θ_O); for higher coverages the reaction becomes poisoned. A similar observation has been made for ammonia dehydrogenation at Cu(110)–O surfaces leading to adsorbed imide species [41]

$$O(a) + NH_3(a) \rightarrow NH(a) + H_2O(g)$$

This suggested a common mechanism which was elucidated in a Monte Carlo simulation of the distribution of different sites on the oxidised copper surface [42] (see Section 4.1), where it was found that the active sites are isolated oxygen adatoms and oxygen adatoms at the ends of Cu–O–Cu chains. Subsequent STM studies confirmed this model, initially for methoxy formation from methanol at Cu(110)–O [43,44] and subsequently for the ammonia oxydehydrogenation reaction on Cu(110) [45]. Similar mechanisms have since been shown to operate at Ni(110) [46] and Rh(110) [47] surfaces. Furthermore, by adsorbing oxygen on Cu(110) at low temperature (100 K), where the reconstructed (2×1)–O phase is inhibited, Sueyoshi *et al.* [48–50] have shown that the rate of CO oxidation is much higher for a surface where the oxygen atoms are not incorporated into Cu–O chains. The high mobility at low temperatures (< 80 K) of these reactive, short Cu–O chains has been demonstrated recently by Bradshaw and co-workers [51]. There appears therefore to be an increase in reactivity with decreasing degree of coordination of the oxygen, and the case of oxygen chemisorbed at face-centred cubic fcc(110) surfaces shows this is not simply a steric effect, since oxygen adatoms with vacant nearest-neighbour sites in the [110] direction are as unreactive as the fully coordinated adatoms.

3.5   *Summary*

Following early scepticism, oxygen activation at metal surfaces is now accepted as a widespread phenomenon [30,52,53] — the clean metal may be unreactive but hydrogen transfer to O(a) leads to the formation of a bond between the remaining molecular fragment and the surface. The surface oxygen may be regarded as a *chemical switch* [30] that selectively opens otherwise inaccessible reaction channels [28].

Of course, chemisorbed oxygen can have an inhibiting effect on surface reactivity, as is found, for example, with nitric oxide adsorption at nickel surfaces. At the clean nickel surface, both chemisorbed nitride (the majority species) and a strongly adsorbed ('bent') molecular species are observed, whereas pre-exposure to oxygen leads to the almost complete suppression of the dissociation reaction channel [54,55], the only species observed being weakly adsorbed ('linear') molecular NO [54]. Similarly, and in contrast with copper, platinum and nickel, pre-oxidised Mg(0001) and Zn(0001) surfaces are unreactive towards ammonia dehydrogenation, activation only occurring through the oxygen transient species [34,56–59] which precede the chemisorbed state.

4   Selected systems

In this section we concentrate on oxygen states at the surfaces of four fcc metals which demonstrate a wide range of reactivities towards oxygen chemisorption: nickel, copper, silver and gold. For a detailed review of the structural aspects of oxygen chemisorption at the low-index single-crystal surfaces of copper, nickel and silver, the reader is directed to reviews by Besenbacher and Norskov [4] and Kiskinova [8]. Oxygen coverages θ_O are quoted with respect to the *saturated* monolayer, for which $\theta_O = 1$. Thus for copper, $\theta_O = 1$ corresponds to one oxygen atom per two surface copper atoms.

4.1   *Oxygen states at copper surfaces*

4.1.1   STRUCTURAL STUDIES

STM has had an enormous impact on our understanding of the structures of chemisorbed oxygen adlayers at copper surfaces and has also provided a unique insight into how the adlayers develop [3–5]. On Cu(110), for example, STM investigations [51,60–62] have established that copper atoms scavenged from step edges are incorporated into added Cu–O–Cu–O rows aligned in the [001] direction. An elegant study by Bradshaw and co-workers [51] has provided the clearest STM images, one of which is reproduced in Fig. 4.2; the structure is illustrated schematically in Fig. 4.3(a). At low coverages ($\theta < 0.5$) short Cu–O chains are mobile even at room temperature but tend to cluster together to form larger immobile islands with an overall p(2×1) structure [61,62]. The equilibrium between the relatively stable islands and the more mobile short chains has been modelled with a Monte Carlo simulation [42]. Interaction energies of 2 kJ mol^{-1} and 7 kJ mol^{-1} in the [110] and [100] directions respectively were found to give a good match between theory and experiment.

After completion of the (2×1) monolayer, the oxygen sticking probability drops dramatically, but higher exposures lead to the development of a c(6×2) overlayer [7,63,64]. The building blocks for the latter structure can still be viewed as Cu–O chains but with added copper atom links between chains every second repeating unit [63,64]. This structure results in the chemisorbed oxygen species occupying two distinct sites at this stage, which has been confirmed by HREELS [65].

The reconstruction of the Cu(100)–O(a) surface occurs by a different mechanism from that of the (110) surface [6,66]. Oxygen chemisorption leads to a squeezing out of copper adatoms from the terraces, the atoms then congregating as separate islands on the terraces. Interestingly the reconstructed (100) surface also consists of pairs of Cu–O

O / Cu(110)

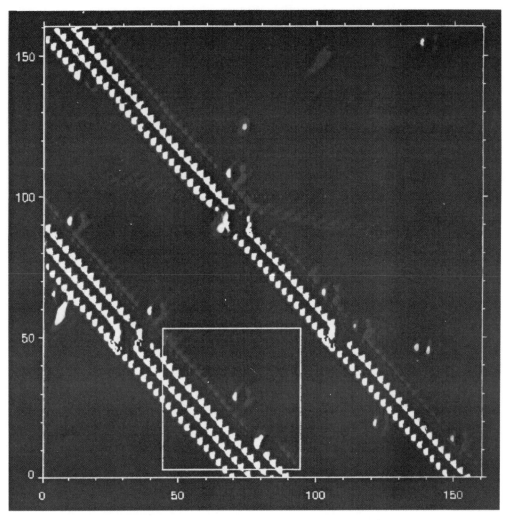

ΔI image, 4K, adsorption at 300K

Figure 4.2. STM image, 160 Å × 160 Å, of a O–Cu–O–Cu chain at a Cu(110) surface dosed with oxygen at 300 K and imaged at 4 K. Note the isolated oxygen species which are stabilised by the very low temperature. Note also that the chain continues across the monoatomic step on the copper terrace. Reproduced with permission from [51].

chains with every fourth row missing, as is illustrated schematically in Fig. 4.3(b). The Cu(111)–O interaction is more complex than the other two low-index planes [67]. Early work reported a disordered adsorption at low coverage, with varying LEED patterns reported for the monolayer coverage [67]. STM has provided a more detailed picture [68,69], in which the reconstructed surface is a coincidence lattice between the (1 × 1) surface and the (111) plane of Cu_2O. Locally the reconstruction is again stabilised by linearly coordinated O–Cu–O units (Fig. 4.3(c)); indeed, according to Jensen *et al.* [69] 'all of the oxygen induced reconstructions of the low index copper surfaces seem locally, with respect to the copper coordination, to be reminiscent of Cu_2O'. However, as we

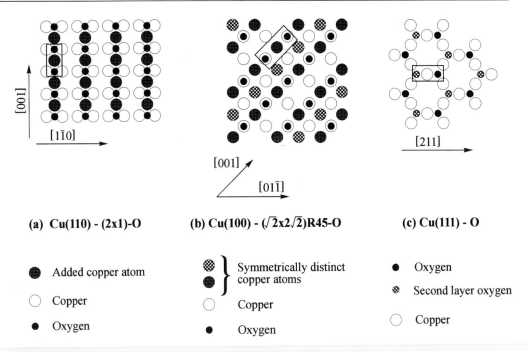

(a) Cu(110) - (2x1)-O

(b) Cu(100) - ($\sqrt{2}$x2$\sqrt{2}$)R45-O

(c) Cu(111) - O

⬢ Added copper atom

○ Copper

● Oxygen

⬡ } Symmetrically distinct
● } copper atoms

○ Copper

● Oxygen

● Oxygen

⊗ Second layer oxygen

○ Copper

Figure 4.3. Schematic diagram of the reconstructions of the three low-index copper planes at low oxygen coverages: (a) Cu(110)(2 × 1)–O; (b) Cu(100)($\sqrt{2}$ × 2$\sqrt{2}$)R45–O; (c) Cu(111)–O. The O–Cu–O linking units are indicated in each case by a rectangular box.

will discuss later, oxygen adatoms at the three low-index copper planes show very different reactivities. This can be attributed to the fact that the coordination with respect to the *oxygen* at the three surfaces is different. At the (110) and (100) surfaces the oxygen is essentially twofold coordinated and at the chain ends only onefold, whereas oxygen adsorbed at the (111) face is essentially threefold coordinated.

4.1.2 SPECTROSCOPIC STUDIES

XPS studies of low coverages of oxygen chemisorbed at clean copper surfaces (polycrystalline and single crystal) show only a single oxygen species characterised by a binding energy of about 530 eV, and labelled (as discussed in Section 1) as O^{2-}(a). As the oxygen coverage is increased towards saturation, an increase in the O(1s) binding energy of about 0.7 eV has been reported at both Cu(110) [65] and Cu(111) [70] surfaces. HREELS spectra of the Cu(110)–(2 × 1)O adlayer [65] show a single vibrational loss at 391 cm^{-1} (at 100 K and 300 K) consistent with the bridge site shown by STM. As the c(6 × 2) adlayer develops, however, two other loss peaks are observed at 330 cm^{-1} and 502 cm^{-1}, both assigned to a second oxygen site. This is confirmed by low energy ion scattering (LEIS) measurements [71] and ellipsometry [72]. This second oxygen state has been suggested to be a subsurface species, the HREELS [65] and STM [64] results suggesting that it lies within the first two atomic layers.

4.1.3 PROBE MOLECULE STUDIES

We described earlier how chemisorbed oxygen and hydroxyl species at a copper surface could be identified through their different reactivities towards CO [33]. This approach has since been applied to explore the nature of oxygen states at copper surfaces, the probe molecules used, including carbon monoxide [48–50,73–75], ammonia [41,42,76–79], methanol [43,44,80–84] and water [85,86]. The latter three, which do not chemisorb at clean copper surfaces, undergo facile hydrogen abstraction reactions in the presence of oxygen and share many common features. In particular, the extent of reaction in all three cases is a sensitive function of the oxygen coverage, reaching a maximum where $\theta_O \approx 0.5$ and becoming poisoned as $\theta_O \to 1$. This aspect has been investigated in detail for the reaction of ammonia with pre-oxidised Cu(110) [42]. The proportion of oxygen adatoms reactive to ammonia at different oxygen coverages was determined by XPS and compared with the concentrations of different types of adsorption site for the oxygen adatoms at the Cu(110)–O(a) surface, calculated from the Monte Carlo simulations mentioned above. Four sites, illustrated in Fig. 4.4, were considered: the centres of the oxygen islands, the [110] and [001] edges of the islands (the latter being the so-called 'chain end' sites) and isolated oxygen adatoms. A strong correspondence was found between the experimentally determined active oxygen species and the total concentration of 'chain ends' and isolated oxygen atoms (Fig. 4.5). This is not simply a steric effect, since the oxygen adatoms at the [110] edges of the oxygen islands are inactive but not sterically hindered in the [110] direction. In other

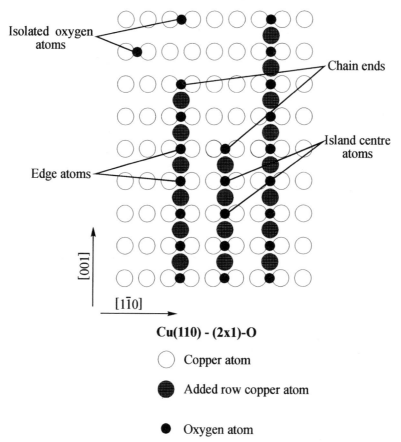

Figure 4.4. The different environments available for chemisorbed oxygen atoms on Cu(110) within an incomplete (2 × 1) added-row structure.

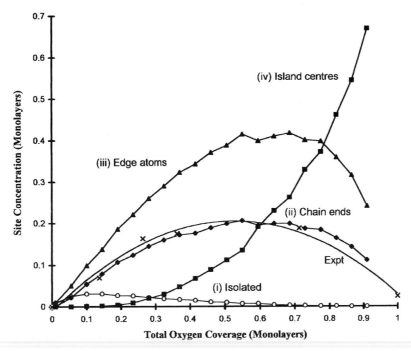

Figure 4.5. A plot of the results of a Monte Carlo simulation [42] of the development of the (2 × 1)–O adlayer on Cu(110). The graph shows the concentration of the different types of chemisorbed oxygen atoms identified in Figure 4.4 as a function of total oxygen coverage. Also plotted is the experimentally determined concentration of 'active' oxygen atoms; the experimental data fit very closely with the sum of the 'chain ends' and isolated oxygen atoms.

words, there is a chemical difference between the oxygen adatoms adsorbed at these different sites. The reactive nature of the 'chains ends' has been confirmed by more recent STM studies of methanol [44], carbon monoxide [74,75] and ammonia [45], and appears to be a common feature of fcc(110)–O surfaces, with similar results having been found on rhodium [47] and nickel surfaces [46].

The oxidation of carbon monoxide at copper surfaces is a much slower process than the hydrogen abstraction reactions, and early studies of the system involved relatively high temperatures, pressures and oxygen coverages [17,72]. A common feature of the studies was the observation of two types of oxygen for $\theta > 1$, correlating well with spectroscopic and structural studies. More recently, Iwasawa and co-workers have investigated the reactivity of carbon monoxide with low coverages of oxygen chemisorbed at Cu(110) surfaces at low temperatures [48–50, 73]. They find that oxygen adsorbed at temperatures below 200 K, where the added row reconstruction of the surface is inhibited, is very much more reactive than oxygen adsorbed at higher temperatures. They concluded that 'even a short Cu–O chain without long range order affected the oxygen reactivity strongly'. This observation agrees well with the results of the coadsorption studies discussed in the next section.

4.1.4   EFFECT OF COADSORBED MOLECULES ON OXYGEN
        AT COPPER SURFACES

There is some compelling spectroscopic evidence [78,79] to suggest that coadsorbates can influence the state of oxygen at copper surfaces. When ammonia and dioxygen were coadsorbed at 100 K at a Cu(111) surface three novel oxygen states were observed: a molecular species, identified by a high resolution electron energy loss (HREELS) peak at 1500 cm^{-1}, and two atomic states characterised by X-ray photoelectron (XP) binding energies of 531.3 and 528.9 eV (Fig. 4.6). The two atomic states showed very different chemistry; the state characterised by the higher binding energy was stable to 400 K, was unreactive towards ammonia and was tentatively assigned to a subsurface state. The other state was unstable above 200 K, reacting with the coadsorbed ammonia to form OH(a) and NH$_2$(a); it was assigned to oxygen adatoms prevented from forming the more stable O^{2-}(a) state by a matrix of adsorbed ammonia.

The important role played by coadsorbate-induced oxygen states in reactions at copper surfaces has been highlighted by a series of coadsorption experiments at room temperature involving mixtures of dioxygen with ammonia [76,77], water [85,86] and methanol [87–89]. In each case coadsorption results in a highly selective and efficient reaction giving a single product in high concentrations; for example, ammonia–dioxygen coadsorption at Cu(110) and Cu(111) surfaces gives rise to close to a monolayer of pure NH(a) or NH$_2$(a) species, depending upon the mixture ratio (Fig. 4.7). This is in sharp contrast to the mixed NH$_2$(a)–NH(a)–N(a) adlayer that results from the reaction of ammonia with *pre-adsorbed* oxygen [41]. Furthermore, at both surfaces the rate of reaction was shown to be equal to the rate of dioxygen chemisorption. Methanol–dioxygen coadsorption on copper also results in a fast reaction at room temperature, and by changing the mixture ratio it is possible to change the products from predominantly formaldehyde to pure formate and back to formaldehyde [88].

The high reactivity of the coadsorption experiments has been attributed [76,77,85, 86] to the very low surface oxygen concentrations maintained throughout the experiment by the continuous reaction, and hence the absence of inhibiting oxygen islands. The result is that every oxygen atom can be considered an active site. However, there is also evidence that the nature of the oxygen species involved is fundamentally different in the coadsorption experiments. At the Cu(110) surface, for example, the formation of formate from the reaction of methoxy species with pre-adsorbed oxygen is activated, but when methanol and dioxygen are coadsorbed the reaction proceeds with 'zero activation energy' at room temperature [88]. In addition, at Cu(111) surfaces, very low concentrations of surface oxygen are unreactive but the rate-determining step when ammonia and dioxygen are coadsorbed is the chemisorption of the dioxygen [77]. This implies a very fast hydrogen abstraction step in the coadsorption experiment, and the involvement of a different oxygen state from that when oxygen is pre-adsorbed.

Clearly, when reactants are coadsorbed at copper surfaces a short-lived and highly reactive oxygen state must be involved. It is possible that this is a molecular species: van Santen's comprehensive quantum-chemical treatment of the NH$_3$–O$_2$ dissociation process identified [90] the O$_2$ + NH$_3$ → OH + NH$_2$ step as the lowest-energy pathway. Alternatively, it may be that 'isolated' atomic oxygen atoms have a distinctly different reactivity from those incorporated into oxygen islands. Support for this point of view comes from recent theoretical calculations [91] and the CO titration experiments [48–50,73] which show much higher reactivity at temperatures where the surface is prevented from reconstructing.

O(1s)

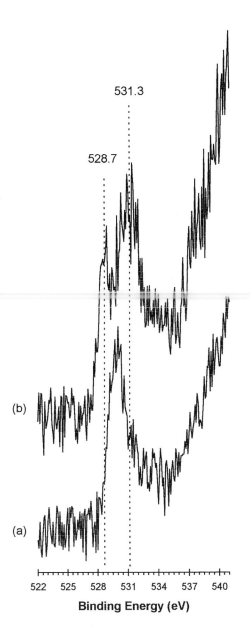

Figure 4.6. O(1s) XP spectra showing the effect of pre-adsorbed ammonia on the state of oxygen at Cu(111) surfaces (reproduced with permission from [78]). Note the presence of at least three oxygen species with binding energies of 528.7, 529.8 and 531.3 eV. (a) Clean surface exposed to 50 L O_2(g) at 298 K; (b) ND_3 precovered surface exposed to 250 L O_2(g) at 100 K.

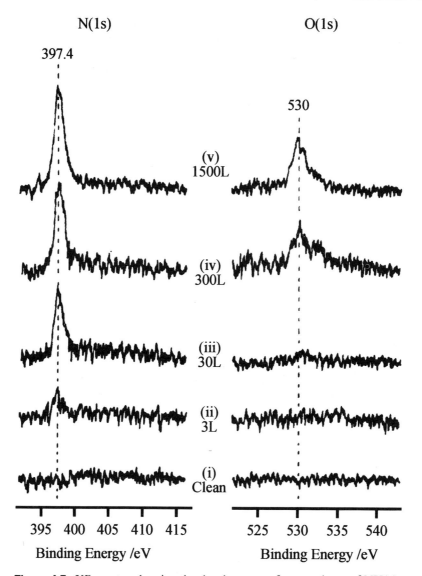

N(1s)   O(1s)

Figure 4.7. XP spectra showing the development of a monolayer of NH(a) at a Cu(110) surface at room temperature following the dosing of a 36 : 1 $NH_3(g)$: $O_2(g)$ mixture at room temperature. See [76] for further details.

4.2   Oxygen states at nickel surfaces

4.2.1   STRUCTURAL STUDIES

For the Ni(111) and Ni(100) surfaces detailed structural information on chemisorbed oxygen is only available for annealed and well-ordered adlayers involving relatively high surface oxygen coverages [6,92,93]. At both surfaces oxygen adsorbs in the highest coordination sites available, causing a relaxation of the nickel lattice, but, in contrast to the case of copper, no further reconstruction occurs. At the Ni(110) surface, on the other hand, oxygen chemisorption results in an added row reconstruction involving [001]-directed Ni–O chains [7]. However, unlike copper, the packing of the chains in the [110] direction changes with coverage, giving rise to a (3×1), (2×1) and (3×1)

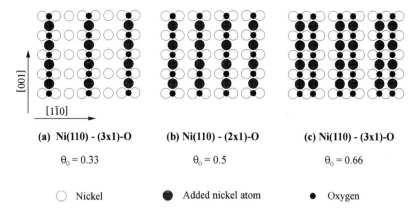

Figure 4.8. Models for the structure of the Ni(110)–O surface at different oxygen coverages: (a) Ni(110)(3 × 1)–O, $\theta_O = 0.33$; (b) Ni(110)(2 × 1)–O, $\theta_O = 0.5$; (c) Ni(110)(3 × 1)–O, $\theta_O = 0.66$.

sequence of structures (Fig. 4.8), with corresponding oxygen coverages of 0.33, 0.5 and 0.66 monolayers. Low coverages of chemisorbed oxygen at Ni(110) surfaces were thought until recently to involve a disordered phase, but this has now been shown by STM [94] to consist of a mixture of [001]-directed Ni–O chains and a different reconstruction involving [110]-directed Ni–O 'strings'. The 'strings' spontaneously 'dissolve' if the oxygen coverage is increased, or if the sample is heated to 373 K, the constituent atoms becoming incorporated into the [001]-directed chains.

4.2.2 SPECTROSCOPIC STUDIES

X-ray photoelectron spectra of oxygen chemisorbed at various nickel surfaces are very consistent [14,95–99]. Two states of oxygen are observed in every case, characterised by O(1s) binding energies of c. 529.6 eV and c. 531 eV (Fig. 4.9). The former is the dominant species at high oxygen coverage, or after low oxygen coverages are annealed, and is clearly related to the $O^{2-}(a)$ oxidic species observed at other metal surfaces. The second oxygen state coexists with the $O^{2-}(a)$ state at low temperatures and where the total oxygen coverage is low, and has been observed in particularly high concentrations at the open (210) surface plane [14]. It converts into the $O^{2-}(a)$ state on heating above 370 K. Because of its low stability, high reactivity (see below) and relatively high binding energy, it has been labelled as $O^{\delta-}(a)$.

   HREELS studies of oxygen chemisorption at nickel surfaces have also been interpreted in terms of more than one oxygen state. On Ni(100), for example, high coverages of oxygen show [100,101] a single dipole active vibrational mode at a frequency of 410 cm^{-1}, but low coverages of oxygen give rise to a further loss peak at 370 cm^{-1}. Similarly, at Ni(110) surfaces [102] low oxygen coverages show two dipole active species with characteristic loss frequencies of 500 cm^{-1} and 388 cm^{-1}, but only one (the latter) at high coverages.

   The $O^{2-}(a)$ state formed at high coverages can be unambiguously assigned to strongly chemisorbed oxygen atoms incorporated into the [001]-directed oxygen chains on Ni(110), and to the adatoms absorbed in three- and fourfold hollows, for Ni(111) and Ni(100), respectively. Despite the different adsorption sites at each surface, the O(1s) binding energy indicates that the nature of the oxygen is similar.

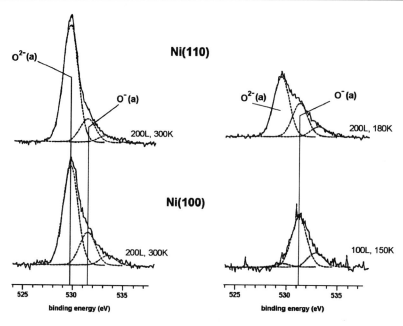

Figure 4.9. O(1s) spectra of oxygen adsorbed on Ni(110) and Ni(100) at the temperatures and exposures indicated on the figure. Two major species, O^{2-}(a) and O^{1-}(a) are identified, the latter being the active species for hydrogen abstraction from coadsorbed ammonia [98].

We note that the occurrence and stability of the $O^{\delta-}$(a) oxygen state correlates well at Ni(110) surfaces with the [110]-directed oxygen strings observed by STM. If this assignment is correct, however, it has interesting implications for the (100) and (111) planes where a similar state is observed by XPS. To the authors' knowledge, there have been no atomically resolved studies of the chemisorption of oxygen at either plane at low coverages, and thus we have no clues as to the structure of this state at these surfaces.

4.2.3   PROBE MOLECULE STUDIES

There are a number of studies in the literature concerning the interaction of chemisorbed oxygen at nickel surfaces with probe molecules such as ammonia [46,103,104], hydrogen sulfide [105], water [106–108], and benzene [101]. Several of these studies involve annealed oxygen adlayers and the probe molecule results are consistent with the presence of a single state of oxygen. For example, the chemisorptive replacement of oxygen at a pre-oxidised Ni(110) surface preannealed to 600 K and exposed to H_2S [105] was facile at all coverages, even at $\theta_O \approx 1$. Similarly, more recent investigations into the interaction of benzene [109] and ammonia [46] involved nickel surfaces dosed with oxygen at 420 K, conditions under which only a single state of oxygen, O^{2-}(a), is present. No direct reaction between benzene and the surface oxygen was observed, but the ammonia reacted with the ends of the Ni–O chains in a similar fashion to the oxydehydrogenation reaction on Cu(110) [42].

Where probe molecules have been used to study pre-oxidised nickel surfaces which have not been annealed, the results are more interesting. Benndorf *et al.*, for example, studied the decomposition of water at nickel surfaces with various pre-coverages of oxygen dosed at 300 K [108]. They found a distinct maximum in the extent of water dissociation for an oxygen coverage $\theta_O \approx 0.3$, with dissociation ceasing by $\theta_O \approx 1$. Carley *et al.* observed [106] that oxygen adsorbed at low temperatures was more reactive than oxygen chemisorbed at higher temperatures, the most reactive oxygen species being present under conditions where the $O^{\delta-}(a)$ state is formed. Direct confirmation that the $O^{\delta-}(a)$ state is the more reactive oxygen species comes from two recent studies which have shown that ammonia [98] and water [99] react preferentially with the $O^{\delta-}(a)$ state at nickel (110) and (100) surfaces.

4.3   *Oxygen states at silver surfaces*

The significant effort devoted since the mid-1980s to understanding the chemisorption of oxygen at Ag surfaces reflects its uniqueness as an effective catalyst for the industrially important oxidation of ethylene to ethylene oxide. Improving the selectivity of the reaction towards partial oxidation rather than complete combustion (the thermodynamically preferred reaction pathway) has been one of the incentives for trying to understand the nature and reactivity of oxygen species formed at silver surfaces.

4.3.1   STRUCTURAL STUDIES OF OXYGEN CHEMISORPTION AT
Ag(100), Ag(110) AND Ag(111)

The strongly bound chemisorbed atomic oxygen state on Ag(110) is associated in the coverage range $0 < \theta_O < 0.5$ with a series of ordered $p(n \times 1)$ LEED patterns, where n decreases from 7 to 2 as the oxygen coverage increases. Initial conclusions were that these corresponded to oxygen overlayer structures, the oxygen atoms forming rows along the [001] direction, but subsequent studies supported a missing row-type reconstruction. More recent STM investigations suggest that an *added-row* reconstruction model [110–117] is more correct. This is similar to the reconstruction observed with Cu(110), but there are differences in their modes of formation [118]. The added rows run along the [001] direction on top of the Ag(110) surface, the oxygen atoms occupying the long bridge sites. Atomic detachment rates from step edges on Ag(110) have been shown [119] to be sufficient to supply the atoms for the added-row reconstruction of the oxidised surface, provided the oxygen partial pressure is low. For higher partial pressures the source of silver atoms is supplemented by vacancy-island generation on the terraces. After high-pressure dosing, an oxygen coverage of $\theta_O = 0.67$ is associated with a c(6×2) structure, as found for Cu(110), which transforms back to the (2×1) structure on heating to 565 K.

Few structural studies exist for either the Ag(100) or Ag(111) surfaces. A c(2×2) structure is observed for dissociative chemisorption on Ag(100) at 180 K, the oxygen atoms occupying fourfold hollow sites [120]. Annealing to room temperature transforms this to a p(1×1) structure. For Ag(111) a p(4×4) LEED pattern is seen for oxygen adsorption above 373 K [121], and is stable up to the desorption temperature of 579 K. A new phase exhibiting long-range order has been reported [122] after reaction of Ag(111) with O_2 at atmospheric pressure and a temperature greater than 800 K.

There has been much debate regarding the identification of the several oxygen species observed at the surfaces of polycrystalline and single-crystal silver surfaces, often in the context of the assignment of the O(1s) photoemission peaks. Four distinct states of adsorbed oxygen have been reported to exist, two molecular species (physisorbed and chemisorbed), and two atomic species (chemisorbed and subsurface or dissolved). It is interesting to note that all the species introduced in the scheme on p. 77 have been isolated on silver, although the population of each species depends strongly on the crystallographic plane.

The problem of assignment is complicated by the high oxygen exposures necessary, particularly with the Ag(111) plane, in order to achieve a significant surface concentration of adsorbed oxygen species. Contamination of the surface by adsorbed water, hydroxyl and carbonate species, due to concomitant reactions with ambient gases [123–125], must always be considered and may go some way towards explaining the disparate binding energy values quoted in the literature. The high reactivity which may be observed in such coadsorption experiments (albeit inadvertent in this case) has been discussed elsewhere [1,2].

As with most fcc metals, it is the (110) surface of silver which has received the most attention, with relatively few studies of the (100) and (111) planes having been reported. Physisorbed oxygen adsorbs at temperatures below 40 K at all the low-index silver surfaces studied, giving rise to an O(1s) peak at a binding energy of 536.7 eV [126]. The vibrational frequency $\nu(O-O)$ of the dioxygen bond [121] at Ag(111) is 1536 cm^{-1}, which is close to the gas phase value (1580 cm^{-1}) [127] and thus consistent with a very weak interaction with the surface. For Ag(110) a peroxo-like species with the O–O bond parallel to the surface has been reported [128].

Desorption of dioxygen from Ag(111) is complete at 40–50 K [121] but at Ag(110) there is a partial conversion to a chemisorbed peroxidic species [128–134], with a low O–O stretching frequency of 620–660 cm^{-1} [129–131,135] indicative of charge transfer to the O_2 antibonding π^* orbital and consequent weakening of the O–O bond. This species gives rise to a narrow O(1s) peak at 529.3 eV, consistent with its peroxo-like character. It is chemisorbed with its O–O bond parallel to the surface and aligned along the grooves in the [110] direction [128,132,136,137], and is stable up to 170 K. Desorption of the molecular species is accompanied by some dissociation [128–131] to give strongly chemisorbed oxygen atoms, the corresponding O(1s) peak having a binding energy of 528.5 eV [129]. The interaction between the oxygen atoms and the surface is highly ionic, with a calculated charge transfer of approximately one electron [138].

Adsorption of dioxygen at an Ag(111) surface at 150 K leads to the formation of a superoxo-like species with its molecular axis perpendicular to the surface plane. The experimental evidence for this assignment comes from XPS. A broad O(1s) peak centred at 531.9 eV is observed [121,139], the high value of the full width at half maximum (2.4 eV) contrasting with that observed for Ag(110). This is consistent with coordination of the molecule through one oxygen atom only, resulting in two non-equivalent oxygen atoms with somewhat different O(1s) binding energies, a conclusion supported by recent calculations [140]. Adsorption occurs with a very low sticking probability, with reported values ranging from 6×10^{-7} [141] to 10^{-4} [86]. Warming to c. 220 K leads to almost complete desorption of dioxygen [121,139], the dissociation process being highly inefficient. Campbell [121] has suggested that this is due to an unfavourable configuration of the molecule, the activation energy for O(a) formation actually being small. This is consistent with theoretical conclusions that dissociation on silver proceeds through the peroxide [142], and not the superoxide species. Adsorption at elevated temperatures is necessary to obtain a significant fraction of a monolayer of chemisorbed oxygen atoms — for example, a surface coverage of 0.4 was obtained by Campbell [121] after exposure of Ag(111) to dioxygen at 490 K. The oxygen species was characterised by an O(1s) binding of 528.5 eV and the estimated dissociative sticking probability was 10^{-6}. It is worth noting in this context that coadsorption from a dioxygen–ammonia mixture results in an increase of several orders of magnitude in the rate of dioxygen bond cleavage at an Ag(111) surface, through the formation of a $[NH_3–O_2]$ complex which exhibits facile dissociation kinetics even below room temperature [56,57,123]. Ertl's group has also reported [143] an energetically more favourable mechanism for O(a) formation than the direct dissociation of O_2, via the hydroxylation of an oxygen-exposed Ag(111) surface using water vapour. Dehydroxylation at higher temperatures then leads to O(a) formation. Chemisorbed oxygen atoms desorb as O_2 at 580–600 K from both the Ag(111) and Ag(110) surfaces [121,144].

The formation of subsurface oxygen states is widely accepted, but at silver surfaces there is some disagreement about the temperature at which subsurface oxygen (O_{sub}) first starts to form, both 300 K [145] and 420 K [130,146,147] having been suggested; dissolution into the bulk occurs for $T > 470$ K [148]. The observed O(1s) binding energy for subsurface oxygen depends on the occupancy of subsurface sites [149], decreasing from 531.6 eV to 529.7 eV as the concentration increases, due to an increase in the extra-atomic relaxation shift. Recombination of O_{sub} followed by molecular desorption is generally (but not unanimously [150]) accepted to occur at 880 K [130,145,147,149], except when the O_{sub} concentration is below a low threshold value — in this case a strong Ag–O interaction results in a significantly higher desorption temperature of 1040 K [149]. A near edge X-ray absorption fine-structure (NEXAFS) study found that subsurface oxygen on Ag(110) had no effect on the orientation of the molecular oxygen species [137]. On both Ag(110) and Ag(111), Raman bands at 803 cm^{-1} and 627 cm^{-1} have been assigned to O^{2-}(a) and O_{sub} respectively, after oxygen treatment at c. 800 K [151].

Ag(100) is the least studied of the low-index faces of silver [120,152]. Low-temperature physisorption of dioxygen followed by warming to 130 K leads to the formation of a species with an O–O stretching frequency (c. 640 cm^{-1}) which by analogy with Ag(110) is assigned to a peroxo-type oxygen [152] bonded parallel to the surface — this is supported by a molecular orbital calculation [153]. Dissociative chemisorption is observed at temperatures above 170 K [120,152], the oxygen–surface bond being largely ionic, with the charge on the atom estimated to be c. – 1.5 electrons [154].

Bare *et al*. [155] found that for both Ag(111) and Ag(110) surfaces, chemisorbed oxygen atoms could be formed by adsorption of NO_2 at 500–520 K at moderate pressures, in comparison to the high pressures required with dioxygen. In particular, with Ag(111) the (4×4)–O and $c(6 \times 2)$–O overlayers generated by NO_2 adsorption exhibited the same structure as those formed less easily from dioxygen — another example of *chimie douce* which has analogies with the ammonia–dioxygen coadsorption studies referred to earlier.

4.3.3   REACTIVITY OF OXYGEN SPECIES AT SILVER SURFACES

Madix's group has made an extensive study of oxygen-activated reactions at silver surfaces — see [30], in which many of them are discussed. Three distinct behaviours for atomic oxygen adsorbed on silver surfaces are identified: (i) it can act as a Brønsted base, extracting hydrogen from molecules such as ammonia and ethylamine in an oxydehydrogenation reaction; (ii) it is nucleophilic, attacking carbon in aldehydes and esters; (iii) it creates partially electron-deficient (Lewis acid) sites, which can lead to stabilisation of donor–acceptor bonding. In the case of ethylamine dehydrogenation, the product desorbing at 370 K is acetonitrile [156], demonstrating highly selective C–H bond-breaking, with no C–C or C–N bond scission.

Carbonate formation at Ag(110)–O [124,157–159] and Ag(111)–O [121] surfaces after reaction with $CO_2(g)$ has been observed. An STM study of Ag(110) found several reaction regimes depending on the initial oxygen coverage [157]. For $\theta_O < 0.5$ the reaction is initiated mainly at steps and defects and proceeds along the [001]-directed added rows, with two oxygen atoms being consumed for each carbonate species formed. For $\theta_O < 0.25$ a complete transformation to carbonate occurs, whereas for $0.25 < \theta_O < 0.5$ the unreacted rows are compressed to a (2×1) structure. The similarities between these observations and those for methanol and ammonia oxidation at Cu(110) surfaces, where oxygen atoms with differing reactivities have been identified (Section 4.1.3), are striking.

Chemisorbed oxygen on silver may be titrated off as $CO_2(g)$ by exposing the surface to CO [121,124,160,161], although the reaction rate depends on oxygen coverage, being higher on Ag(110) for the $c(6 \times 2)$–O adlayer compared with the $p(2 \times 1)$ surface [124]. The peroxo species formed on Ag(110) after dioxygen adsorption at 95 K reacts with CO to give both adsorbed CO_2 and surface carbonate [129]; the formation of the latter is facilitated by the increased lifetime and population of CO_2 on the surface compared with the room-temperature reaction.

4.3.4   ETHYLENE EPOXIDATION – CONTROLLING THE REACTIVITY
OF ADSORBED OXYGEN

An ideal reaction for illustrating the subtleties of oxygen states present on metals is provided by the selective oxidation of ethylene at silver surfaces. A crucial question to be answered is the nature of the oxygen species active for epoxidation — a Raman spectroscopic study [162] of an industrial silver catalyst found that both atomic and molecular oxygen species existed on the surface at 870 K in a flowing oxygen atmosphere. The active species was thought for some time to be molecular (implying a maximum theoretical selectivity for ethylene oxide formation of 86%), but as experimental and theoretical evidence accumulated [163,164] it became apparent that atomic oxygen is responsible for epoxidation, and 100% selectivity is attainable in principle.

The significance of subsurface oxygen was first noted by Backx *et al.* [158], who found that ethylene adsorption was promoted on a silver surface subjected to many oxygen adsorption–desorption cycles, HREELS measurements [165] suggesting that the molecule is π-bonded to an $Ag^{\delta+}$ site, with the $C=C$ axis parallel to the surface. Grant and Lambert [166], investigating ethylene epoxidation at Ag(111), further proposed that complete oxidation involved charge transfer *from* the O^{2-}(a), whereas partial oxidation required charge transfer *to* the oxygen; thus, selectivity is related to the charge on the chemisorbed oxygen. Subsurface oxygen controls this charge by competing with O^{2-}(a) for metal electrons and both O_{sub} and O^{2-}(a) are necessary for epoxidation. Coadsorbed chlorine atoms also reduce the charge on O^{2-}(a) [167], which explains the beneficial effect on selectivity of incorporating traces of chlorinated hydrocarbons in the industrial gas feed. The effect of subsurface oxygen on the interaction of ethylene with Ag(111)–O has been modelled by van Santen and co-workers [168], who found that O_{sub} performs two crucial roles: it reduces the bond energy between the surface atoms and O^{2-}(a), and also transforms the interaction between O(a) and ethylene from a repulsive one into an attractive one. A previous theoretical study [164] had suggested the involvement of an $O^{\delta-}$-like oxyradical anion in selective oxidation, but had not considered the possible effect of subsurface oxygen.

Bukhtiyarov *et al.* have also concluded from XPS and isotope studies [147,169,170] that two oxygen species are necessary for selective oxidation, which they termed 'ionic' and 'covalent', with O(1s) binding energies and desorption temperatures of 528.4 eV and 580 K, and 530.5 eV and 800 K, respectively. Their interpretation is that the 'ionic' oxygen is responsible for creating $Ag^{\delta+}$ sites which absorb ethylene without activation of the C–H bond, while the 'covalent' oxygen is electrophilic ($O^{\delta-}$-like) and inserts in the $C=C$ bond to form ethylene oxide. The 'ionic' oxygen, when present on its own, is responsible for the total combustion of ethylene, whereas both species are necessary for ethylene epoxidation. The two species are produced by treating the silver foil with a mixture of C_2H_4 and O_2 at 470 K, which also leads to the formation of elemental carbon, both C(a) and C_{sub}, via oxydehydrogenation of the ethylene. It is suggested that the carbon prevents the development of islands of 'ionic' and 'covalent' oxygen — adjacent pairs of the two species are needed for the epoxidation reaction. In contrast, generation of the two oxygen species from pure oxygen results in a surface which is inactive for ethylene oxide formation. An angle-resolved XPS study [171], with a claimed depth resolution of one atomic layer, placed the 'ionic' oxygen between the top and second silver layers, and the 'covalent' oxygen on or within the first layer. This is in conflict with other studies [172,173] which assigned the 530.5 eV species as subsurface oxygen.

The adsorption of oxygen at silver clusters deposited on a carbon foil substrate has been studied [174] as a model system for the industrial epoxidation catalyst. For small clusters only the covalent oxygen was observed after treatment in oxygen (100 Pa, 470 K, 10 min), the ionic species only appearing as the cluster size increased (Fig. 4.10). This observation correlates very well with the known dependence of the rate of ethylene epoxidation on cluster size.

The addition of nitric oxide to the C_2H_4–O_2 gas mixture has been reported to give rise to 'ultra-selective' ethylene epoxidation at potassium-doped Ag(110) and Ag(111) surfaces [175]. Nitrate and nitrite species are formed at the alkali metal sites and

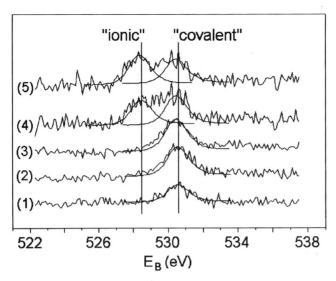

Figure 4.10. O(1s) spectra for oxygen adsorption (100 Pa, 470 K, 10 mins) on silver deposited on a graphite substrate [174], for increasing silver coverage (1)–(5). The mean particle size for (3) is estimated to be *c.* 100 Å and for (4) *c.* 300 Å. Note that the 'ionic' oxygen species only appears for particle sizes approximately equal to or greater than 300 Å.

decompose at 500 K, the temperature at which epoxidation occurs — the decomposing species will be a continuously regenerated source of reactive 'nascent' oxygen atoms which may play a role in the epoxidation reaction. CO_2 in the gas feed poisons the reaction through the preferential formation of the more stable carbonate.

The effect of caesium doping on the epoxidation reaction is twofold: the presence of Cs^+ enhances ethylene adsorption [164] but also inhibits the isomerisation of ethylene oxide to acetaldehyde [176], the latter being susceptible to further oxidation — acetate formation has been suggested as the intermediate step to total combustion [177]. The control of surface oxygen species on a silver–alumina catalyst has been achieved through alloying with gold [178], which led to an enhancement in the population of a molecular species, a weakening of the Ag–O bond and a reduction in the concentration of the subsurface oxygen state.

4.4   *Oxygen states at gold surfaces*

4.4.1   BACKGROUND

The oxidation of gold has been the subject of some controversy, which was only resolved in the mid-1980s. Chesters and Somorjai reported in 1975 [179] that oxidation of Au(111) began at 500°C in an ambient oxygen pressure of 6×10^{-7} torr and proceeded at an accelerated rate at 700°C. Significantly, they took pains to ensure that the surface of the crystal was free of segregated calcium, a common impurity in gold and known to enhance greatly the dissociation of oxygen [180]. The oxidation reaction was found to be structure-insensitive and the adsorbed oxygen was remarkably stable, desorbing only slowly at 1073 K. In contrast, bulk Au_2O_3 decomposes at 410 K.

However, this high-temperature state was shown unequivocally by Madix's group [181] to be due to oxidation of Si which segregates to the gold surface. A surface compound with a stoichiometry of SiO_2 was identified by Auger electron spectroscopy and moreover, the Si(KVV) Auger feature exhibited a chemical shift reflecting SiO_2 formation. No shifted component was observed in the Au(4f) photoemission peaks, again arguing against oxide formation on the gold. Oxygen adsorption was only observed in the presence of a silicon surface impurity, a crystal cleaned for several weeks to deplete it of silicon showing no detectable adsorption at 600 K and an oxygen pressure of 10^{-6} torr. A subsequent study [182] extended measurements to oxygen pressures of 1400 torr; no oxygen adsorption could be detected, despite earlier claims in the literature.

Although dissociation is not favoured on gold, oxygen atoms generated at a hot platinum filament placed close to the sample do chemisorb on the gold surface with a saturation coverage of about two monolayers [181]. A chemically shifted Au(4f) peak is now observed, indicating a formal oxidation state of $+1$ if relaxation effects are ignored; furthermore, the surface oxide has a stoichiometry of $O : Au = 2 : 1$. The oxygen desorbs at 670 K, which is lower than the previously claimed high-temperature state but still significantly higher than the decomposition temperature for the bulk oxide. The oxygen adatoms do not form an ordered structure on Au(110), and indeed oxygen adsorption results in the destruction of the Au surface crystallinity [182].

Ozone treatment has also been used to generate coverages of atomic oxygen of up to one monolayer on Au(111) [183]. Interestingly, the latter authors also reported that coadsorption of H_2O and NO_2 resulted in the formation of chemisorbed oxygen, despite both molecules being unreactive when adsorbed individually. The reaction mechanism is likely to involve the formation of a surface complex similar to that proposed for the activation of dioxygen by coadsorbed ammonia at a Zn(0001) surface [57]. In contrast, attempts to generate surface oxygen atoms through the chemisorption of nitrous oxide, at pressures of up to 5 torr at 450 K, proved unsuccessful [182].

Chemisorption of oxygen on gold may also be achieved by 'doping' the surface with barium atoms [184]; the oxygen species which are adsorbed migrate from the active site on to the metal — an example of spillover.

4.4.2   REACTIVITY OF CHEMISORBED OXYGEN ON GOLD

The reactivity of oxygen atoms on gold has been investigated by Madix's group. Although the maximum surface oxide coverage attained via the dosing of oxygen atoms was found to be two monolayers, all of the chemisorbed oxygen could be removed by exposure to formic acid [181] by means of a chemisorptive replacement reaction (Section 3.1.1). This indicated that the oxygen had not penetrated the surface very deeply.

The surface oxygen can be titrated away [182] using hydrogen (7.6 torr, 400 K) with a reaction probability ($P_r \approx 6 \times 10^{-7}$) similar to that observed for Cu(110)–O and Cu(111)–O surfaces but much smaller than that measured for surfaces of Pt($P_r \approx 0.5$) and Pd ($P_r \approx 0.1$). This is due to the much lower dissociative sticking probability for H_2 on gold compared with platinum and palladium.

CO oxidation proceeds at an Au(110)–O surface [185] with an activation energy of 10 kJ mol^{-1}, the gas phase reaction product being CO_2. Isotopic labelling experiments show that CO dissociation does not take place. The reaction rate depends on both the instantaneous and initial oxygen coverages, the latter dependence arising from the heterogeneity of active oxygen sites on the surface. This is exactly analogous with the observations made for the oxidation of methanol and ammonia at copper surfaces (Section 4.1), higher oxygen coverages leading to larger adatom islands, only the periphery of these islands being reactive. On Au(111) CO oxidation occurs readily below room temperature, the oxidation rate being first-order in oxygen coverage [183]. CO_2 does not react with adsorbed oxygen on either Au(110) [185] or Au(111) [183] to form adsorbed carbonate species, in contrast with observations made for silver surfaces.

Madix has also studied the interaction of formic acid [186], formaldehyde [186], methanol [187], acetylene [187], water [187] and ethylene [187] with Au(110)–O surfaces, with the aim of exploring the relationship between reactivity and gas phase acidity. With formic acid, reactions analogous to those on Cu and Ag are observed, with the exception that no H_2 evolution occurs. This was suggested to be a consequence of the weakness of the Au–H bond, which allows the adsorbed hydrogen atoms to be scavenged by HCOO(a) or O(a) species. Formaldehyde is oxidised to formate, as reported for copper surfaces. Methyl formate formation via a methoxy intermediate is observed for the methanol reaction at Au(110)–O. Acetylene undergoes total combustion to CO_2 and H_2O, whereas ethylene is unreactive. Water is more strongly bound to the surface due to hydrogen bonding with the adsorbed oxygen.

On Au(111) combustion of methanol to CO_2 and H_2O was reported [183], but no reaction was observed between chemisorbed oxygen and either H_2O or C_2H_4 [183].

5   General discussion

We can classify oxygen states at metal surfaces into at least three main categories by their different chemical and, to a lesser extent, spectroscopic characteristics. The first of these, the $O^{2-}(a)$ state, is often associated with reconstruction of the metal surface and was discussed in some detail in Section 4. We now consider the other types of surface oxygen.

5.1   *Metastable O^{2-}-like species*

As we have noted, some metal–oxygen systems do not exhibit chemisorptive replacement activity, at least as far as the stable, oxide-like $O^{2-}(a)$ species is concerned. However, reactivity can be induced by controlling the conditions under which oxygen is adsorbed, the aim being to trap the oxygen in a metastable state from which the barrier to further reaction is lower. One approach is to adsorb the oxygen at low temperature. Thus, an aluminium surface pre-oxidised at 295 K is unreactive towards CO at 80 K, whereas pre-oxidation at 80 K leads to a surface which reacts with CO leading to C–O bond cleavage *at 80 K* and the formation of surface carbonate and other products [188]. Similarly, a lead surface pre-oxidised at 295 K is unreactive at 80 K towards both H_2O [106] and C_2H_4 [189], but pre-exposure to oxygen at 80 K results in a surface $O^{\delta-}$-like species characterised by an O(1s) feature at 531.0 eV. This reacts to form OH(a) species with both molecules. A similar $O^{\delta-}$-like species was also reported on Zn(0001), hydroxylation being observed when water was adsorbed on the surface pre-exposed to

O_2 at 80 K [190,191]. On Cu(110) the CO oxidation reaction was used [48–50,73] to probe the reactivities of oxygen adsorbed at 100 K and at 300 K, the low-temperature species being found to be 25 times more active for CO_2 formation. At nickel surfaces a reactive $O^{\delta-}(a)$ species, with a characteristic O(1s) binding energy of 531 eV, is formed at room temperature but is favoured by low temperatures and low coverages [98,99].

In most studies, therefore, the reactive $O^{\delta-}(a)$ state seems to be associated with an O(1s) binding energy of approximately 531 eV, consistent with a lower charge on the oxygen atom. The exception is copper, where two studies have indicated that the $O^{\delta-}(a)$ state has an O(1s) binding energy of 528.9 eV [79,192]

An alternative approach to studying metastable oxygen states is to coadsorb probe molecules with dioxygen with the aim of intercepting short-lived species before the $O^{2-}(a)$ state is formed. Thus adsorption of oxygen simultaneously with water, ammonia and methanol at copper [41,76,77,85,87–89], magnesium [34,59], and nickel [98,99] surfaces has been shown to result in very efficient and selective reactions, consistent with the involvement of a reactive oxygen species. This species has been interpreted as an oxygen transient, although distinguishing between a reactive, isolated oxygen adatom at a non-reconstructed surface and either a molecular or 'hot' oxygen atom transient is not always straightforward.

5.2 *Subsurface oxygen*

Place exchange, resulting in the penetration of oxygen ions below metal surfaces, is an essential step in the development of multilayer oxide films, but it has been suggested that in some cases this process begins before the completion of the first oxygen monolayer. The experimental discrimination between 'surface' and 'subsurface' oxygen species is not straightforward, however. At aluminium surfaces, for example, HREELS spectra for low oxygen coverages [193,194] were initially interpreted in terms of a subsurface oxygen state, in agreement with early work-function measurements on evaporated aluminium films [195]. Recent medium energy ion scattering (MEIS) results [196] and STM images [197], however, strongly suggest subsurface oxygen is not formed until the monolayer is complete.

Work-function measurements are perhaps the most straightforward indication of the presence of subsurface oxygen, the surface dipole being reversed as the oxygen penetrates the surface, thus decreasing the work function. The presence of subsurface oxygen at Ni(100) [198] and Pt(100) [199–201] surfaces has thus been established by work-function measurements. The case of silver, where subsurface and dissolved oxygen are well characterised, has been discussed already (Sections 4.3.2 and 4.3.4).

Two characteristics shared by most of the oxygen states that have been assigned to subsurface oxygen are low reactivities towards probe molecules [200,202] and high O(1s) binding energies (greater than 531 eV) [79,149]. Both observations can, however, be misleading, the high temperatures and pressures often required to form the subsurface state increasing the chances of contamination with species such as carbonate and nitrate.

An interesting new development is the suggestion that coadsorbates may increase the rate of formation of subsurface oxygen. At a Pt(100) surface [199–201], for example,

CO and H_2 both increased the rate of formation of subsurface oxygen by controlling the reconstruction of the surface, and a similar observation has been reported for Cu(111) surfaces [79].

5.3   *Oxygen dimerisation*

There are now several reported studies where dissociated oxygen species have been observed to recombine and form an adsorbed molecular species. In the case of Bi(0001)–O, a second oxygen species is observed to form at 145 K at the expense of the chemisorbed oxygen species [203]. The second species was unstable and warming the surface to 295 K caused it to transform back to its precursor. The initial conversion $O^{2-}(a) \rightarrow O_2^-(a)$ was found to be induced by the X-rays used for the XPS study.

Oxygen dimerisation at a Zn(0001)–O surface was induced by depositing several monolayers of barium — the O(1s) spectra showed changes reflecting the formation of BaO and also a peroxo-like species associated with both the Zn and Ba sites [204]. The latter was identical to the oxygen species observed during the oxidation of a Ba-doped Zn(0001) surface [205] and an assignment as BaO_2 was ruled out.

Oxygen adatom recombination followed by desorption, the complete reversal of the reaction sequence shown on p. 77, was reported in a study designed to prepare a model catalyst surface by depositing copper clusters on a Zn(0001)–O surface where the surface concentration of oxygen atoms was about a monolayer [206]. Deposition at 295 K resulted in a 75% decrease in the intensity of the O(1s) peak due to $O^{2-}(a)$ and the disappearance of the Zn(LMM) chemically shifted Zn^{2+} component. The copper diffused into the substrate to form a brass-like alloy which had little affinity for oxygen. By preparing the Zn(0001)–O surface under somewhat more severe conditions of temperature and oxygen pressure, a stable (presumably more oxide-like) surface could be prepared.

6   Reactivity and the charge on the oxygen species

The charge on an adsorbed oxygen species will depend on the local atomic environment, but as a rule of thumb the higher the coordination of an oxygen species the higher its relative charge. Since an O^{2-}-like species is isolectronic with neon and an O^--like species with fluorine, this picture fits very well with the observed reactivity of surface oxygen species, which decreases sharply as the coordination increases. Although there is as yet no reliable method for measuring the charge on an adsorbed atom, in many situations XPS can provide useful information: thus, highly reactive $O^{\delta-}(a)$-like species have binding energies of 531 eV consistent with a charge lower than the chemisorbed oxygen state. However, final-state effects can exert a strong influence on the observed binding energy, as in the case of subsurface oxygen states: these are expected to be highly charged yet also exhibit O(1s) binding energies of *c*. 531 eV.

The relationship between charge and reactivity is the subject of theoretical calculations [91], which will be useful in determining, for example, how the charge varies between the different oxygen states (with different chemical reactivities) which were identified in the Cu(110)–O system (Section 4.1).

7 Conclusions

Our understanding of the nature of oxygen states at metal surfaces has advanced considerably since the days in which oxygen was often regarded as a poison for reactions over supported metal catalysts. This has occurred largely through the synergistic combination of spectroscopic and chemical reactivity measurements. We can now identify a range of oxygen species with distinct characteristics. At one extreme is the 'oxide'-like state characterised by low XP binding energies and, in general, little reactivity towards probe molecules. Spectroscopically distinct from the latter, though equally unreactive, are the adatoms in the centres of chemisorbed oxygen islands. The perimeters of the oxygen islands are more reactive than the centres, though generally the two cannot be distinguished spectroscopically. The most reactive of the chemisorbed oxygen states is the species formed under conditions where the metal surface is prevented from reconstructing. These are stabilised either by the presence of a coadsorbate or by low temperatures. Finally, though not discussed here in any detail, there are the highly reactive oxygen transients, both molecular and atomic species, which may play key roles in many surface reactions, in particular the selective oxidation of hydrocarbons.

8 Acknowledgements

We would like to thank Prof. G. Ertl for providing a preprint of [39], and Professor A.M. Bradshaw for permission to reproduce Fig. 4.2. We are also grateful to Professor M.W. Roberts both for encouraging us to write this review and for many stimulating discussions.

9 References

1   Roberts MW. *Appl. Surf. Sci.* 1991; **52**: 133.
2   Roberts, MW. *Surf. Sci.* 1994; **300**: 769.
3   Besenbacher F, Stensgaard I. In King DA, Woodruff DP (eds) *The Chemical Physics of Solid Surfaces.* London: Elsevier 1996; 573.
4   Besenbacher F, Norskov JK. *Progr. Surf. Sci.* 1993; **44**: 5.
5   Stensgaard I, Ruan L, Besenbacher F, Jensen F, Lægsgaard E. *Surf. Sci.* 1992; **270**: 81.
6   Woodruff DP. In King DA, Woodruff DP (eds) *The Chemical Physics of Solid Surfaces.* London: Elsevier 1996; 465.
7   Barnes CJ. In King DA, Woodruff DP (eds) *The Chemical Physics of Solid Surfaces.* London: Elsevier 1996; 501.
8   Kiskinova M. *Chem. Rev.* 1996; **96**: 1431.
9   Quinn CM, Roberts MW. *Nature* 1963; **200**: 648.
10 Quinn CM, Roberts MW. *Trans. Faraday Soc.* 1964; **60**: 899.
11 Brundle CR, Carley AF. *Chem. Phys. Lett.* 1975; **31**: 423.
12 Carley AF, Roberts MW. *Proc. Roy. Soc., London A* 1978; **363**: 403.
13 Carley AF. *Catal. Today* 1992; **12**: 413.
14 Carley AF, Chalker PR, Roberts MW. *Proc. Roy. Soc. London A* 1985; **399**: 167.
15 Carley AF, Chalker PR, Riviere JC, Roberts MW. *J. Chem. Soc. Faraday Trans. I* 1987; **83**: 351.
16 Madey TE, Yates JT, Erickson NE. *Chem. Phys. Lett.* 1973; **19**: 487.
17 Jupille J, Dolle P, Besançon M. *Surf. Sci.* 1992; **260**: 271.

18 Pettenkofer C, Pockrand I, Otto A. *Surf. Sci.* 1983; **135**: 52.
19 Vandenhoek PJ, Baerends EJ. *Surf. Sci.* 1989; **221**: L791.
20 McKee CS, Perry DL, Roberts MW. *Surf. Sci.* 1973; **39**: 176.
21 Kulkarni GU, Laruelle S, Roberts MW. *J. Chem. Soc., Chem. Commun.* 1996; 9.
22 Kulkarni GU, Laruelle S, Roberts MW. *J. Chem. Soc. Faraday Trans.* 1996; **92**: 4793.
23 Diehl RD, McGrath R. *Surf. Sci. Rep.* 1996; **23**: 43.
24 Kishi K, Roberts MW. *J. Chem. Soc., Faraday Trans. I* 1975; **71**: 1721.
25 Joyner RW, Kishi K, Roberts MW. *Proc. Roy. Soc. London A* 1977; **358**: 223.
26 Saleh JM, Wells BR, Roberts MW. *Trans. Faraday Soc.* 1964; **60**: 1865.
27 Carley AF, Hegde MS, Roberts MW. *Chem. Phys. Lett.* 1982; **90**: 108.
28 Blake PG, Carley AF, Roberts MW. *Surf Sci.* 1982; **123**: L733.
29 Blake PG, Carley AF, di Castro V, Roberts MW. *J. Chem. Soc. Faraday Trans. I* 1986; **82**: 723.
30 Madix RJ. *Science* 1986; **233**: 1159.
31 Au CT, Breza J, Roberts MW. *Chem. Phys. Lett.* 1979; **66**: 340.
32 Fisher GB, Sexton BA. *Phys. Rev. Lett.* 1980; **44**: 683.
33 Au CT, Roberts MW. *Chem. Phys. Lett.* 1980; **74**: 472.
34 Au CT, Li XC, Tang JA, Roberts MW. *J Catal.* 1987; **106**: 538.
35 Au CT, Roberts MW. *J. Chim. Phys.* 1981; **78**: 921.
36 Falconer J, McCarty J, Madix RJ. *Surf. Sci.* 1974; **42**: 329.
37 Madix RJ. *Surf. Sci.* 1994; **299–300**: 785.
38 Amorelli TS, Carley AF, Rajumon MK, Roberts MW, Wells PB. *Surf. Sci.* 1994; **315**: L900.
39 Wintterlin J, Schuster R, Ertl G. *Phys. Rev. Lett.* 1996; **77**: 123.
40 Madix RJ, Wachs IE. *J. Catal.* 1978; **53**: 208.
41 Afsin B, Davies PR, Pashusky A, Roberts MW, Vincent D. *Surf. Sci.* 1993; **284**: 109.
42 Carley AF, Davies PR, Roberts MW, Vincent DJ. *Topics Catal.* 1994; **1**: 35.
43 Leibsle FM, Francis SM, Davis R, Xiang N, Haq S, Bowker M. *Phys. Rev. Lett.* 1994; **72**: 2569.
44 Leibsle FM, Francis SM, Haq S, Bowker M. *Surf. Sci.* 1994; **318**: 46.
45 Madix RJ. Private communication, 1996.
46 Ruan L, Stensgaard I, Læsgaard E, Besenbacher F. *Surf. Sci.* 1994; **314**: L873.
47 Leibsle FM, Murray PW, Francis SM, Thornton G, Bowker M. *Nature* 1993; **363**: 706.
48 Sueyoshi T, Sasaki T, Iwasawa Y. *Chem. Phys. Lett.* 1995; **241**: 189.
49 Sueyoshi T, Sasaki T, Iwasawa Y. *Surf. Sci.* 1995; **343**: 1.
50 Sueyoshi T, Sasaki T, Iwasawa Y. *J. Phys. Chem.* 1996; **100**: 1048.
51 Buisset J, Rust HP, Schweizer EK, Cramer L, Bradshaw AM. *Surf. Sci.* 1996; **349**: L147.
52 Roberts MW. In Morterra C, Zecchina A, Costa G (eds) *Structure and Reactivity of Surfaces.* Amsterdam: Elsevier 1989; 787.
53 Roberts MW. In Nowotny J, Dufour LC (eds) *Surface and Near-Surface Chemistry of Oxide Materials.* Amsterdam: Elsevier 1988, 219.
54 Carley AF, Rassias S, Roberts MW, Wang TH. *J. Catal.* 1979; **60**: 385.
55 Vattuone L, Yeo YY, King DA. *J. Chem. Phys.* 1996; **104**: 8096.
56 Carley AF, Roberts MW, Yan S. *J. Chem. Soc., Chem. Commun.* 1988; 267.
57 Carley AF, Yan S, Roberts MW. *J. Chem. Soc. Faraday Trans.* 1990; **86**: 2701.
58 Au CT, Roberts MW. *J. Chem. Soc. Faraday Trans. I* 1987; **83**: 2047.
59 Au CT, Roberts MW. *Nature* 1986; **319**: 206.
60 Kuk Y, Chua FM, Silverman PJ, Meyer JA. *Phys. Rev. B* 1990; **41**: 12393.
61 Wintterlin J, Schuster R, Coulman DJ, Ertl G, Behm RJ. *J. Vac. Sci. Technol. B* 1991; **9**: 902.
62 Jensen F, Besenbacher F, Lægsgaard E, Stensgaard I. *Phys. Rev. B* 1990; **41**: 10233.
63 Feidenhansl R, Grey F, Nielsen M *et al. Phys. Rev. Lett.* 1990; **65**: 2027.
64 Coulman D, Wintterlin J, Barth JV, Ertl G, Behm RJ. *Surf. Sci.* 1990; **240**: 151.
65 Baddorf AP, Wendelken JF. *Surf. Sci.* 1991; **256**: 264.

66 Jensen F, Besenbacher F, Lægsgaard E, Stensgaard I. *Phys. Rev. B* 1990; **42**: 0926.

67 Judd RW, Hollins P, Pritchard J. *Surf. Sci.* 1986; **171**: 643.

68 Jensen F, Besenbacher F, Lægsgaard E, Stensgaard I. *Surf. Sci.* 1991; **259**: L774.

69 Jensen F, Besenbacher F, Stensgaard I. *Surf. Sci.* 1992; **270**: 400.

70 Tillborg H, Nilsson A, Hernnas B, Martensson N. *Surf. Sci.* 1992; **270**: 300.

71 Hupkens TM, Fluit JM, Niehaus A. *Surf. Sci.* 1986; **165**: 327.

72 Vanpruissen OP, Dings MMM, Gijzeman OLJ. *Surf. Sci.* 1987; **179**: 377.

73 Sasaki T, Sueyoshi T, Iwasawa Y. *Surf. Sci.* 1994; **316**: L1081.

74 Crew WW, Madix RJ. *Surf. Sci.* 1996; **349**: 275.

75 Crew WW, Madix RJ. *Surf. Sci.* 1994; **319**: L34.

76 Afsin B, Davies PR, Pashuski A, Roberts MW. *Surf. Sci.* 1991; **259**: L724.

77 Boronin AI, Pashusky A, Roberts MW. *Catal. Lett.* 1992; **16**: 345.

78 Davies PR, Shukla N, Vincent DJ. *J. Chem. Soc. Faraday Trans.* 1995; **91**: 2885.

79 Davies PR, Roberts MW, Shukla N, Vincent DJ. *Surf. Sci.* 1995; **325**: 50.

80 Carley AF, Owens AW, Rajumon MK, Roberts MW, Jackson SD. *Catal. Lett.* 1995; **37**: 79.

81 Barnes C, Pudney P, Guo QM, Bowker M. *J. Chem. Soc. Faraday Trans.* 1990; **86**: 2693.

82 Francis SM, Leibsle FM, Haq S, Xiang N, Bowker M. *Surf. Sci.* 1994; **315**: 284.

83 Bowker M, Madix RJ. *Surf. Sci.* 1980; **95**: 190.

84 Russell JN, Gates SM, Yates JT. *Surf. Sci.* 1985; **163**: 516.

85 Carley AF, Davies PR, Roberts MW, Shukla N, Song Y, Thomas KK. *Appl. Surf. Sci.* 1994; **81**: 265.

86 Carley AF, Davies PR, Roberts MW, Thomas KK. *Surf. Sci.* 1990; **238**: L467.

87 Davies PR, Mariotti GG. *J. Phys. Chem.* 1996; **100**: 19975.

88 Davies PR, Mariotti GG. *Catal. Lett.* 1997; **43**: 261.

89 Davies PR, Mariotti GG. *J. Chem. Soc., Chem. Commun.* 1996; 2319.

90 Neurock M, van Santen RA, Biemolt W, Jansen APJ. *J. Am. Chem. Soc.* 1994; **116**: 6860.

91 Davies PR, van Santen RA, Mingdan C. In preparation 1996.

92 Oed W, Lindner H, Starke U, Heinz K, Muller K, Pendry JB. *Surf. Sci.* 1989; **224**: 179.

93 Oed W, Lindner H, Starke U, Heinz K *et al. Surf. Sci.* 1990; **225**: 242.

94 Eierdal L, Besenbacher F, Lægsgaard E, Stensgaard I. *Surf. Sci.* 1994; **312**: 31.

95 Carley AF, Grubb SR, Roberts MW. *J. Chem. Soc., Chem. Commun.* 1984; 459.

96 Rajumon MK, Prabhakaran K, Rao CNR. *Surf. Sci.* 1990; **233**: L237.

97 Rao CNR, Vijayakrishnan V, Kulkarni GU, Rajumon MK. *Appl. Surf. Sci.* 1995; **84**: 285.

98 Kulkarni GU, Rao CNR, Roberts MW. *J. Phys. Chem.* 1995; **99**: 3310.

99 Kulkarni GU, Rao CNR, Roberts MW. *Langmuir* 1995; **11**: 2572.

100 Rocca M, Lehwald S, Ibach H. *Surf. Sci.* 1985; **163**: L738.

101 Franchy R, Wuttig M, Ibach H. *Surf. Sci.* 1989; **215**: 65.

102 Voigtlander B, Lehwald S, Ibach H. *Surf. Sci.* 1990; **225**: 162.

103 Netzer FP, Madey TE. *Surf. Sci.* 1982; **119**: 422.

104 Madey TE, Benndorf C. *Surf. Sci.* 1985; **152**: 587.

105 Huntley DR. *Surf. Sci.* 1990; **240**: 24.

106 Carley AF, Rassias S, Roberts MW. *Surf. Sci.* 1983; **135**: 35.

107 Madey TE, Netzer FP. *Surf. Sci.* 1982; **117**: 549.

108 Benndorf C, Nobl C, Madey TE. *Surf. Sci.* 1984; **138**: 292.

109 Stensgaard I, Ruan L, Lægsgaard E, Besenbacher F. *Surf. Sci.* 1995; **337**: 190.

110 Hashizume T, Taniguchi M, Motai K, Lu H, Tanaka K, Sakurai T. *Jpn J. Appl. Phys. Part 2* 1991; **30**: L1529.

111 Taniguchi M, Tanaka K, Hashizume T, Sakurai T. *Surf. Sci.* 1992; **262**: L123.

112 Taniguchi M, Tanaka K, Hashizume T, Sakurai T. *Chem. Phys. Lett.* 1992; **192**: 117.

113 Hashizume T, Taniguchi M, Motai K, Lu H, Tanaka K, Sakurai T. *Ultramicroscopy* 1992; **42**: 553.

114 Hashizume T, Taniguchi M, Motai K, Lu H, Tanaka K, Sakurai T. *Surf. Sci.* 1992; **266**: 282.

115 Sakurai T, Hashizume T, Lu H. *Vacuum* 1992; **43**: 1107.

116 Shimizu T, Tsukada M. *J. Vac. Sci. Technol. B* 1994; **12**: 2200.

117 Shimizu T, Tsukada M. *Solid State Commun.* 1993; **87**: 193.

118 Shimizu T, Tsukada M. *Surf. Sci.* 1993; **295**: L1017.

119 Pai WW, Bartelt NC, Peng MR, Reuttrobey JE. *Surf. Sci.* 1995; **330**: L679.

120 Fang CSA. *Surf. Sci.* 1990; **235**: L291.

121 Campbell CT. *Surf. Sci.* 1985; **157**: 43.

122 Bao X, Barth JV, Lehmpfuhl G *et al. Surf. Sci.* 1993; **284**: 14.

123 Yan S. Spectroscopic and kinetic studies of the adsorption and coadsorption of some small molecules on Zn(0001) and Ag(111) surfaces. *PhD thesis, University of Wales* 1989.

124 Campbell CT, Paffett MT. *Surf. Sci.* 1984; **143**: 517.

125 Barteau MA, Madix RJ. *Surf. Sci.* 1981; **103**: L171.

126 Eickmans J, Otto A, Goldmann A. *Surf. Sci.* 1985; **149**: 293.

127 Schmeisser D, Demuth JE, Avouris P. *Chem. Phys. Lett.* 1982; **87**: 324.

128 Prince KC, Paolucci G, Bradshaw AM. *Surf. Sci.* 1986; **175**: 101.

129 Capote AJ, Roberts JT, Madix RJ. *Surf. Sci.* 1989; **209**: L151.

130 Backx C, Degroot CPM, Biloen P. *Surf. Sci.* 1981; **104**: 300.

131 Sexton BA, Madix RJ. *Chem. Phys. Lett.* 1980; **76**: 294.

132 Vandenhoek PJ, Baerends EJ. *Surf. Sci.* 1989; **221**: L791.

133 Outka DA, Stohr J, Jark W, Stevens P, Solomon J, Madix RJ. *Phys. Rev. B* 1987; **35**: 4119.

134 Upton TH, Stevens P, Madix RJ. *J. Chem. Phys.* 1988; **88**: 3988.

135 Barteau MA, Madix RJ. *J. Electron Spectrosc. Relat. Phenom.* 1983; **31**: 101.

136 Selmani A, Sichel JM, Salahub DR. *Surf. Sci.* 1985; **157**: 208.

137 Pawelacrew J, Madix RJ, Stohr J. *Surf. Sci.* 1995; **339**: 23.

138 Ricart JM, Torras J, Clotet A, Sueiras JE. *Surf. Sci.* 1994; **301**: 89.

139 Campbell CT. *Surf. Sci.* 1986; **173**: L641.

140 Broomfield K, Lambert RM. *Mol. Phys.* 1989; **66**: 421.

141 Demongeot FB, Valbusa U, Rocca M. *Surf. Sci.* 1995; **339**: 291.

142 Nakatsuji H, Nakai H. *J. Chem. Phys.* 1993; **98**: 2423.

143 Bao X, Muhler M, Pettinger B *et al. Catal. Lett.* 1995; **32**: 171.

144 Grant RB, Lambert RM. *Surf. Sci.* 1984; **146**: 256.

145 Bao X, Deng J, Dong SH. *Surf. Sci.* 1985; **163**: 444.

146 Ayre CR, Madix RJ. *Surf. Sci.* 1994; **303**: 297.

147 Bukhtiyarov VI, Boronin AI, Savchenko VI. *Surf. Sci.* 1990; **232**: L205.

148 Bukhtiyarov VI, Boronin AI, Savchenko VI. *J. Catal.* 1994; **150**: 262.

149 Rehren C, Muhler M, Bao X, Schlog R, Ertl G. *Z. Phys. Chem.* 1991; **174**: 11.

150 Rehren C, Isaac G, Schlogl R, Ertl G. *Catal. Lett.* 1991; **11**: 253.

151 Pettinger B, Bao X, Wilcock IC, Muhler M, Ertl G. *Phys. Rev. Lett.* 1994; **72**: 1561.

152 Garfunkel EL, Ding X, Dong G, Yang S, Hou X, Wang X. *Surf. Sci.* 1985; **164**: 511.

153 Mehandru SP, Anderson AB. *Surf. Sci.* 1989; **216**: 105.

154 Torras J, Ricart JM, Illas F, Rubio J. *Surf. Sci.* 1993; **297**: 57.

155 Bare SR, Griffiths K, Lennard WN, Tang HT. *Surf. Sci.* 1995; **342**: 185.

156 Thornburg DM, Madix RJ. *Surf. Sci.* 1990; **226**: 61.

157 Stensgaard I, Lægsgaard E, Besenbacher F. *J. Chem. Phys.* 1995; **103**: 9825.

158 Backx C, Degroot CPM, Biloen P, Sachtler WMH. *Surf. Sci.* 1983; **128**: 81.

159 Backx C, Degroot CPM, Biloen P. *Appl. Surf. Sci.* 1980; **6**: 256.

160 Burghaus U, Conrad H. *Surf. Sci.* 1995; **338**: L869.

161 Liu YH, Wang WC, Zheng JB. *Phys. Rev. B* 1993; **47**: 3929.

162 Deng JF, Xu XH, Wang JH, Liao YY, Hong BF. *Catal. Lett.* 1995; **32**: 159.

163 van Santen RA, Kuipers HPC. *Adv. Catal.* 1986; **35**: 265.

164 Carter EA, Goddard WA. *Surf. Sci.* 1989; **209**: 243.

165 Backx C, Degroot CPM. *Surf. Sci.* 1982; **115**: 382.

166 Grant RB, Lambert RM. *J. Catal.* 1985; **92**: 364.

167 Hawker S, Mukoid C, Badyal JPS, Lambert RM. *Surf. Sci.* 1989; **219**: L615.

168 Vandenhoek PJ, Baerends EJ, van Santen RA. *J. Phys. Chem.* 1989; **93**: 6469.

169 Bukhtiyarov VI, Prosvirin IP, Kvon RI. *Surf. Sci.* 1994; **320**: L47.

170 Bukhtiyarov VI, Boronin AI, Prosvirin IP, Savchenko VI. *J. Catal.* 1994; **150**: 268.

171 Baschenko OA, Bukhtiyarov VI, Boronin AI. *Surf. Sci.* 1992; **271**: 493.

172 Joyner RW, Roberts MW. *Chem. Phys. Lett.* 1979; **60**: 459.

173 Bao X, Muhler M, Pettinger B, Schlogl R, Ertl G. *Catal. Lett.* 1993; **22**: 215.

174 Bukhtiyarov VI, Carley AF, Dollard LA, Roberts MW. *Surf. Sci.* (accepted for publication).

175 Ormerod RM, Peat KL, Wytenburg WJ, Lambert RM. *Surf. Sci.* 1992; **270**: 506.

176 Grant RB, Lambert RM. *J. Catal.* 1985; **93**: 92.

177 Sault AG, Madix RJ. *Surf. Sci.* 1986; **172**: 598.

178 Kondarides DI, Verykios XE. *Studies Surf. Sci. Catal.* 1994; **82**: 471.

179 Chesters MA, Somorjai GA. *Surf. Sci.* 1975; **52**: 21.

180 Legare P, Hilaire L, Sotto M, Maire G. *Surf. Sci.* 1980; **91**: 175.

181 Canning NDS, Outka D, Madix RJ. *Surf. Sci.* 1984; **141**: 240.

182 Sault AG, Madix RJ, Campbell CT. *Surf. Sci.* 1986; **169**: 347.

183 Lazanga MA, Wickham DT, Parker DH, Kastanas GN, Koel BE. In Oyama ST, Hightower JW (eds) *Catalytic Selective Oxidation.* Washington DC: American Chemical Society, 1996; 90.

184 Akhlaque MS, Carley AF, Davies PR, Read S, Roberts MW. In preparation.

185 Outka DA, Madix RJ. *Surf. Sci.* 1987; **179**: 351.

186 Outka DA, Madix RJ. *Surf. Sci.* 1987; **179**: 361.

187 Outka DA, Madix RJ. *J. Am. Chem. Soc.* 1987; **109**: 1708.

188 Gallagher DE. Photoelectron spectroscopic studies of aluminium and chromium surfaces. *PhD thesis, University of Wales* 1987.

189 Carley AF, Roberts MW. Unpublished results.

190 Au CT, Roberts MW. *Proc. Roy. Soc. London A* 1984; **396**: 165.

191 Au CT, Roberts MW, Zhu AR. *J. Chem. Soc., Chem. Commun.* 1984; 737.

192 Usuki N. *Vacuum* 1990; **41**: 1683.

193 Erskine JL, Strong RL. *Phys. Rev. B* 1982; **25**: 5547.

194 Crowell JE, Chen JG, Yates JT. *Surf. Sci.* 1986; **165**: 37.

195 Roberts MW, Wells BR. *Surf. Sci.* 1969; **15**: 325.

196 O'Connor DJ, Wouters ER, Vandergon AWD *et al. Surf. Sci.* 1993; **287**: 438.

197 Brune H, Wintterlin J, Trost J, Ertl G, Wiechers J, Behm RJ. *J. Chem. Phys.* 1993; **99**: 2128.

198 Pope TD, Bushby SJ, Griffiths K, Norton PR. *Surf. Sci.* 1991; **258**: 101.

199 Rotermund HH, Lauterbach J, Haas G. *Appl. Phys. A* 1993; **57**: 507.

200 Lauterbach J, Asakura K, Rotermund HH. *Surf. Sci.* 1994; **313**: 52.

201 Vonoertzen A, Mikhailov A, Rotermund HH, Ertl G. *Surf. Sci.* 1996; **350**: 259.

202 Ladas S, Imbihl R, Ertl G. *Surf. Sci.* 1993; **280**: 14.

203 Joyner RW, Roberts MW, Singhboparai SP. *Surf. Sci.* 1981; **104**: L199.

204 Carley AF, Rajumon MK, Roberts MW, Wang FC. *J. Solid State Chem.* 1994; **112**: 214.

205 Carley AF, Rajumon MK, Roberts MW, Wang FC. *Solid State Commun.* 1994; **91**: 791.

206 Carley AF, Roberts MW. *Topics Catal.* 1996; **3**: 91.

5 Chemical Reactivity of Metal Clusters at Solid Surfaces

V.I. BUKHTIYAROV

Boreskov Institute of Catalysis, Prospekt Akademika Lavrentieva, 5, Novosibirsk 630090, Russia

1 Introduction

Nanometre metal clusters are an intriguing class of research objects due to their unusual electronic and structural properties. In consequence, their chemical reactivity can be different from that of bulk metals. The variation of chemical properties of small clusters is of great importance for catalysis, since a great number of real catalysts represent metal particles supported on an inert carrier. The economic incentive for decreasing the metal particle size, especially for catalysts in which noble or rare metals serve as an active component, explains the long-standing interest of researchers in the field of catalysis. Moreover, the chemical reactivity of metal clusters on a solid is becoming one of the most important fields of study in the development of fundamental catalytic science.

During the 1990s the empirical approach to the development of new catalysts or the improvement of existing ones has begun to be replaced by a fundamental approach to the study of the mechanisms of catalytic reactions in order to understand the catalytic action of various systems. This transformation of catalytic science correlates with the appearance of commercially designed equipment using surface-sensitive techniques such as Auger electron spectroscopy (AES), low-energy electron diffraction (LEED), X-ray and ultraviolet photoelectron spectroscopies (XPS and UPS), electron energy-loss spectroscopy (EELS), scanning tunnelling microscopy (STM), etc. The 30th anniversary issue of the journal *Surface Science* [1] not only represents an excellent review of the history of and modern developments in these techniques, but also contains many interesting examples of their application to the study of catalytic and other systems. However, despite the fact that information can now be obtained at the atomic and molecular level, there are two limitations on the application of surface-science techniques to real catalysts: the *pressure gap* and *material gap* problems. The pressure gap problem originates from the difference between the pressures at which almost all surface-science methods can work ($P < 10^{-4}$ Pa) and those used in real catalytic processes ($P > 10^5$ Pa). The other peculiarity of surface-science studies is the use of well-ordered single crystals as model samples, while real catalysts represent dispersed metal as an active component supported on an inert carrier. The discrepancy in the nature of the bulk and dispersed metals is the basis for the material gap problem.

Since the mid-1980s researchers have paid much attention to tackling the pressure gap problem. The integration of high-pressure cells or microreactors into high-vacuum equipment has made it possible to measure the reaction rates at higher pressure followed by analysis of a surface composition, without exposing the sample to the atmosphere. A good example is found in the works of Campbell [2,3] who has used a spectrometer with very fast transfer (less than 20 s) from a high-pressure microreactor

109

to the vacuum chamber to study ethylene epoxidation on Ag(111) and Ag(110) single crystals. The successful development of a high-pressure electron spectrometer (VG Scientific) should also be noted. This spectrometer enables one to record XPS and UPS spectra *in situ* at pressures up to 100 Pa, which is six orders of magnitude higher than that used in conventional spectrometers [4]. Two of these spectrometers have been used effectively at the University of Wales College of Cardiff (Cardiff, UK) and at the Boreskov Institute of Catalysis (Novosibirsk, Russia) [5,6].

The material gap problem has been also studied for a long time. Investigations of the dependence of the catalytic properties of supported metal catalysts on particle size (see, for example, the review of Che [7]) should be recognised as a first step in this direction. It has been shown that for many catalytic reactions their rate, normalised to one surface atom of a metal, varies with the metal particle size. Contrasting with these reactions, termed by Boudart [8] structure-sensitive, there are also structure-insensitive reactions, the rate of which does not depend on the supported particle size. Obviously, to explain these phenomena, the electronic, structural and adsorption properties of small metal clusters should be precisely studied and compared with the bulk metal properties. Real supported catalysts, which are usually prepared by impregnation of a high-area support by an active component salt solution followed by calcination and reduction to the metallic state, should be related to colloidal systems. The behaviour of the latter is known to be described by a statistical approach. This, together with the possible appearance of uncontrolled impurities introduced both with the support and at the stages of the catalyst preparation, can explain the discrepancies in the data obtained for similar catalytic systems by different authors [7]. It is clear that many repetitions in the preparation of similar samples are necessary to avoid irreproducibility in the results. Furthermore, researchers should take into account that the preparation of a series of catalysts with various metal cluster sizes using impregnation is itself the time-consuming step.

There are two ways in which we can decrease the time necessary to produce reliable and convincing data: first, the preparation of catalysts with various particle sizes by step-by-step ageing (sintering) of the freshly prepared sample under reaction conditions; and second, the use of ultrahigh-vacuum (UHV) metal deposition for the controlled preparation of metal clusters. Despite the evident advantages, neither method is free of limitations. Thus, the treatment of the fresh catalyst by a reaction mixture at high temperatures, used as a sintering procedure, can change the nature and composition of the samples in a series. Furthermore, this preparation method retains all the drawbacks characteristic of real catalysts: undesirable impurities, insufficiently complete reduction of metallic clusters, etc. The other method of metal cluster preparation, based on resistive evaporation of metal, allows one to avoid these difficulties. Indeed, the realisation of this method inside the high-vacuum environment makes it possible, first, to clean the support and anneal the evaporation system before metal deposition, and second, to transfer the prepared sample to the analyser chamber without contact with the atmosphere. The mechanisms of nucleation and growth of clusters deposited in UHV have been critically reviewed by Poppa [9]. The data presented in Poppa's review have shown that UHV deposition provides a means of controlled, reproducible preparation of small supported metal clusters on to a planar support. These aspects explain the great interest in studying clusters prepared by this method using various surface-science techniques. Many interesting observations about the influence of metal cluster sizes on their electronic and structural properties have been revealed. However, the catalytic properties have not been tested because of the low surface area of the planar supports used in those model studies.

It should, however, be noted that the replacement of a real support by a model one can also have positive advantages, since it can remove a limitation on the application of some physical methods for studying the reactivity of clusters: for example, the use of graphite instead of an oxidic support for studying oxygen adsorption on metal clusters by XPS. Nevertheless, it is obvious that the subsequent combining of surface science and catalytic data is necessary, and studying the chemical reactivity of metal clusters towards reagents and reaction products can help to achieve this. Summarising the above-mentioned discussion, a fundamental approach to studying a catalytic system can be represented by the following scheme.

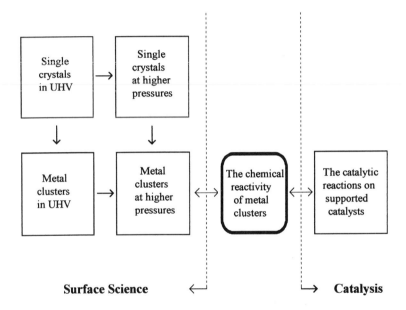

One can see that the proposed scheme consists of two parts: the field of surface science and the field of catalysis. Starting from the investigation of well-ordered single crystals in UHV, one should move both from left to right (the pressure gap problem) and from top to bottom (the material gap problem). Studying the nature of adspecies produced on the surface of a catalytically active metal and their variation in the sequence

single crystal → polycrystalline foils → powders → supported particles

should help to explain the peculiarities of the chemical reactivity of metal clusters and to make the final step into the field of catalysis.

In this paper I will demonstrate the validity of such an approach, paying special attention to the analysis of the chemical reactivity of metal clusters, which is the active component for different catalytic systems. Investigations devoted to the electronic and structural properties of small metal clusters will also be reviewed in brief.

2 Electronic properties

The electronic properties of metal clusters and their variation with cluster size have been extensively studied since the mid-1980s. It has been shown that as the size decreases, the electronic properties of small clusters deviate from those of the bulk. In many studies [10–16], this conclusion has been based on the variation of the core-level binding energy (E_B) of a metal measured by XPS. In general, the variation of the observed E_B of a given core level in XPS spectra (ΔE_B) is made up of the following contributions:

$$\Delta E_B = \Delta E + \Delta E_R + \Delta E_F \tag{1}$$

ΔE is the 'true' chemical shift, i.e. reflects the difference between the energies of the various initial states of the studied element. This contribution is responsible for the other name for XPS, which is well known to chemists — electron spectroscopy for chemical analysis (ESCA). The last two contributions in (1) refer to final-state effects, i.e. those which are caused by X-ray photoelectron emission, and transform the name 'ESCA' to 'XPS'. ΔE_R represents the variation of the relaxation energy. This contribution corresponds to the reorganisation of the electrons of the solid that provides the screening of the photoelectron hole remaining after electron emission. The last contribution, ΔE_F, represents the possible variation of the Fermi level in the samples studied with respect to the spectrometer Fermi level. If the conductivity of the sample is large enough, it is in equilibrium with the spectrometer. In this case, the positive charges induced by photoelectron emission are effectively neutralised by the electron flux from earth to the sample, and this contribution to the binding energy variation is negligible. Such a situation is realised if a bulk metal is studied. If the conductivity of the sample is not high enough to compensate for the charge induced by photoemission, then a significant positive charge accumulates at the surface. The electrons passing through this charged layer lose kinetic energy that results in an increase in apparent binding energy. This means that the Fermi levels of the sample and spectrometer are no longer in equilibrium, providing an unknown surface potential ϕ — the charging value.

The increase in E_B of small metal clusters (2–3 nm) with respect to the bulk metal has been discussed in terms of the initial state or relaxation effects. However, the studies of Mason on alloys and other systems [11] and of Rao *et al.* who have used high-energy spectroscopy [12,13] have shown that the relaxation effect cannot completely explain the observed binding energy shift. This has allowed the authors [12,13] to conclude that a transition from metal to non-metal takes place with a decrease in cluster size below 2–3 nm. This conclusion is of great significance since the changes in the initial state of metal atoms in a small cluster should affect their chemical reactivity towards adsorbed gases.

Unlike the initial state and relaxation effects, many authors who have studied the electronic properties of metal clusters have not taken into account the final contribution to E_B. This is justified in the case of homogeneous insulating samples, when the charging potential (ϕ) is the same for all core levels of the XPS spectrum, and it can be easily excluded by the internal reference method. A more complex situation appears in heterogeneous systems, where different phases possess different conductivities. In consequence, the surface charges will be different, and the so-called differential charging of a more conducting phase compared with a less conducting one will appear.

Returning to metal clusters on solid surfaces and taking into account the variation of internal conductivity of metal clusters with their size, one can expect the appearance of differential charging in two cases: large metal particles on such dielectric supports as alumina or silica; and small clusters on such conductive support as carbon. Indeed, Barr [17,18] has shown that metal crystallites of 50 to several hundred angstroms would appear to possess the solid-state properties of a conductive metal, providing easier compensation of the charge on their surface than on the surface of a dielectric support. In consequence, the charging potential will be less on the metal particles compared with the dielectric support and the internal reference method results in an apparently negative shift of the metal XPS lines. On the other hand, weakening the conductive properties by decreasing the metal cluster sizes below 50–100 Å should result in less effective compensation of their surface charge induced by photoemission compared with a conductive support. The latter, along with the initial state and relaxation effects, can be responsible for the increase in binding energies of XPS lines of a metal observed for small metal particles [10–16]. Indeed, Wertheim and co-workers [14–16] have suggested that such a surface positive charge is responsible for the increase in binding energies with particle size for Ag and Pd particles supported on graphite.

XPS parameters which do not depend on charging should be used to discriminate the effects associated with differential charging. One of these is the modified Auger parameter α that represents the sum of a core-level binding energy and the kinetic energy of the corresponding Auger peak [19,20]. Since the charging potential increases the core level binding energy and decreases the Auger kinetic energy by the same value, so their sum is not changed with charging.

The application of the modified Auger parameter has allowed us to show that it is the change in the conduction properties of the supported silver and nickel particles which is responsible for the variation of the $Ag(3d_{5/2})$ and $Ni(2p_{3/2})$ core-level binding energy measured for the fresh and treated $Ag–Al_2O_3$ and $Ni–C$ catalysts for ethylene epoxidation and selective butadiene hydrogenation, respectively [21–24]. The former catalytic system will be repeatedly cited in the present review, since it has been investigated using both the surface-science and catalytic approaches, i.e. in full agreement with the proposed scheme (Section 1).

A continuous increase in the $Ni(2p_{3/2})$ core-level binding energies up to 1.1 eV has been observed for Ni particles supported on filamentous carbon, together with a decrease in the Ni/C ratio, as result of their treatment in a reaction mixture [24]. The constancy of α and the absence of a shake-up satellite in the $Ni(2p_{3/2})$ spectra indicate that both the fresh and the treated samples contain metallic nickel, and the shift of the $Ni(2p_{3/2})$ spectra arises from the variation of the differential charging value due to a change in the properties of Ni conduction electrons. It should be noted that to observe these results we have used beryllium ceramics to mount the catalysts on to the sample holder [24]. This unusual material has a high thermal conductivity that is necessary to reduce the Ni surface by hydrogen at $T = 600$ K inside the spectrometer and disrupts the electrochemical equilibrium of a sample with the spectrometer due to its low electrical conductivity. The absence of electrochemical contact resulted in the appearance of total charging that stimulated the differential charging effect for the supported Ni compared with filamentous carbon.

The dielectric properties of alumina allowed us to observe a similar phenomenon for

Ag–Al$_2$O$_3$ catalysts without using such special mounting material [21,22]. It has been shown that the treatment of the freshly prepared Ag catalysts with a mean size of silver particles of 160 Å by C$_2$H$_4$ + O$_2$ reaction mixtures at T = 500 K increases the Ag(3d$_{5/2}$) binding energy by 0.4 eV, and this is accompanied by the same shift in the Ag(MNN) Auger spectrum. As a consequence, the Auger parameter remains unchanged and equals 726.2 eV. The comparision of this value of α with the known data for metallic silver (726.0–726.3 eV) and silver oxides (724.0–724.5 eV) [20] indicates that the samples contain metallic silver. This conclusion is in agreement with X-ray diffraction of synchrotron radiation (XRSDR) and transmission electron microscopy (TEM) data obtained on the same catalysts, which indicate that their phase composition corresponds to α-alumina and metallic silver [25]. Thus, the variation of E_B(Ag(3d$_{5/2}$)) caused by the reaction mixture treatment originates from the change of silver internal conductivity which in turn depends on Ag particle size [22]. The latter conclusion is based on the study of a number of Ag–Al$_2$O$_3$ catalysts with various metal particle sizes — from 100 to 1000 Å (Fig. 5.1). The E_B values have been measured after shifting all XPS lines by the charging value determined for each sample, with the Al(2p) line from the support being taken as an internal reference with a binding energy of 74.5 eV [20]. One can see that Ag particles with sizes above 300 Å are characterised by an E_B(Ag(3d$_{5/2}$)) value of 367.8 eV, which is lower by 0.3 eV than that for the bulk metal (368.2 eV) [20]. According to Barr [17,18] this means that this size range is characteristic of the silver differential charging compared with alumina, due to the appearance of silver internal conductivity. To check this suggestion we have studied dispersed silver deposited on graphite in UHV [26,27]. If the suggestion about the influence of the conductive electron properties of silver on the variation of the Ag(3d$_{5/2}$) binding energy is correct, then the use of conductive support such as graphite should change the trend in the E_B variation, increasing its magnitude above the bulk level (368.2 eV) for the particles less than 300 Å in diameter. This system allowed us, furthermore, to follow the electronic properties of supported silver over a wider range of particle sizes than for the Ag–Al$_2$O$_3$ system: from small clusters (10 Å) to large crystallites.

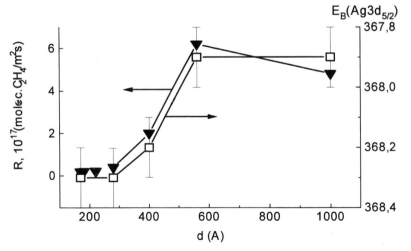

Figure 5.1. Variation of ethylene epoxidation rate (▼) and E_B(Ag(3d$_{5/2}$)) values (□), with mean sizes of metal particles measured for supported Ag–Al$_2$O$_3$ catalysts.

Figure 5.2 shows the variation of $Ag(3d_{5/2})$ core-level binding energies with $Ag(3d)/C(1s)$ intensity ratios, measured after Ag deposition on a graphite surface at room temperature. The estimation of the mean sizes of Ag particles presented also in Fig. 5.2 has been made using the data of Vijaykrishnan *et al.* [12] who have studied metal deposition on graphite by a combination of XPS and STM. The existence of two size ranges where the electronic properties of the supported silver change before they become close to that of the bulk metal can be seen from the data in this diagram. Recording the corresponding Auger spectra, followed by determination of the modified Auger parameter [27], has allowed us to conclude that the first decrease in binding energy of $Ag(3d_{5/2})$ from 368.9 to 368.4 eV originates from the transition from metal to non-metal, while the change in the conductivity of the deposited silver is responsible for the further decrease in binding energy which occurs at sizes of 200–300 Å. The fact that, unlike other metals [10–16], the increase in silver cluster sizes up to 100 Å is not enough to achieve the bulk value of the $Ag(3d_{5/2})$ binding energy, is in accordance with the data of previous studies [12,15], where the electronic properties of silver clusters have been analysed.

The coincidence of the size ranges where the $E_B(Ag(3d_{5/2}))$ values become lower than the bulk value for large silver particles on dielectric alumina, and higher for small silver particles on conductive graphite, proves that 300–500 Å is the range in which the properties of silver conduction electrons deviate from the bulk level. Although this contribution appears as a result of photoelectron emission (final-state effect), its study is of great significance for the analysis of the chemical reactivity of metal particles. Participation of the conduction electrons in the formation of chemical bonds between the metal surface and chemisorbed species was proposed as long ago as 1937 [28]. Quantum-chemical justification of this phenomenon has been provided by Newns [29]. The involvement of the metal conduction electrons in the surface chemical bond has been experimentally confirmed by the observation of the Knight shift in the ^{13}C nuclear magnetic resonance (NMR) resonance for CO adsorbed on supported metal particles

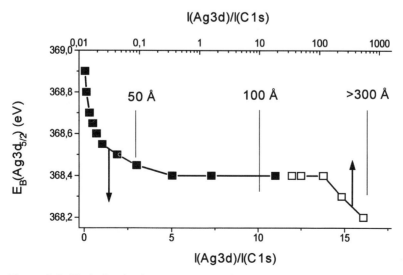

Figure 5.2. Variation in the $E_B(Ag(3d_{5/2}))$ core-level energies with coverage of silver deposited on a graphite surface.

[30,31]. Our catalytic data on the rate of silver-catalysed ethylene epoxidation superimposed in Fig. 5.1 also confirms this suggestion. One can see a very good correlation between catalytic activity and XPS data: the lower the binding energy of $Ag(3d_{5/2})$, the higher the rate of ethylene oxide formation.

It should be noted that the observed change in the internal conductivity of the silver takes place at much larger particle sizes (in excess of 300 Å) than the well-known transition from metal to non-metal (less than 50 Å). The following reasons can be suggested for this: sintering of Ag particles; modification of their subsurface layers by the reaction mixture; and changes to their morphology and shape. TEM analysis of the fresh and treated samples has not shown any considerable change in the mean sizes of the Ag particles under the conditions used [25], hence this reason can be excluded. The preparation of dispersed silver by its high-vacuum deposition on the thoroughly cleaned surface of graphite [26,27] allows us to suggest that the variation of the properties of the silver conduction electrons occurring around 100 Å is most likely to depend not only on the influence of dissolved oxygen and carbon atoms, formed from the $C_2H_4 + O_2$ reaction mixture, but also on changes in particle morphology. The influence of the morphology of supported particles on their conduction properties has been emphasised by Barr [17,18] who studied the differential charging of metal clusters compared with a dielectric support. I now consider the structural properties of metal clusters and their variation with particle size.

3   Structural properties

The surface structure of metal clusters has been studied both theoretically and experimentally [32–35]. The theoretical approach analyses the equilibrium shape of a crystal which is determined [32] by the Gibbs free energy of the system. The latter may be written as

$$G = \sum_{\substack{\text{phases}}} G_i + \sum_{\substack{\text{surface} \\ \text{and} \\ \text{interfaces}}} \gamma_i A_i + \sum_{\substack{\text{edges}}} \varepsilon_i l_i, \tag{2}$$

where the first term is associated with bulk thermodynamic phases, the second term contains flat surface contributions, and the third term is the energy originating from intersections between planes and corresponds to the curved surface region. It is clear that the surface structure of clusters is determined by the ratio between the flat and curved contributions, which have been suggested to depend on the cluster size. For the large particles, edge and curvature effects are negligible and the equilibrium shape of a crystal is determined by the relation $\gamma_i A_i = $ minimum, where the γ_i is the surface free energy per unit area A_i of exposed surface. At zero temperature the surface is characterised by planes at all rational Miller indices [32,33]. On raising the temperature only high surface atom density crystallographic orientations such as (111), (110) and (100) will be present, i.e. under real conditions. A decrease in cluster size will increase the curved contribution. Thus, depending on its size, a cluster can consist presumably of flat facets, of facets and curved regions, or of totally curved regions.

Experimentally, studying the surface structure of small clusters is not a straightforward matter, since atomic resolution should be inherent in a physical method used to tackle this problem. One of the most effective methods is TEM, which has allowed the existence of the flat and curved surface regions to be proved and their variation with cluster size to be studied [33–35]. Drechsler, who has studied clusters of different metals using TEM analysis combined with computer modelling of the surface structure [35], has shown that the surface of a Pt particle (c. 80 Å radius) on graphite consists of c. 37% (111) planes, c. 28% (100) faces and c. 35% curved area. Moreover, this improved TEM analysis indicates that the curved region is mainly composed of high-index planes such as (001), (113), (012) and (133). The results of computer structure modelling of the curved region and a computer image of the whole particle are presented in Fig. 5.3.

The other method for the analysis of surface structure consists of the adsorption of a molecule, the spectral characteristics of which are sensitive to surface structure. To use this method effectively, attention needs to be paid to two problems: first, the choice of the test molecule suitable for the metal studied; and second, the measurement of its spectral characteristics on well-defined single crystals with different surface structure. Obviously, to tackle the latter problem, data from the literature can also be used. It should be noted that this method of surface structure analysis leads to *in-situ* information, as opposed to TEM, which is an *ex-situ* method. This is of great significance in studying catalytic reactions, especially if a test molecule is one of the reagents. This affirmation is based on the well-known fact that the adsorption of reagents can reconstruct the surface not only of clusters, but even single crystals [36].

For examples of this method, the reader should consult references [37–40], where the adsorption of CO on silica-supported Pd clusters was studied by NMR or infrared absorption spectroscopy (IRAS). It has been shown that a fraction of the ^{13}CO adsorbed on supported Pd catalysts is bound in a linear as opposed to a bridging fashion. Raising the Pd dispersion from 16–19% to 56% increases the fraction of linear CO. The authors explain this effect by postulating a larger fraction of low-coordination Pd surface sites as the particle sizes decrease. This explanation is in agreement with the above-mentioned TEM observations that small clusters are characterised by a large proportion of curved surface [35]. Similar conclusions follow from the works of Goodman and co-workers, who have shown that on small palladium particles, more CO is adsorbed on the top

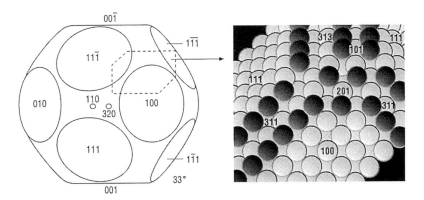

Figure 5.3. Profiles and faces obtained for the 80 Å Pt particle by computer calculation [35]. A model of the structure for the region indicated by broken lines is also shown.

sites than on the bridging ones [38]. This fact is consistent with there being more edge/defect sites on the small particles than on the large ones.

To study the surface structure of supported silver we have proposed the use of low-temperature O_2 adsorption followed by analysis of O(1s) core-level spectra [41]. This approach arises from the fact that oxygen dissociation on silver is determined by the surface structure: the more defective the silver surface, the lower the O_2 dissociation temperature [42–45]. Thus, the appearance of atomic oxygen after O_2 adsorption at low temperature correlates with the defectiveness of the supported silver surface structure. Alternatively, molecular oxygen will be indicative of a less defective surface. We have found that the nature of the chemisorbed oxygen depends on the Ag coverage or silver particle size [41]. The absence of molecularly chemisorbed oxygen for Ag particles with sizes below 100 Å is consistent with the more defective surface of small particles compared with that of the large ones. The appearance of molecular oxygen is observed as the Ag particle sizes exceed 200–400 Å, indicative of the formation of low-index planes such as (111) and (110). These data confirm again our suggestion about the considerable influence of silver-particle morphology on the properties of silver conduction electrons (see Section 2).

Summarizing the results cited in this section, it should be noted that despite the various approaches of different groups, the following general conclusion may be made: the surfaces of small metal clusters cannot be represented by low-index, more stable planes such as (111), (110) or (100) only; curved regions are also of great significance. This is one of the factors which limits the use of single-crystal data to real catalytic systems. Indeed, edges, kinks, steps and other defective sites, which are characteristic of the curved region, most probably have distinctive properties towards reagents compared with the flat surface. Clearly, this should be taken into account when the chemical reactivity of small metal particles is studied.

4   The chemical reactivity of metal clusters

The deviation of the electronic and structural properties of metal clusters from those of bulk metals, which is observed as their sizes decrease, suggests that their reactivity towards gas molecules should also depend on particle size. The first examples confirming this suggestion were obtained from catalytic studies [7], which demonstrate that the rate of many reactions catalysed by metal clusters varies in the same range of cluster sizes as that characteristic of the changes in electronic and structural properties [7,10–16,39]. However, the real nature of the supported catalysts used in those studies has not allowed the authors to correlate the variation of catalytic activity with the changes in the cluster reactivity. To achieve progress in this direction we must turn to the preparation of model samples by means of the high-vacuum deposition of metals on planar supports.

The first studies analysing the reactivity of metal clusters were devoted to CO oxidation on Pt-group metals [46–49]. Apparently, this fact is explained by the existence of a great number of precise investigations of this reaction on bulk Pt-group metals [50–53], the results and conclusions of which have been used to discuss the cluster data, i.e. the authors have followed the approach being discussed in this paper (see Section 1).

To explain the features of CO oxidation on Pd clusters deposited on a flat α-Al_2O_3 support, Ladas and co-workers [46] used the reaction mechanism established by Engel and Ertl for Pd(111) [50]. They proposed that the threefold increase in the rate of CO_2 formation with decreasing Pd cluster size from 4.9 to 1.5 nm observed at $T > 500$ K, originates from the increase in the impingement rate of CO on edge and corner Pd atoms, the amount of which increases for the small clusters [46]. However, in a later work [47] the authors changed the explanation by introducing an effective collection zone on the alumina which surrounds each Pd particle. Molecules of CO impinging on the support within this zone can diffuse to the Pd clusters where they are chemisorbed and oxidised [47]. Nevertheless, this novel explanation was not confirmed by Stara and Matolin [54] who studied the reactivity of Pd clusters towards CO by temperature programmed desorption (TPD). They showed the TPD spectrum of CO adsorbed on 27 nm Pd particles to be identical to that for the Pd(111) surface, while the spectrum for 2.5 nm particles is characterised by an additional state of CO with lower desorption temperature. The measurement of similar TPD spectra of CO in the case of Pd particles smaller than 5 nm deposited on mica [55] indicates that the formation of the weakly bonded CO does not depend on the nature of the support and is a general property of small Pd particles. It should be also noted that, besides the appearance of the weakly bonded CO, the data of Stara and Matolin [54] indicate that dissociation of CO occurs on small Pd particles in contrast with large particles and single crystals. This conclusion arises from the appearance of a CO_2 TPD peak produced via disproportionation of CO ($2CO \rightarrow C + CO_2$), leading to the deposition of carbon on the surface. This conclusion about CO dissociation was also reached by Gillet and co-workers [56] who studied CO adsorption on Pd–mica samples using TEM, TPD and static secondary ion mass spectrometry (SSIMS). These observations suggest that the strength of the CO–Pd bond at the surfaces of small particles can both increase and decrease compared with large particles. Without discussing the reasons for such different effects, I wish to note that the appearance of the weakly bonded state of CO is likely to determine the increasing rate of CO_2 formation observed in the works of Ladas *et al.* [46,47] with decreasing Pd cluster sizes.

The distinctive behaviour of the TPD peaks of CO has been observed for Pt particles supported on Al_2O_3 [57] and on mica [58]: decreasing the Pt cluster size below 4 nm reduces the intensity of the low-temperature CO peak at $T = 400$ K. This has allowed the authors to conclude that the CO molecule adsorbed on the small particles is more strongly bonded than the CO adsorbed on the large clusters. This observation is in accordance with the data of Herskowitz and co-workers, who reported that a decrease in the rate of CO oxidation occurs for the small Pt particles (less than 2.5 nm) [48].

A comprehensive study of CO adsorption on supported Pt and Rh particles has indicated that the relative population of the low- and high-temperature desorption states of CO is a much weaker function of particle size on Rh than on Pt [59]. The latter fact seems to determine the structure insensitivity of CO oxidation catalysed by rhodium defined by Oh and Eickel [49].

Thus, the established data demonstrate that despite essential differences in the size effects in CO + O_2 reaction on various Pt-group metals, they can be easily correlated with changes of strength of the CO–metal bond. The success in explaining such features of CO + O_2 catalytic reactions as structure insensitivity for Rh and sympathetic or antipathetic structure sensitivity for Pd and Pt, respectively, resulted in attempts by many researchers to use this approach for other catalytic systems [49,60–62]. Thus, to understand the reasons for the 45-fold decrease in the rate of CO + NO reaction with decreasing size of Rh particles supported on α-alumina from 676 to 10 Å, Oh and Eickel [49] applied the NO TPD results of Altman and Gorte [60]. The latter showed that for large particles most N_2 desorbs in a low-temperature peak at 190°C, with a relatively small amount of N_2 desorbing above 280°C, while the decrease in the particle size enhances the high-temperature peak at the expense of the low-temperature one. The slower removal of N_2 from the surface of small Rh particles is responsible for the structure sensitivity of the CO + NO reaction on rhodium [49]. It should be also noted that Oh and Eickel have observed that the support material (α-Al_2O_3, θ-Al_2O_3, SiO_2) had a small effect on the activity of Rh for the CO + NO reaction [49].

Another interesting example of this type of investigation is the work of Briot and co-workers [61], who themselves studied both the $CH_4 + O_2$ reaction and O_2 adsorption on Pt–Al_2O_3. To analyse the reactivity of Pt particles towards O_2, the authors determined the heat of oxygen chemisorption and oxygen reactivity towards H_2. Two samples with different mean Pt particle sizes (< 2 nm and 20 nm) were used for this study. They revealed that the sample with large particles is characterised by an eightfold increase in catalytic activity in methane oxidation, by a lower initial heat of oxygen adsorption (250 kJ mol^{-1} versus 280 kJ mol^{-1}) and by a higher reactivity of adsorbed oxygen towards hydrogen compared with the sample with small particles [61]. The latter observation is based on the fact that oxygen adsorbed on the Pt–Al_2O_3 catalyst with large particles is almost fully titrated by hydrogen at 195 K, while for small Pt clusters the oxygen is not titrated up to $T = 276$ K. All these facts point to a decrease in the strength of the surface platinum–oxygen bond as the Pt particle sizes increase, and this is correlated with the increase in catalytic activity of the corresponding samples [61].

Many other examples could be presented which demonstrate the usefulness of studying the chemical reactivity of metal clusters as a means of understanding the behaviour of supported catalysts. However, we consider now the attempts of many researchers to study the fundamental reasons for the variation of cluster reactivity, with both electronic and structural factors being taken into account.

The fact that CO desorption from Pt is strongly dependent on the crystal plane orientation [63–65] has led to discussion of the influence of structural factors on the variation of the heat of adsorption of CO with cluster size [59]. The closely packed Pt(111) surface is characterised by a main CO TPD peak at about 400 K [63], while an additional state at 510 K appears in the case of the more open Pt(110) face and Pt foil, with the relative intensity of the high-temperature peak increasing with increasing density of step sites [64,65]. The comparison of CO TPD spectra for single crystals with those for Pt–Al_2O_3, as a function of particle size, has allowed Altman and Gorte [57] to conclude that the weakening of the CO–Pt bond for large particles originates from the preferable presence of Pt(111) terraces, while the formation of defective sites with lower coordination is responsible for the opposite tendency on small particles. Such low-coordination sites have been associated by Stara and Matolin [54] with the dissociation of CO on small Pd clusters. They modelled their structure as a fourfold site on the (100) steps separating the (111) planes [54]. Repulsive interactions between carbon atoms with CO molecules coadsorbed on Pt(111) terraces has been suggested to weaken the CO–metal bond observed in the same study (see above). The structural approach is confirmed by the permanence of CO desorption spectra from crystal-plane orientations [66,67] in the case of Rh that is in agreement with structure insensitivity of the $CO + O_2$ reaction catalysed by rhodium. The formation of flat crystal planes, as the sizes of Pt particles supported on alumina increase, has been postulated by Briot *et al.* [61] in order to describe the decrease in the heat of oxygen adsorption and the increase in its reactivity.

The electronic approach to explaining the unusual reactivity of small metal clusters has been followed by those who have used physical methods (XPS, NMR, etc.) sensitive to both the nature of adspecies and the electronic properties of the metals [37,68–70]. The XPS study of CO adsorption on Pd, Cu and Ni clusters on graphite carried out by Santra and co-workers [68] showed that both the transition from metal to non-metal and the strengthening of the CO–metal bond caused by the decrease in cluster size occur in the same size range. The latter conclusion is based on the observation that higher temperatures are necessary to reduce the C(1s) signal assigned to the adsorbed CO (binding energy of 285.9 eV) in the case of small particles compared with larger ones [68]. Moreover, strengthening the CO–metal bond in the case of Ni results in CO dissociation that is confirmed by the appearance of a C(1s) line at 284.2 eV character-istic of elementary carbon, after warming the sample up to 300 K. The authors have suggested that the increase in the chemical reactivity of small nickel clusters towards CO arises from the closeness of their Ni(3d) level to the $2\pi^*$ level of CO.

The loss of metallicity for small Pd and Ni clusters (less than 3 nm) has been correlated with an increase in their reactivity towards O_2 [69] and hydrogen sulfide [70], respectively. In the former case, the conclusion about the strengthening of the O–Pd bond has been made on the basis of microcalorimetric data, which show the practical constancy of the heat of oxygen adsorption ($Q \approx 50$ kcal mol^{-1}) in the size range from 1000 to 3 nm, and a sharp increase up to $Q = 80$ kcal mol^{-1} for smaller particles [69]. No obvious influence of the support was observed for Pd dispersed on SiO_2, η-Al_2O_3 and SiO_2–Al_2O_3. As for H_2S adsorbed at Ni, Rao and co-workers have demonstrated the higher extent of H_2S dissociation on small Ni clusters than on larger ones, determined by the intensity of the S(2p) signal at 162.0 eV (referred to as S^{2-}) relative to the signal at 164.5 eV characteristic of the molecularly adsorbed H_2S [70].

It should, however, be recalled that the same range of metal cluster size (less than 3–5 nm) is characterised by the appearance of curved regions with large amounts of low-coordination sites such as steps, kinks, edges, etc. (see Section 3). This suggests that the electronic and structural factors are interrelated and cannot be discriminated between in their influence on the chemical reactivity of metal clusters. This conclusion is in agreement with the data of Zilm and co-workers [37] who have analysed the ^{13}C NMR spectra of CO adsorbed on silica-supported Pd particles as a function of Pd cluster size. The authors have concluded that the tighter bonding of adsorbed CO molecules with small Pd clusters is responsible for the decrease in the rate of CO diffusion on their surface, which is *c.* 10^6 times slower than that on the samples with larger particles [37]. This conclusion is based on the absence of motional narrowing of the ^{13}C resonance. Comparison of magic angle spinning (MAS) spectra, both with and without suppression of spinning sidebands (TOSS), in combination with T_1 relaxation time measurement, indicates that this tightly bonded CO is adsorbed in a linear form on low-coordination sites such as edges, corners and kinks, the fraction of which increases as the Pd particle size decreases. Furthermore, the comparison of the Knight shift for linear and bridge-bonded CO has allowed the authors to draw conclusions about the electron deficiency of these low-coordinated sites due to decoupling of their Fermi level from the bulk metallic level.

It should be noted that the deviation of the properties of the conduction electrons from the bulk properties observed by us for supported Ni [23,24] and Ag [22,26,27] particles is also correlated with changes in the surface structure, which becomes more defective due to roughening of the surfaces of Ni and Ag under treatment by reaction mixtures. In the case of the Ni–C catalysts for butadiene hydrogenation, this conclusion is based on the TEM data [24], while changes in the O(1s) binding energies of adsorbed oxygen have been used to study the surface structure of silver catalysts for ethylene epoxidation (see Section 3).

Whether electronic or structural factors are important in determining the variation of the metal cluster reactivity, the data indicate that as a rule decreasing cluster size results in increased reactivity towards gas. Adsorbate–metal bond strengthening means that most catalytic systems exhibit the antipathetic size effect. Moreover, the synthesis of supported iridium clusters consisting of four or six atoms allowed Xu and co-workers [71] to show that even for such structure-insensitive reactions as hydrogenation of toluene and cyclohexene, the reaction rate expressed as a turnover frequency is decreased by one order of magnitude on these smallest clusters. This study, regarded by Boudart as a milestone [72], together with the other data presented here, allows one to conclude that the tendency for a decrease in catalytic activity for small clusters is a general one. This supports Boudart in his opinion that selectivity as opposed to activity should become a driving force in future investigations devoted to the reactivity of metal clusters. This advice is justified by the already existing examples, which demonstrate the variation of selectivity with particle size for a number of catalytic reactions [23,39,73]. Thus, Bond and Slaa [73] showed that the composition of reaction products for n-butane hydrogenolysis on Ru–Al$_2$O$_3$-supported catalysts depends upon the size of the Ru clusters. The sample with small Ru particles is characterised by high ethane selectivity, whereas methane becomes a main product as the Ru particle sizes increase as a result of high-temperature treatment [73].

The selectivity for butadiene hydrogenation to butene has also been observed to increase from 10% to 95% with a decrease in Ni/C ratio [23]. This effect is accompanied by the roughening of the surface of the Ni particles supported on filamentous carbon, and a weakening of the nickel internal conductivity [24].

The influence of the electron-deficient, low-coordination sites on NO adsorption has been used by Xu and Goodman [39] to explain changes in the pathways for NO decomposition and CO + NO interaction with a decrease in the sizes of Pd particles supported on silica. The strengthening of the NO–Pd bond on these edge and corner sites, the presence of which has been proved by the comparison of infrared absorption spectra (IRAS) of CO adsorbed on Pd particles, Pd(111) and Pd(110) single crystals [74,75], results in dissociation of NO to nitrogen and oxygen atoms. The absence of molecularly adsorbed NO makes impossible a reaction channel for N$_2$O formation, as shown by temperature-programmed reaction (TPR) spectroscopy.

It is likely that the number of studies which demonstrate the variation of the selectivity of products for metal clusters as opposed to the bulk metals will increase sharply in the near future. This will be achieved by focusing efforts on studying the chemical nature of adspecies realised on the metal clusters in comparison with those on bulk metals, and its variation with cluster structure and electronic properties. An NMR study of CO adsorption on silica-supported Pd, carried out by Zilm and co-workers [37] and already cited in this review, can be considered as the first step in this direction. Indeed, the authors could show that the small Pd particles are characterised by two types of CO_{ads} with considerably different extents of mixing of their electron levels with the metal electronic bands.

The importance of studying this aspect of the chemical reactivity of metal clusters is confirmed by recent investigations which revealed the correlation between the nature of the adspecies and the catalytic behaviour of supported Ni and Ag clusters, with selectivity being also analysed [21–24,26,27]. Figure 5.4 presents the H_2 TPD spectra recorded after hydrogen adsorption for 2 min at $P = 1$ Pa and room temperature on two Ni–C catalysts for selective butadiene hydrogenation to butene. The samples differ in the time of treatment in the reaction conditions which in turn decreases the Ni/C atomic ratio due to an increase in the length of carbon filaments [23]. The sample with the lower concentration of carbon (Ni/C = 0.4 wt%) exhibits a selectivity towards butene formation of 45%, whereas a much higher selectivity of 95% is observed for the sample with the higher carbon content (Ni/C = 0.1 wt%) [23]. From Fig. 5.4, it can be seen that the more selective catalyst is characterised by one TPD peak at 320 K, while an additional TPD peak of H_2 at about 360 K appears for the non-selective sample. To assign these peaks to different hydrogen species we analysed the data of references [76,77] where hydrogen adsorption on samples of bulk Ni such as single crystals and foils has been studied using both TPD and measurement of work function. Exposure-dependent experiments have shown that the Ni surfaces are first covered by the high-temperature state that is accompanied by a sharp increase in the work function.

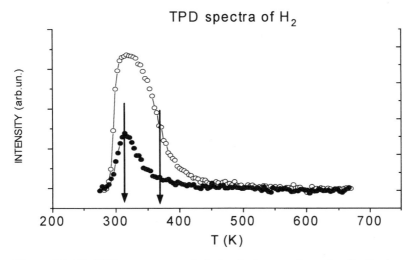

Figure 5.4. H_2 TPD spectra recorded after hydrogen adsorption for 2 min at $P(H_2) = 10$ Pa and $T = 270$ K on Ni–C catalysts with different Ni/C ratios: 0.1 wt% (solid circles) and 0.4 wt% (open circles).

Such behaviour allowed the authors to assign this TPD peak to the desorption of negatively charged hydrogen ions: H⁻ [76,77]. Development of the low-temperature state was after saturation of the high-temperature one. The invariability of the work function in this range of H_2 exposures suggests the uncharged nature of this hydrogen adspecies [76]. The well-known high reactivity of the hydride ions produced on the sample with a high Ni/C ratio of $c.$ 0.3 can be a reason for their non-selective hydrogenation, while its disappearance (Fig. 5.4) on the other sample is correlated with the increase in selectivity. Unfortunately, since the application of surface-sensitive methods such as XPS to study hydrogen adsorption is impossible, more substantiated conclusions are not feasible.

Oxygen adsorption on silver is free of this limitation. Furthermore, substituting the commercially used support (α-Al_2O_3) for the model one (graphite) has allowed us to avoid the screening effect of the O(1s) signal from the alumina and for the first time to apply XPS to study oxygen adsorption on silver clusters as a function of their sizes [26,27]. Figure 5.5 shows the O(1s) spectra recorded after O_2 adsorption for 10 min at $T = 470$ K and $P = 100$ Pa on the Ag–C samples prepared by the high-vacuum deposition of silver on the cleaned surface of graphite [27]. It can be seen that the surface of the small silver clusters (up to 10 nm) is characterised by one O(1s) peak with a binding energy of 530.5 eV, while the appearance of a second line with E_B of 528.5 eV is observed for large particles (Fig. 5.5). Comparison of the measured values with those for oxygen adsorbed at bulk silver samples [44,45,78,79] allows us to conclude the formation of two atomically adsorbed oxygen species with different ionicities for the O–Ag bond: the component with lower E_B is assigned to an ionic oxygen species (O^{2-}), and the second line to a covalent one ($O^{\delta-}$).

As shown in our previous studies [78,79], to produce ethylene oxide both the ionic and the covalent oxygen should coexist at the silver surface. The 'covalent' oxygen incorporates into ethylene providing the formation of ethylene oxide, while the 'ionic'

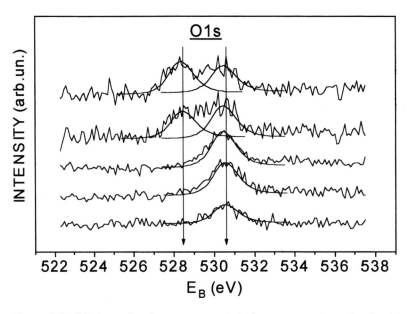

Figure 5.5. O(1s) core-level spectra recorded after oxygen adsorption for 10 min at $P(O_2) = 100$ Pa and $T = 470$ K on Ag deposited on graphite for various silver coverages (1–5) measured at $T = 470$ K. (1) $I_{Ag}/I_C = 0.35$; (2) 1.5; (3) 10; (4) 28; (5) 65.

oxygen is necessary to create ions of Ag^+ (or surface silver oxide) as adsorption centres for ethylene molecules [79]. According to this, the key stage of the ethylene epoxidation can be presented as follows:

$$O^{\delta-}_{ads} + C_2H_4{}_{ads} (Ag^+) \rightarrow C_2H_4O$$

Taking into account this mechanism and the disappearance of the ionic oxygen on small particles (Fig. 5.5), it becomes possible to explain the sharp decrease in the epoxidation rate for catalysts with Ag particles less than 50 nm [22]. Indeed, less-effective adsorption of ethylene on the small Ag particles due to the absence of Ag^+ sites will decrease the rate of epoxidation. The fact that the ionic oxygen is active only in total oxidation of C_2H_4 to CO_2 and H_2O [78,79] enables one to predict the variation of selectivity with the Ag particle sizes. Indeed, the disappearance of the ionic oxygen with decrease in particle size should decrease the rate of total oxidation and, hence, increase the selectivity towards ethylene oxide formation. This suggestion is in agreement with the data of Verykios *et al.* [80] who have observed an increase in selectivity as the Ag particle size decreases, the same range of sizes (200–500 Å) having been studied. These data, along with our results on the variation of activity with Ag particle size, suggest an optimum size range of Ag particles of 200–500 Å: the smaller particles exhibit a decrease in activity, and the larger ones are characterised by a decrease in selectivity.

5 Summary

The data reviewed and, in particular, our investigation of the reasons for the catalytic action of silver in ethylene epoxidation, demonstrate that the chemical reactivity of metal clusters: (i) correlates with the activity and selectivity of the corresponding supported catalysts; (ii) can be explained using the data obtained on the corresponding bulk metals, both electronic and structural properties, as well as their variation with cluster size, having to be considered; and hence (iii) can serve as a bridge which connects surface-science studies with real catalysis. This all confirms the validity of the proposed approach linking surface science to real catalysis which, in the author's opinion, will be applied more and more widely to the design of new catalysts. This affirmation is based on the success of the Boreskov Institute of Catalysis in the synthesis of a new generation of silver catalysts which comprise silver particles with mean sizes of about 500 Å supported on a specially designed carbon support. Preliminary lab-scale testing of their catalytic properties shows that they exhibit a selectivity close to 90%.

It should, however, be noted that some limitations connected with the influence of support, reaction medium, promoters, etc. must be considered when this approach is applied. The fact that I have not taken into account the first and the most serious limitation is justified by the use of the data obtained using such supports as alumina, silica or carbon, i.e. when a strong metal–support interaction is negligible. Indeed, the invariability of the results obtained for the same metal on different supports has been demonstrated throughout this review. Furthermore, there are studies which indicate that, like supported clusters, both the electronic properties and activity with selectivity of free ones depend on cluster sizes [81–84]. Nevertheless, the participation of a support in catalytic reactions, including secondary ones, as well as the influence of promoters and reaction mixtures on the reactivity of metal clusters, should be studied to produce a precise and full understanding of each specific catalytic system.

6 Acknowledgements

I am grateful to M.W. Roberts for the opportunity to make a contribution to this book and to A.F. Carley for checking and correcting my English. Furthermore, I would like to thank B.S. Balzhinimaev, A.I. Boronin, A.F. Carley, L.A. Dollard, R.I. Kvon, I.P. Prosvirin, M.W. Roberts and V.I. Savchenko for helpful discussions and technical assistance in the study of silver catalysts for ethylene epoxidation.

7 References

1 *Surf. Sci.* 1994; **299–300**.
2 Campbell CT. *Surf. Sci.* 1984; **139**: 396.
3 Campbell CT. *J. Catal.* 1985; **94**: 436.
4 Joyner RW, Roberts MW, Yates K. *Surf. Sci.* 1979; **87**: 501.
5 Joyner RW, Roberts MW. *Chem. Phys. Lett.* 1979; **60**: 459.
6 Boronin AI, Bukhtiyarov VI, Vishnevskii AL, Boreskov GK, Savchenko VI. *Surf. Sci.* 1988; **201**: 195.
7 Che M, Bennett CO. *Adv. Catal.* 1989; **36**: 55.
8 Boudart M. *Adv. Catal.* 1969; **20**: 153.
9 Poppa H. *Catal. Rev. Sci. Engg.* 1993; **35**: 359.
10 Mason MG. *Phys. Rev. B* 1983; **27**: 748.
11 Mason MG. In Pacchioni G (ed.) *Cluster Model for Surface and Bulk Phenomena.* New York: Plenum Press, 1992.
12 Vijayakrishnan V, Chainani A, Sarma DD, Rao CNR. *J. Phys. Chem.* 1992; **96**: 8679.
13 Aiyer HA, Vijayakrishnan V, Subbanna GN, Rao CNR. *Surf. Sci.* 1994; **313**: 392.
14 Wertheim GK. *Z. Phys. B* 1987; **66**: 53.
15 Wertheim GK, Dicenzo SB, Buchnan DNE. *Phys. Rev. B* 1986; **33**: 5384.
16 Wertheim GK. *Phase Transition* 1990; **24–26**: 203.
17 Barr TL. *J. Vac. Sci. Technol. A* 1989; **7**: 1677.
18 Barr TL. *Critical Rev. Anal. Chem.* 1991; **22**: 229.
19 Wagner CD. *Anal. Chem.* 1975; **47**: 1201.
20 Wagner CD. In Briggs D, Seach MP (eds) *Practical Surface Analysis*, Vol. 1, 2nd edn. New York: Wiley, 1990; 595.
21 Bukhtiyarov VI, Prosvirin IP, Kvon RI. *J. Electron. Spectrosc. Relat. Phenom.* 1996; **77**: 7.
22 Bukhtiyarov VI, Prosvirin IP, Kvon RI, Goncharova SN, Balzhinimaev BS. *J. Chem. Soc. Faraday Transac.* (accepted).
23 Chesnokov VV, Molchanov VV, Zaitseva NA *et al.* In *Book of Abstracts of 2nd International Conference—Modern Trends in Chemical Kinetics and Catalysis.* Novosibirisk, Russia, 1995, part II(1), 191.
24 Molchanov VV, Chesnokov VV, Buyanov RA *et al. Kinetika i Kataliz* (submitted).
25 Tsybulya SV, Kryukova GN, Goncharova SN, Shmakov AN, Balzhinimaev BS. *J. Catal.* 1995; **154**: 194.
26 Bukhtiyarov VI, Prosvirin IP, Carley AF, Dollard LA, Roberts MW. In *Book of Abstracts of 2nd Intern. Conference on Catalysis — EUROPACAT-2.* Maastricht, Netherlands, 1995.
27 Bukhtiyarov VI, Carley AF, Dollard LA, Roberts MW. *Surf. Sci.* (submitted).
28 Lennard-Jones J-E, Goodwin RT. *Proc. Roy. Soc. A* 1937; **163**: 101.
29 Newns DM. *Surf. Sci.* 1986; **171**: 600.
30 Ansermet J-P, Wang P-K, Slichter CP, Sinfelt JH. *Phys. Rev. B* 1988; **37**: 1417.
31 Rudas SL, Ansermet J-P, Wang P-K, Slichter CP, Sinfelt JH. *Phys. Rev. Lett.* 1985; **54**: 71.
32 Winterbottom WL. *Acta Met.* 1967; **15**: 303.
33 Wang T, Lee C, Schmidt LD. *Surf. Sci.* 1985; **163**: 181.
34 Drechsler M. *Surf. Sci.* 1985; **162**: 755.

35 Drechsler M, Dominguez JM. *Surf. Sci.* 1989; **217**: L406.
36 Somorjai GA, van Hove MA. *Progr. Surf. Sci.* 1989; **30**: 201.
37 Zilm KW, Bonneviot L, Haller GL, Han OH, Kermarec M. *J. Phys. Chem.* 1990; **94**: 8495.
38 Xu X, Szanyi J, Xu Q, Goodman DW. *Catal. Today* 1994; **21**: 57.
39 Xu X, Goodman DW. *Catal. Lett.* 1994; **24**: 31.
40 Goodman DW. *Surf. Sci.* 1994; **299–300**: 837.
41 Bukhtiyarov VI, Carley AF, Dollard LA, Roberts MW (in preparation).
42 Backx C, de Groot CPM, Biloen P. *Surf. Sci.* 1981; **104**: 300.
43 Backx C, de Groot CPM, Biloen P, Sachtler WMH. *Surf. Sci.* 1981; **128**: 81.
44 Campbell CT. *Surf. Sci.* 1984; **143**: 517.
45 Campbell CT. *Surf. Sci.* 1985; **157**: 43.
46 Ladas S, Poppa H, Boudart M. *Surf. Sci.* 1981; **102**: 151.
47 Rumpf F, Poppa H, Boudart M. *Langmuir* 1988; **4**: 722.
48 Herskowitz M, Holliday R, Cutlip MB, Kenney CN. *J. Catal.* 1982; **74**: 408.
49 Oh SH, Eickel CC. *J. Catal.* 1991; **128**: 526.
50 Engel T, Ertl G. *J. Chem. Phys.* 1978; **69**: 1267.
51 Engel T, Ertl G. *Adv. Catal.* 1979; **22**: 2.
52 Savchenko VI. *Uspekhi Khimii* 1986; **55**: 462.
53 Campbell CT, Shi S-K, White JM. *Appl. Surf. Sci.* 1979; **2**: 382.
54 Stara I, Matolin V. *Surf. Sci.* 1994; **313**: 99.
55 Gillet E, Channakhone S, Matolin V, Gillet M. *Surf. Sci.* 1985; **152–153**: 603.
56 Gillet MF, Channakone S. *J. Catal.* 1986; **97**: 448.
57 Altman EI, Gorte RJ. *Surf. Sci.* 1986; **172**: 71.
58 Doering DL, Poppa H, Dickinson JT. *J. Vac. Sci. Technol.* 1982; **20**: 827.
59 Altman EI, Gorte RJ. *Surf. Sci.* 1988; **195**: 329.
60 Altman EI, Gorte RJ. *J. Catal.* 1988; **113**: 185.
61 Briot P, Auroux A, Jones D, Primet M. *Appl. Catal.* 1990; **59**: 141.
62 Tardy B, Noupa C, Leclercq C *et al. J. Catal.* 1991; **129**: 1.
63 McCabe RW, Schmidt LD. *Surf. Sci.* 1977; **66**: 101.
64 Hofman P, Bare SR, King DA. *Surf. Sci.* 1982; **117**: 245.
65 Ko CS, Gorte RJ. *J. Catal.* 1984; **90**: 59.
66 Baird RJ, Ku RC, Wynblatt P. *Surf. Sci.* 1980; **97**: 346.
67 Root TW, Shmidt LD, Fisher GB. *Surf. Sci.* 1985; **150**: 173.
68 Santra AK, Ghosh S, Rao CNR. *Langmuir* 1994; **10**: 3937.
69 Chou P, Vannice MA. *J. Catal.* 1987; **105**: 342.
70 Rao CNR, Vijayakrishnan V, Santra AS, Prins MWJ. *Angew. Chem. Int. Ed. Engl.* 1992; **31**: 1062.
71 Xu Z, Xiao F-S, Purnell SK *et al. Nature* 1994; **372**: 346.
72 Boudart M. *Nature* 1994; **372**: 320.
73 Bond GC, Slaa JC. *Catal. Lett.* 1994; **23**: 293.
74 Kuhn WK, Szanyi J, Goodman DW. *Surf. Sci.* 1992; **274**: L611.
75 Ortega A, Hoffman FM, Bradshaw AM. *Surf. Sci.* 1982; **119**: 79.
76 Savchenko VI, Boreskov GK. *Kinetika i Kataliz* 1968; **9**: 142.
77 Christmann K, Schober O, Ertl G, Neumann M. *J. Chem. Phys.* 1974; **60**: 4528.
78 Bukhtiyarov VI, Boronin AI, Prosvirin IP, Savchenko VI. *J. Catal.* 1994; **150**: 262.
79 Bukhtiyarov VI, Prosvirin IP, Kvon RI. *Surf. Sci.* 1994; **320**: L47.
80 Verykios XE, Stein FP, Coughlin RW. *J. Catal.* 1980; **66**: 368.
81 Meiwes-Broer KH. *Appl. Phys. A* 1992; **55**: 430.
82 Martin TP, Bergmann T, Gölich H, Lange T. *J. Phys. Chem.* 1991; **95**: 6421.
83 El-Sayed MA. *J. Phys. Chem.* 1991; **95**: 3898.
84 Kalenik Z, Ladna B, Wolf EE, Fehlner TP. *Chem. Mater.* 1993; **5**: 1247.

6 Kinetics of Clustering on Surfaces

M. TOMELLINI and M. FANFONI*

*Dipartimento di Scienze e Tecnologie Chimiche, Università di Roma 'Tor Vergata', Via della Ricerca Scientifica, 00133 Rome, Italy

*Dipartimento di Fisica, Università di Roma 'Tor Vergata' and Istituto Nazionale di Fisica della Materia, Via della Ricerca Scientifica, 00133 Rome, Italy

1 Introduction

In order to obtain an intimate contact between two solids, one of them, the overlayer, is often grown on the other, the substrate, via vapour deposition. Depending on the relative magnitudes of the adsorbate, substrate and interfacial surface free energies, an overlayer can grow in three different modes: (i) layer by layer (Frank–van der Merve); (ii) several ordered layers followed by three-dimensional (3D) islands (Stranski–Krastanov); and 3D islands on the bare substrate (Volmer–Weber) [1,2]. Besides being an interesting subject *per se*, the morphology of growth of an overlayer is of considerable importance for technological applications. For example, several metal clusters dispersed on oxide films might be preferred to a continuous metal film when preparing a supported metal catalyst [3–5]. Conversely, a continuous diamond film is the ultimate aim in the field of materials coating, as in the production of cutting tools [6].

The subject of the present paper is the Volmer–Weber growth mechanism in which nucleation and growth processes occur at the substrate surface. There is increasing interest in the fundamentals of clustering on surfaces in connection with both the kinetics and the thermodynamic behaviour of the systems. As a matter of fact, depending on temperature and deposition flux, different morphologies of the deposit can be obtained.

The kinetics of clustering on surfaces is usually modelled according to three different approaches: (i) rate equations; (ii) scaling theories; and (iii) kinetic Monte Carlo simulations [7–10]. Modelling allows the evaluation of important quantities of the deposition process such as the particle size distribution function, the growth rate of the islands, the nucleation kinetics and the time dependence of the surface fraction covered by the islands.

Several reviews of clustering are available [4,7,11–16] and the present one will mainly be devoted to an analysis of kinetic solutions in closed form based on rate equations. These solutions provide, under certain restrictive assumptions, a straightforward description of experimental data. Film-growth kinetics will be examined from nucleation and early stages up to film closure. In the regime of high surface coverages, modelling of collisions among islands becomes compulsory, since clusters are not well isolated on the surface. In order to evaluate the time dependence of the surface portion covered by islands, the problem of cluster impingement will be tackled in detail.

This paper is subdivided as follows. In Section 2 the thermodynamic aspects of a cluster aggregate on a solid surface, under equilibrium conditions, are revisited. The atomistic theory of nucleation, based on rate equations, is presented in Section 3, together with the modelling of the particle size distribution function based on the cluster growth law. Section 4 is devoted to the analytical and phenomenological model which allows the description of the entire kinetics of film growth. In Section 5 the potentialities of some surface techniques for studying the kinetics of clustering on solid surfaces have been outlined. Quantitative measurements of energy parameters, such as the formation energy of clusters and the activation energy of adatom diffusion, may serve as inputs for kinetic models and enable one to verify the validity of such a modelling process. Particular attention is given to both the latest-generation synchrotron-radiation sources and scanning tunnelling microscopy (STM), which are becoming very promising tools for studying clustering on surfaces in real time.

2 Equilibrium thermodynamics of a cluster aggregate

Let us consider a homogeneous and isotropic surface where adatoms and clusters are distributed at random. At equilibrium the homogeneous mixture of clusters and adatoms can be described by means of chemical thermodynamics by simply studying the surface reactions for the formation of clusters from admonomers:

$$j\mathrm{A}_1 \rightleftarrows \mathrm{A}_j \tag{1}$$

where A_j stands for a surface cluster constituted by j monomers. The chemical potential of a cluster made of j monomers can be written in the usual form [17]:

$$\mu_j = \mu_j^0 + kT \ln\left[\gamma_j \frac{N_j}{N_0}\right] \tag{2}$$

where N_j and γ_j are, respectively, the surface density of clusters and the activity coefficient, and N_0 is the density of surface sites available for adsorption, the superscript '0' indicating the standard state. As usual, the standard state of adatoms (clusters) is chosen as follows: (i) ideal behaviour of the thermodynamic system with respect to the interaction among adatoms (adatoms and clusters); (ii) the ratio N_j/N_0 is equal to 1. The standard state of the gas phase is at $P = 1$ atm. The logarithmic term on the right-hand side of equation (2) stems from the configurational entropy of mixing. Considering M lattice points, the conservation of the number of sites is given by:

$$M = N_0 + N_1 + \sum_{i>1} \xi_i N_i \tag{3}$$

where ξ_i is the number of surface sites occupied by the ith island. The configurational contribution to the entropy of mixing is:

$$S_{\mathrm{c}} = k \ln \frac{M!}{\prod_j N_j!(M - N_1 - \sum_{j>1} N_j)!} \tag{4}$$

where the approximation $\xi_j = 1$ is employed, i.e. each island occupies a single lattice site. This is a reasonable approximation as long as $(\Sigma_{i>1} N_i)/M \ll 1$. A low value of the cluster-covered surface fraction does not necessarily imply the ideal behaviour of the homogeneous mixture formed by admonomers and clusters. In fact, at higher adatom coverages, interactions among adatoms and between adatoms and clusters may be no longer negligible, leading to a non-ideal behaviour. Under these circumstances equations (2) are expected to hold, with the simplest expressions for the γ_j given by:

$$\gamma_1 = e^{z_1\theta_1\varepsilon/kT} \tag{5}$$

$$\gamma_j = e^{z_j\theta_1\tilde{\varepsilon}/kT} \tag{6}$$

where z_j is the number of neighbouring lattice sites in a cluster made up of j monomers, ε and $\tilde{\varepsilon}$ the couple interaction energies between free adatoms and between adatoms and island monomers at the periphery of the cluster, respectively, and $\theta_1 = N_1/(N_1 + N_0)$ the adatom surface coverage of that portion of the surface uncovered by clusters. For reaction (1) the law of mass action is given by:

$$\left(\frac{N_1}{N_0}\right)^j = \frac{\gamma_j}{\gamma_1^j}\left(\frac{N_j}{N_0}\right)e^{\Delta G_j^0/kT} \tag{7}$$

where

$$\Delta G_j^0 = (\mu_j^0 - j\mu_1^0) = \Delta G_j' - j(\mu_1^0 - \mu_g) \tag{8}$$

is the standard free energy change for cluster formation, $\Delta G_j' = \mu_j^0 - j\mu_g$ and μ_g is the chemical potential of monomers in the gas phase. Moreover,

$$\Delta G_j^0 = \Delta G_j' - j(\Delta G_a^0 - kT \ln P) \tag{9}$$

in which $\Delta G_a^0 = \mu_1^0 - \mu_g^0$ is the standard free energy change for monomer adsorption and P is the gas pressure. For γ_1 and γ_j equal to one equation (7) becomes:

$$\left(\frac{N_1}{N_0}\right) = \left(\frac{N_j}{N_0}\right)^{1/j}Pe^{-[\Delta G_a^0 - (\Delta G_j'/j)]/kT} \tag{10}$$

which is a quite general relationship between the surface coverages of adatoms and clusters at equilibrium. Equation (10) holds for an ideal mixture of clusters and adatoms without any restriction on the size of the cluster. In the specific case of macroscopic islands ($j \gg 1$) the cluster itself is considered to be a thermodynamic system and quantities such as the surface free energy and the wetting angle can be used to evaluate $\Delta G_j'$ [18]. As an example, in the case of disc-shaped islands the following expression has been established for the surface density of adatoms in equilibrium with clusters made up of j monomers [7,18]:

$$\left(\frac{N_1}{N_0}\right) = \left(\frac{N_1}{N_0}\right)_\infty e^{\Gamma/j^{1/2}kT} \tag{11}$$

$$\left(\frac{N_1}{N_0}\right)_\infty = e^{-[\Delta G_a^0 + \Delta G_f^0 - (\Delta\sigma/h\rho)]/kT}$$

$$\Gamma = 2\sigma_e(h\pi/\rho)^{1/2}$$

$$\Delta\sigma = \sigma_{c,v} + \sigma_{s,c} - \sigma_{s,v}$$

h is the cylinder height, ρ the cluster density and σ indicates the surface free energy with the subscripts c, v, s and e standing for 'cluster', 'vacuum', 'substrate' and 'edge',

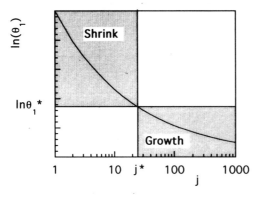

Figure 6.1. Typical plot of the adatom surface coverage against cluster sizes as described by the Gibbs–Thomson equation. At $\theta = \theta^*$ clusters with sizes larger than j^* experience a θ value higher than the equilibrium value: they are expected to grow. Conversely, for $j < j^*$, islands are expected to shrink.

respectively. ΔG_f^0 is the standard free energy change for the formation of a gas monomer. Equation (11) is a restatement of the Gibbs–Thomson equation [7] in which $(N_1/N_0)_\infty$ is the surface coverage of the adatoms in equilibrium with the solid ($j \to \infty$).

A typical plot of the equilibrium function $\theta_1 = (N_1/N_0)$ against j is shown in Fig. 6.1, as calculated from equation (11). Let us consider a *non-equilibrium* state of the cluster aggregate with *any size* distribution function for the islands and an *average value* for the adatom surface coverage equal to θ_1^*. According to Fig. 6.1, clusters with sizes larger than j^* will experience an adatom concentration higher than that expected for the sizes considered at equilibrium; these clusters are expected to grow. On the other hand, clusters having sizes smaller than j^* are expected to shrink. As a consequence, for a conservative system, i.e. where only adatom redistribution occurs at the surface, a net flux of monomers from smaller ($j < j^*$) to larger ($j > j^*$) clusters is expected at $\theta_1^* = $ constant. This phenomenon is usually called 'coalescence ageing' or Ostwald ripening [16,19–23].

Experimental data and thermodynamic computation for the adatom concentration between clusters as a function of temperature have been reported for the Ge–Si system [24,25]. This study was carried out using Auger electron spectroscopy by monitoring the changes in the attenuation signal coming from the substrate with temperature. When some adatoms coalesce in large 3D clusters they do not contribute to the attenuation of the substrate signal and the Si Auger emission is expected to increase. On the other hand, when clusters release atoms the Auger emission from the substrate decreases. An equilibrium value of the adatom surface coverage up to 0.1 ML was measured [24,26].

3   Atomistic nucleation theory based on rate equations

The theory of homogeneous nucleation was developed during the 1930s and 1940s thanks to the work of Volmer and Weber (VW), Becker and Döring (BD) and Zeldovich [18,27,28]. They derived an expression for the rate of formation of nuclei in the case of vapour–liquid 3D phase transitions. Volmer worked out the theory of heterogeneous nucleation (nucleation on an impurity surface) by straightforwardly extending the VW work [27,28]. There is not enough space here to discuss the theory in detail, but it is none the less worth remembering the historical evolution from the work of VW to that of BD.

Both the theories deal with macroscopic nuclei of the order of 10^2–10^3 atoms in size. So as to determine the density of condensed-phase critical nuclei, the latter were assumed, by VW, to be in equilibrium with the supersaturated phase. Moreover, they assumed that the continuous removal of supercritical nuclei from the distribution did not distort the equilibrium by formally replacing each eliminated nucleus by an equivalent number of single atoms. In this way, the vapour remains in a stationary state. This is a reasonable approximation when the density of nuclei is very low. Once the concentration of critical nuclei is determined, the nucleation rate, J^e, is easily obtained by multiplication:

$$J^e = N_n^e K_n^+ \tag{12}$$

where N_n^e is the concentration of critical nuclei at equilibrium and K_n^+ is the rate at which single atoms join the critical cluster of size n. Instead of facing the question from a thermodynamical viewpoint, BD, more correctly, developed the theory through an entirely kinetic argument. They considered not only the attachment of vapour atoms to the surface of drops, but also the reverse process. Like VW, they considered a stationary state and eventually were able to obtain:

$$J^s = N_n^s K_n^+ = N_n^e Z K_n^+ \tag{13}$$

in which the superscript s stands for 'steady state', while Z, called the Zeldovich factor, takes into account the departure from equilibrium caused by nucleation. In order to clarify its meaning we will solve a very simple, but we think instructive, nucleation model.

Let us consider the following two-step kinetic model for surface nucleation:

$$\text{surface site } (N_0) \underset{K_n^-}{\overset{K_0^+}{\rightleftharpoons}} \text{critical nucleus } (N_n) \overset{K_n^+}{\rightarrow} \text{stable cluster } (N)$$

Surface densities are indicated in parentheses, while K are the rate constants. The rate equation for the critical nucleus is given by:

$$\dot{N}_n = N_0 K_0^+ - N_n (K_n^+ + K_n^-) \tag{14}$$

The steady-state condition reduces to:

$$N_n^s = N_0^s \frac{K_0^+}{K_n^+ + K_n^-} \tag{15}$$

and the nucleation rate becomes:

$$J^s = K_n^+ N_n^s = N_0^s \frac{K_n^+}{\left(1 + \dfrac{K_n^+}{K_n^-}\right)} \frac{K_0^+}{K_n^-} = N_0^s e^{-\Delta G^*/kT} K_n^+ Z = Z K_n^+ N_n^e \tag{16}$$

where ΔG^* is the free energy change for critical nucleus formation and $Z = (1 + K_n^+/K_n^-)^{-1}$ is the Zeldovich factor of the process. Furthermore, the ratio K_0^+/K_n^- has been expressed, in terms of ΔG^*, by detailed balancing. For $K_n^+ \ll K_n^-$, $Z \approx 1$ and a quasi-equilibrium condition is reached. Thus, in order for quasi-equilibrium to be applicable, the nucleation barrier has to be high or, in other words, the forward rate for supercritical cluster production must be much lower than the backward rate.

In 1962, Walton [29] derived an expression for the surface nucleation rate following a mathematical pathway similar to the classical approaches described above. At variance with the classical approach, Walton's theory is suitable for treating nuclei of extremely small size, the latter being, in turn, related to supersaturation (which is defined as the ratio between the vapour pressure of the atom stream and the vapour tension of the condensate). As a matter of fact, for clustering on surfaces, experiments are usually performed under extremely high supersaturation conditions, up to 10^{21}. Accordingly, on account of the very small size of the critical nucleus that follows (it ranges between 1 and 10 atoms [11,29–33]), the classical thermodynamic approach becomes inapplicable. It would in fact be unjustifiable to assign continuum thermodynamic properties, such as volume energies, surface free energies and contact angles to nuclei of such small size.

When the critical nucleus contains only a few atoms, atomistic theories must be employed. They are to be formulated, as Walton's model does, in terms of atomic quantities such as the adsorption energy, the activation energy for surface diffusion and the pair binding energy of monomers in the islands. Among these theories, approaches based on rate equations have been gaining more and more attention for describing experimental data for clustering on surfaces [7,8,14,15,34–43].

3.1   Rate equations

A set of differential equations describes the changing rate of the surface densities of clusters of each size. Clusters can increase their size by attachment of both adatoms and gas monomers, as well as by diffusive encounters with other clusters. Similarly, clusters shrink by detachment of monomers. Our analysis is restricted to the case of immobile clusters, the only species diffusing at the surface being adatoms. Coalescence between clusters is not considered here but will be discussed briefly in the next paragraph. See [12,44–47] for detailed treatments of both coalescence and cluster mobility phenomena.

The kinetic rate equations are written as:

$$
\left\{
\begin{aligned}
&\dot{N}_1 = J_0 N_0 + \nu_{-2} N_2 - [\nu_{-1} + \nu_1]N_1 + 2\kappa_{-2} N_0 N_2 - 2\kappa_1 N_1^2 + \sum_{i=2}^{\infty}[\kappa_{-(i+1)}N_0 N_{i+1} - \kappa_i N_1 N_i] \\
&\dot{N}_2 = -[\kappa_2 N_1 + \kappa_{-2} N_0 + \nu_{-2} + \nu_2]N_2 + [\kappa_1 N_1 + \nu_1]N_1 + [\kappa_{-3}N_0 + \nu_{-3}]N_3 \\
&\vdots \\
&\dot{N}_j = -[\kappa_j N_1 + \kappa_{-j}N_0 + \nu_{-j} + \nu_j]N_j + [\kappa_{j-1}N_1 + \nu_{j-1}]N_{j-1} + [\kappa_{-(j+1)}N_0 + \nu_{-(j+1)}]N_{j+1} \\
&\vdots
\end{aligned}
\right.
$$

$$(17)$$

where $\dot{N}_j$ is the time derivative of the average surface density of clusters formed by j

monomers, J_0 (v_{-1}) is the rate constant for gas-monomer adsorption (desorption), $\kappa_{\pm j}$ are the rate coefficients for adatom attachment to (+) and detachment from (–) a cluster of size j. Similarly, the rate constants for the direct attachment and detachment of gas monomers to and from the islands are indicated by $v_{\pm j}$. A schematic view of the kinetic terms of equation (17) is shown in Fig. 6.2.

In principle, the equations in system (17) are infinite in number. Therefore, once the rate coefficients are known, system (17) can be solved numerically, provided an adequate number of equations is used to ensure the conservation of the number of particles [47].

An alternative solution of this system can be sought by introducing the physical assumption that there exist a critical size and a stable cluster. These are defined as follows: the critical nucleus is 'that cluster whose probability of growing is less than or equal to one-half, which upon receipt of a single atom in the appropriate position in the cluster acquires a probability of growing which is greater than or equal to one-half. The cluster which is produced by adding the single atom to the critical nucleus will be called

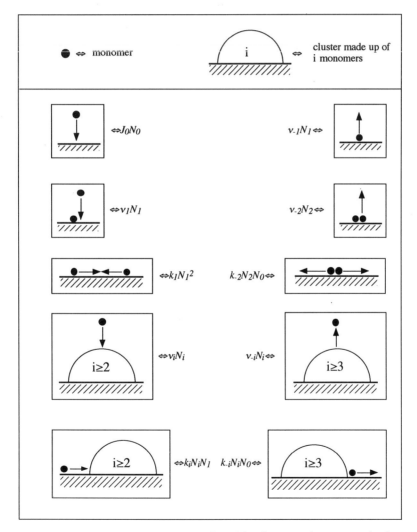

Figure 6.2. Graphic representation of the kinetic terms that appear in the system (17) of rate equations.

the smallest stable cluster' [29]. Under this assumption system (17) reduces to a system of differential equations whose number is just equal to the size of the smallest stable cluster.

$$
\begin{cases}
\dot{N}_1 = J_0 N_0 + \nu_{-2} N_2 - [\nu_{-1} + \nu_1] N_1 + 2\kappa_{-2} N_0 N_2 - 2\kappa_1 N_1^2 + \sum_{i=3}^{n} \kappa_{-i} N_0 N_i \\[2mm]
\qquad - \sum_{i=2}^{n} \kappa_i N_1 N_i - \sum_{i=n+1}^{\infty} \kappa_i N_1 N_i \\[2mm]
\vdots \\[1mm]
\dot{N}_j = -[\kappa_j N_1 + \kappa_{-j} N_0 + \nu_{-j} + \nu_j] N_j + [\kappa_{j-1} N_1 + \nu_{j-1}] N_{j-1} + [\kappa_{-(j+1)} N_0 + \nu_{-(j+1)}] N_{j+1} \\[2mm]
\vdots \\[1mm]
\dot{N}_n = -[\kappa_n N_1 + \kappa_{-n} N_0 + \nu_{-n} + \nu_n] N_n + [\kappa_{n-1} N_1 + \nu_{n-1}] N_{n-1} \\[2mm]
\dot{N} = \kappa_n N_1 N_n + \nu_n N_n
\end{cases}
\tag{18}
$$

where n is the critical size and N is the surface density of *stable* clusters, $N = \sum_{i=n+1}^{\infty} N_i$. The rate equation for N, which is the last in system (18), directly stems from system (17) by simply summing the $\dot{N}_{i>n}$ terms and by performing the appropriate cancellations between similar contributions of the reaction chain. While the system is composed of $n+1$ equations, due to the last term in the $\dot{N}_1$ expression, the number of unknowns is infinite. It is evident that some hypothesis must be set up in order to achieve closure of the system.

It is worth recalling that a sharply defined critical nucleus has a substantial effect on the kinetics. The presence of an irreversible step in the kinetics continuously shifts the reaction towards the formation of stable nuclei. Once the flux of gas monomers is turned off ($J_0 = 0$) the redistribution of adatoms proceeds until the coverage value $\theta_1 = 0$ is reached. The assumption that stable clusters do not release monomers limits the modelling to describing systems where equilibrium between clusters and adatoms is established for non-negligible values of θ_1. The validity of this approximation, however, can be checked with reference to the Gibbs–Thomson equation [48].

On the basis of the above-mentioned kinetic argument the critical size, n, is defined by the following contraints [29]:

$$
\begin{cases}
\kappa_j N_1 < \kappa_{-j} N_0 & 1 < j \leqslant n \\
\kappa_j N_1 > \kappa_{-j} N_0 & j \geqslant n+1
\end{cases}
\tag{19}
$$

where, for the sake of simplicity, only the contribution of the adatoms to nucleation has been taken into account. The rate coefficients for monomer attachment are of the form:

$$
\kappa_j = \sigma_j D
\tag{20}
$$

$$
\nu_j = s_j J_0
\tag{21}
$$

where σ_j and s_j are the capture factors and D the surface diffusion coefficient of

adatoms. From now on the condition $v_{\pm j} = 0$, except for v_{-1}, will be assumed to hold*
[49]. In the case of capture factors independent of the surface density of clusters, and
therefore of time, the detailed balancing can be used to relate forward and backward
rate coefficients:

$$\frac{\kappa_{-(j+1)}}{\kappa_j} = e^{(\Delta G^0_{j+1} - \Delta G^0_j)/kT} \qquad (22)$$

Generally speaking, all the information about the island structure and the spatial
correlation between clusters is included in the time- and size-dependent capture factor,
σ_j, through its dependence on the $N_1, N_2, \ldots, N_j, \ldots$ average densities. Two analytical
approaches have been proposed in the literature, aimed at evaluating the σ_j factors, the
'uniform depletion approximation' and the 'lattice approximation' [10,13,32–33,38–
42,50]. The estimates give a lower and an upper bound limit, respectively. On account
of adatom diffusion at the surface, in order to evaluate σ_j, it is necessary to solve a
transport equation for the *local* surface density of adatoms, $n_1(r,t)$. For a circular cluster
the boundary conditions are: $n_1(r,t) = 0$ for $t > 0$ at the periphery of the cluster ($r = R_j$)
and $n_1(r,t)|_{r \to \infty} = J_0 N_0 \tau$, τ being the constant lifetime of the adatoms. (The diffusion
equation for the local density is solved for an isolated cylindrical cluster. Therefore, the
asymptotic value of n_1 coincides with the average concentration of admonomers
$N_1 = J_0 N_0 \tau$.) The expression for the capture factor is then given by:

$$\sigma_j = \left(\frac{2\pi R_j}{N_1}\right) \frac{\partial n_1(r,t)}{\partial r}\bigg|_{R_j}$$

Two limiting solutions have been obtained, for $t > \tau$ [12,38]:

$$\begin{cases} \sigma_j = \dfrac{2\pi}{\ln \dfrac{\sqrt{D\tau}}{R_j}} & R_j \ll \sqrt{D\tau} \\[20pt] \sigma_j = \dfrac{2\pi R_j}{\sqrt{D\tau}} & R_j \gg \sqrt{D\tau} \end{cases} \qquad (23)$$

*It is possible to show that in the nucleation stage the inequalities $N_1 \kappa_j \gg v_j$ and $N_0 \kappa_{-j} \gg v_{-j}$ are
satisfied. As an example, we consider capture factors $\sigma_j = 2\pi r_j / a$ and $s_j = \pi r_j^2 / a^2$, where r_j is the
cluster radius and a the lattice space. At room temperature the D value is of the order
10^{-13} cm^2 s^{-1} in magnitude when an activation energy of 0.5 eV is used. For the typical value of
$J_0 = 10^{-2}$ s^{-1} and $r_j = a$, equations (20) and (21) lead to: $N_1 \kappa_j = 10^4 \theta_1 (s^{-1})$ and $v_j = 10^{-2}(s^{-1})$. Since
θ_1 is of the order of 0.01 ML, the inequality $N_1 \kappa_j \gg v_j$ is expected to be fulfilled [48]. On the other
hand, the detailed balancing leads to:

$$\frac{\kappa_j N_1}{\kappa_{-(j+1)}} = \frac{v_j N_1}{v_{-(j+1)}} e^{\Delta G^0_a}$$

Since adsorption is exothermic and $N_0 > N_1$ the inequality

$$\frac{\kappa_j N_1}{v_j} < \frac{\kappa_{-(j+1)} N_0}{v_{-(j+1)}}$$

holds. Therefore, in the nucleation stage both the forward and the backward rate constants, $v_{\pm j}$,
may be neglected.

The first application of the rate equations to nucleation kinetics is due to Zinsmeister in the case of stable dimers and for both constant and size-dependent rate coefficients [34–36]. In the former case, $\kappa_j = \kappa = $ constant, the closure of the system is easily obtained as $\sum_{j=n+1}^{\infty} \kappa_j N_j = \kappa N$. For size-dependent rate coefficients, the closure can be achieved on the basis of the argument that follows. For stable clusters, $j > n$, the rate equations are:

$$\dot{N}_j = N_1(\kappa_{j-1}N_{j-1} - \kappa_j N_j) \tag{24}$$

where the contribution of cluster growth due to the direct impingement of gas monomers has been neglected. Multiplying equation (24) by κ_j and summing over the whole population of stable islands easily yields the following:

$$\sum_{j=n+1}^{\infty} \kappa_j \dot{N}_j = \kappa_{n+1}\kappa_n N_n N_1 + \sum_{j=n+1}^{\infty} (\kappa_j \kappa_{j+1} - \kappa_j^2)N_j N_1 \approx \kappa_{n+1}\kappa_n N_n N_1 + \kappa_0 N N_1 \tag{25}$$

where the term $(\kappa_j \kappa_{j+1} - \kappa_j^2)$, being weakly dependent on j, can be assumed to be constant (κ_0) [36]. Integration of equation (25) just gives the term $\sum_{j=n+1}^{\infty} \kappa_j N_j$, which appears in the first equation of the system (18), as a function of N_1, N and N_n. Closure has been achieved, although a system of integral differential equations now has to be solved.

Three regions were identified for the time dependence of the adatom concentration: a linear change in N_1 according to $N_1 \approx J_0 N_0 t$ in the first region; and a nearly constant density of the adatoms in the second region equal to $N_1 \approx J_0 N_0 / \nu_{-1}$ followed by a decrease in the number of admonomers in the last region according to a power law. The relative extension of each region is strictly dependent on temperature, as shown in Fig. 6.3, where typical plots of the functions N_1 and N are shown in the limit of high and low temperatures and for size-independent collision factors. The asymptotic time dependence of both N and N_1 has been computed analytically in each region [34–36]. This kinetic model also allows the size distribution of the aggregates and the $N(t)$ density to be evaluated at any time, the former being a mirror image of the adatom concentration as a function of time [35]. This result is easily generalised whatever the size of the critical island is, provided a growth law exists which is the same for all clusters [51]:

$$\begin{cases} F(r,t) = -\dfrac{\partial N[x(r,t)]}{\partial r} & 0 \leqslant r \leqslant r(0,t) \\ F(r,t) = 0 & r > r(0,t) \end{cases} \tag{26}$$

where $F(r,t)$ is the size distribution of clusters at running time t, $r = r(x,t)$ is the growth law of the islands, and x is the time at which the *stable* island starts growing ($x < t$). In the case of stable dimers with no desorption and for size-independent rate coefficients, system (18) can be rewritten using dimensionless variables $p_1 = N_1(\kappa_1/N_0 J_0)^{1/2}$, $p_x = N(\kappa_1/N_0 J_0)^{1/2}$ and $q = (J_0 N_0 \kappa_1)^{1/2} t$ as follows [37]:

$$\begin{cases} \dfrac{dp_1}{dq} = 1 - p_1(2p_1 + p_x) \\ \dfrac{dp_x}{dq} = p_1^2 \end{cases} \tag{27}$$

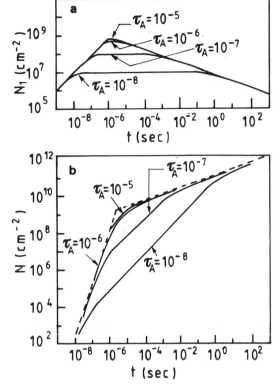

Figure 6.3. The surface densities of (a) single adatoms and (b) stable islands. The numerical solution of equations (18) has been obtained for $n = 1$, $J_0 N_0 = 10^{15}$ cm^{-2} s^{-1} and $\kappa_i = 2 \times 10^{-3}$ cm^2 s^{-1} for all i. From [34], with the kind permission of Elsevier Science SA, Lausanne, Switzerland.

where N_0 is considered to be constant and $\kappa_j = \kappa_1$ for all j. Although no analytic solution is available for this system, it can be solved numerically and solutions, for the N and N_1 densities, can be obtained for all values of the κ_1 and J_0 terms. On the basis of the numerical solution of system (27) a reasonably well-defined point can be identified where the increase in p_x becomes relatively small. A 'saturation point', for $q = q_s$, has therefore been defined, which satisfies the following requirements: (i) the loss of adatoms due to the growth of stable nuclei is equal to the loss of adatoms due to pair formation; and (ii) the population of adatoms is decreasing with time, i.e. $dp_1/dq < 0$. Requirement (i) implies

$$p_x = 2p_1 \bigg|_{q = q_s} = 2\left(\frac{dp_x}{dq}\right)^{1/2}\bigg|_{q = q_s}$$

which leads, after numerical evaluation of system (27), to the conditions $q_s = 3.5$ and $p_x(q_s) = 1$. The saturation value of the N density, attained at time $t = t_s$, is given by [37]:

$$N_s \approx N_0\left(\frac{J_0}{v_d}\right)^{1/2} e^{E_d/2kT} \tag{28}$$

$$t_s \approx 3.5 \frac{N_s}{N_0 J_0} \tag{29}$$

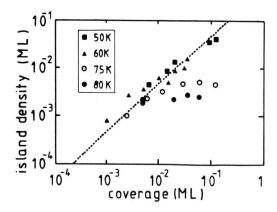

Figure 6.4. Island density versus coverage as deduced from STM images performed at 50, 60, 75 and 80 K. Reproduced with permission from [54].

where E_d and ν_d are the activation energy and the vibrational frequency for the surface diffusion of adatoms, respectively. It is interesting to note that in this case the average size of clusters at the saturation point is approximately equal to or less than $J_0 N_0 t_s / N_s \approx 3.5$, which implies microscopic clusters made up of three or four atoms. An expression similar to equation (28) has been obtained [31,33] for the case $\nu_{-1} \neq 0$:

$$N_s \approx N_0 \left(\frac{\nu_a}{\nu_d}\right) e^{(E_d - E_a)/kT} \tag{30}$$

E_a and ν_a being the adsorption energy of adatoms and the vibrational frequency for desorption, respectively. Equation (30) can be derived by considering a depletion region for the adatoms, around each cluster considered as a point target, with area equal to the square of the diffusion length, $\lambda^2 = 4D\tau = a^2 \nu_d e^{-E_d/kT}\tau = a^2(\nu_d/\nu_a)e^{-(E_d-E_a)/kT}$, a being the lattice spacing. The maximum density of these regions available on a surface with N_0 adsorption sites per unit area is in fact $N_s = N_0 a^2/\lambda^2$.

Saturation of the island density in the nucleation stage of thin-film formation has been experimentally measured by means of variable-temperature STM [52–54]. For Ag deposition on a Pt(111) surface, dimers are the stable nuclei at $T < 120$ K, i.e. adatoms are the critical clusters. The measured nucleation kinetics indicates that the coverage value at saturation decreases as the temperature rises. This is shown in Fig. 6.4 where the nucleation kinetics, as a function of the total coverage, are reported for several temperatures [54]. In Fig. 6.5 the kinetics are shown for $T = 75$ K and compared with theoretical curves computed by solving the rate equations (solid lines). In particular, the rate coefficients were computed in curve (i) according to the expression $\sigma_j \approx 2 + j^{1/1.7}$, and, in curve (iii) by using the uniform depletion approximation which leads to better agreement with the data. It is worth pointing out that in Fig. 6.5 the saturation occurs within 0.05 ML coverage and the nucleation rate can reasonably be expressed by a Dirac δ function.

3.2   Quasi-equilibrium approximation

A physical assumption, due to Venables [14,15,55], can be introduced which consider-

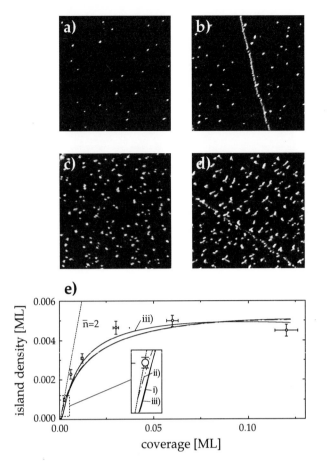

Figure 6.5. STM images showing the evolution of the island shape and density at 75 K for four values of surface coverage θ: (a) 0.0024 ML; (b) 0.006 ML; (c) 0.03 ML; (d) 0.06 ML. In (e) the nucleation kinetics, in terms of θ, are shown for $T = 75$ K. Points on the dotted line are representative of an average island size equal to $\bar{n} = 2$ monomers. Curve (i) gives the solution of rate equations for $n = 1$ and $\sigma_i = 2 + i^{1/1.7}$; curve (ii) gives the solution of rate equations ($n = 1$) for the nucleation density which had developed 20 s after interrupting the flux. For curve (iii) the capture rate had been interpreted as the gradient of adatom concentrations. Reproduced with permission from [54].

ably simplifies the mathematical complexity of system (18). Let us consider the kinetic scheme for nucleation:

$$A_1 + A_1 \underset{\kappa_{-2}}{\overset{\kappa_1}{\rightleftarrows}} A_2 + A_1 \underset{\kappa_{-3}}{\overset{\kappa_2}{\rightleftarrows}} A_3 + \dots + A_1 \underset{\kappa_{-n}}{\overset{\kappa_{n-1}}{\rightleftarrows}} A_n + A_1 \overset{\kappa_n}{\rightarrow} A_{n+1} + A_1 \overset{\kappa_{n+1}}{\rightarrow} \dots \tag{31}$$

In the kinetics the $\nu_{\pm j}$ rate constants have been omitted, except for ν_{-1}, since in the nucleation regime they are numerically negligible. For the sake of simplicity, monomer desorption has not been indicated in the scheme. As was shown at the beginning of this section in connection with the meaning of the Zeldovich factor, if the backward rates are much greater than the forward ones and N_1 is nearly constant in time, equilibrium between each term of the kinetic chain, up to the critical size, is established. Strictly speaking, this is a quasi-equilibrium state, because of the irreversible steps which continuously shift the reaction to the right. The condition $N_1 = $ constant is achieved either at high temperature, through equilibrium between adatoms and gas monomers, or at low temperatures, in case of total condensation ($v_{-1} = 0$), through the adatoms' capture by the growing islands. System (18) reduces to:

$$\begin{cases} \dfrac{dN_1}{dt} = J_0N_0 - v_{-1}N_1 - D\sigma_xN_1N \\[2mm] \dfrac{dN_j}{dt} = 0, \qquad 2 \leqslant j \leqslant n \\[2mm] \dfrac{dN}{dt} = \sigma_nDN_1N_n - 2N\dfrac{dS}{dt} \end{cases} \tag{32}$$

The first equation is for the adatom density and the third one for the density of stable islands. S is the portion of the surface occupied by stable clusters, σ_x the capture factor for the average size. The last term on the right-hand side of the dN/dt equation is the contribution of the coalescence between stable clusters according to Vincent's model [56]. The steady state for the surface density of adatoms leads to:

$$N_1 = \frac{J_0N_0}{v_{-1} + D\sigma_xN} \tag{33}$$

The maximum number of stable clusters, N_{max}, can be determined from the last equation of the system by solving $dN/dt = 0$ [39,40]:

$$2N_{max} = \sigma_nDN_1N_n/(dS/dt)|_{S = S_0} \tag{34}$$

which states that for $S = S_0$ the rate of formation of supercritical clusters is equal to their depletion rate due to coalescence.

Specialising to 2D cluster growth, the changing rate of S can be written as:

$$\frac{dS}{dt} = \frac{1}{N_c}(DN_1\sigma_xN + J_0N_aS) \tag{35}$$

where both the surface diffusion and the direct impingement from the gas-phase contributions are considered. N_c is the atom density in the islands and N_a is the surface density of adsorption sites at the bare substrate. Since, for $N_0 \gg N_1$, $N_0 \cong N_a(1 - S)$, by virtue of equation (33), equation (35) gives $\dot{S} = J_0N_a/N_c$, which holds in the case of complete condensation of admonomers ($v_{-1} = 0$). Using the law of mass action [29] the following expression is eventually obtained:

$$\left(\frac{N_{max}}{N_a}\right) = \left(\frac{\sigma_nN_c}{2\sigma_x^{n+1}N_a}\right)^{1/(n+2)}(1 - S_0)^{2/(n+2)}\left(\frac{J_0}{N_aD}\right)^{n/(n+2)}\exp[-\beta\Delta G_n^0/(n+2)] \tag{36}$$

where $\beta = 1/kT$ and ΔG_n^0 is the standard free energy change for the formation of the critical cluster ($\Delta G_1^0 = 0$).

It can happen, as shown in Fig. 6.6, that the nucleation regime terminates when $S \ll S_0$, i.e. well before the onset of coalescence [43,54,57]. In this case the monomer diffusion length is of the same order of magnitude as the average distance among clusters. The surface density N is nearly constant and equal to the saturation value, N_s, up to S_0 when the coalescence term becomes the dominant one and forces N to

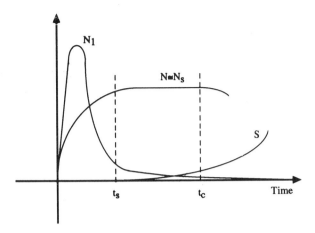

Figure 6.6. Graphic representation of $N_1(t)$, $N(t) = \Sigma_{i>n} N_i(t)$ and $S(t)$ kinetics. The saturation point, $t = t_s$, and the onset of the coalescence region at $t = t_c$, are also indicated.

decrease. Consequently, on account of the conservation of the number of nuclei, N_{max} in equation (34) is equal to N_s.

Equation (36) allows us to estimate material parameters such as activation energy for surface diffusion, critical size and formation energy of critical cluster [14,15,54,55 (and references therein), 58,59].

3.3 *Analytical solution under restrictive assumptions*

An analytical solution for system (18) can be obtained by assuming nucleation to be promoted only at special surface sites (nucleation sites) in the case of constant values for both admonomer and adsorption site surface densities. Under these circumstances system (18) reduces to a system of linear differential equations whose solution, for the surface density of stable clusters, is given by [51]:

$$N(t) = N_\infty \left\{ 1 + \sum_{j=1}^{n+1} (-1)^{(n+j)} \frac{\prod_{i=1}^{n+1} \lambda_i}{\lambda_j \prod_{1 \leq k < j} (\lambda_k - \lambda_j) \prod_{j < m \leq n+1} (\lambda_j - \lambda_m)} e^{-\lambda_j t} \right\} \tag{37}$$

where $(-\lambda_i)$ are the eigenvalues of the associated tri-diagonal kinetic matrix C, whose spectrum is considered to be real and not multiple ($\lambda_i \neq \lambda_j$ $i \neq j$), a condition which is satisfied in the limit of a sharp nucleation barrier [60,61]. The kinetic matrix is given by:

$$C = \begin{pmatrix}
-u_0 & u_{-1} & & & & \\
u_0 & -(u_1 + u_{-1}) & u_{-2} & & 0 & \\
& u_1 & -(u_2 + u_{-2}) & u_{-3} & & \\
& & & \cdot & & \\
0 & & & & \cdot & \\
& & & & \cdot & \\
& & & u_{n-1} & -(u_n + u_{-n}) & 0 \\
& & & & u_n & 0
\end{pmatrix} \tag{38}$$

where $u_{\pm i}$ denotes the probabilities for monomer attachment to (+) and detachment (–) from a cluster made up of i atoms, the index $i = 0$ referring to the nucleation site. The following relationships are derived for the sum and the product of the $n + 1$ eigenvalues (different from zero) in terms of the rate constants:

$$\sum_i \lambda_i = u_0 + \sum_{i=1}^{n} (u_i + u_{-i}) \qquad (39)$$

$$\prod_i \lambda_i = \prod_{i=0}^{n} u_i \qquad (40)$$

For the kinetic problem under consideration the eigenvalues are always less than or equal to zero ($\lambda_i \geq 0$) [60,61]. By virtue of equation (40), the Arrhenius plot of the product of the eigenvalues should be representative of the apparent activation energy for nucleation.

The kinetics reported in equation (37) has extensively been used to process experimental data on diamond deposition at the Si surface by the technique of hot filament chemical vapour deposition (HFCVD) where diamond nucleation is found to be mainly promoted at active sites induced by mechanical treatment of the surface [62]. In this system the particle size distribution function can be derived by scanning electron microscopy (SEM) image analysis performed after a deposition run. The nucleation kinetics, $N(t)$, is successively derived by the size distribution curves applying equation (26) in its integral form and using the growth law of the nucleus [63,64]. The correctness of this method has been verified by analysing experimental data on the bias-enhanced plasma-assisted chemical vapour deposition (PACVD) of diamond, where both the $F(r,t)$ and the experimental $N(t)$ functions were available [65]. The result is shown in Fig. 6.7, where the $N(t)$ curve, given by equation (26), is compared to that obtained by the direct counting of the nuclei (for depositions of different duration). The good agreement between the two sets of data indicates that equation (26) gives a good modelling of the $N(t)$. According to the experimental findings, a linear growth law was used with a typical rate of 0.63 μm h^{-1}. Typical nucleation kinetics on the Si(100) substrate are shown in Fig. 6.8 (open symbols). The solid lines are the best fit of the data using equation (37) for different sizes of the critical cluster, n. The kinetic analysis leads to $n = 2$–3, which indicates the critical nucleus to be microscopic, i.e. made up only of a few carbon atoms [51]. The Arrhenius plot of the eigenvalues (equation (40)), in the case $n = 2$, leads to an activation energy of 52 kJ mol^{-1}. This term should represent the sum of the activation energies for hydrogen abstraction and methylation of the CH_3 radicals [66].

This kinetic analysis shows that the nucleation stage might not be completed at low values of S. Under these circumstances the effect of the capture of both active sites and subcritical clusters by the growing phase has to be taken into account in the description of the experimental kinetics. All the experimental data in [51,63] have been 'corrected' for this effect by making use of Avrami's theory, which will be outlined in the following section.

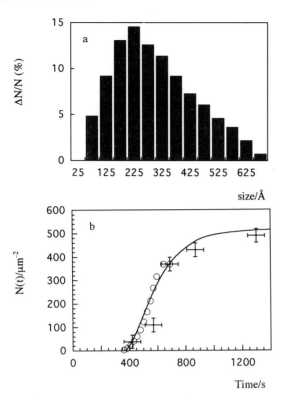

Figure 6.7. Particle size distribution function of diamond crystallites deposited on the Si(100) surface by PACVD (a). The nucleation kinetics extracted by the $F(r,t)$ of panel (a) is shown as open symbols in (b) and compared to the experimental kinetics. The continuous line is just a guide for the eye. Reproduced with permission from [65].

4   High coverage regime

In Section 3.1 rate equations were written on the basis of a 'chemical approximation', namely, by ignoring the spatial variation of the local density of the admonomers caused by the atom diffusion in proximity to the capturing cluster. Accordingly, N_1 is equal to the local density value, $N_1 = n_1(r)|_{r \gg \lambda}$. Since in kinetic equations the N_1 density has been referred to the whole surface, rather than to its uncovered portion, the κ_j coefficient, defined in equation (20), should be divided by the factor $1 - S$. In this respect, rate coefficients are usually computed in the limit $S \rightarrow 0$ because the nucleation stage often ends when the portion of the surface covered is much lower than one. However, when this approximation is not valid, knowledge of the kinetic behaviour of the portion of surface coverage becomes indispensable. For surface coverage greater than about 10% [67,68] overlap among islands cannot be neglected and must be taken into account in the description of the $S(t)$ kinetics.

The $S(t)$ function can be modelled on the basis of Avrami's general theory on the kinetics of phase change [69–71]. This model has successfully been used for studying the formation of thin oxide films on the parent metal [72], the growth of diamond films by the CVD process [73,74] and modelling the growth kinetics of interfaces free of screw dislocations [75].

The following sections are dedicated to Avrami's model, its application to the late stage of cluster growth and its inclusion in the system of rate equations.

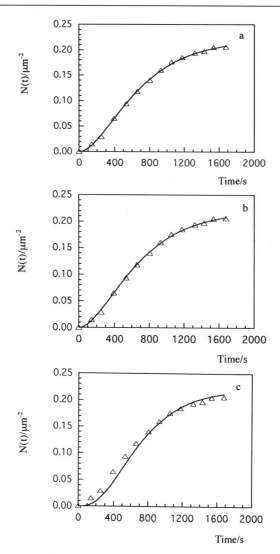

Figure 6.8. Nucleation kinetics of diamond particles at Si(100) substrate. Deposition conditions were: surface temperature $T = 1173$ K, CH_4/H_2 ratio 0.02, total pressure 76 torr. Full lines are the best fit of equation (37) to the data for (a) $n = 1$, (b) $n = 2$, (c) $n = 3$.

4.1 Avrami's model

The basic assumptions of the model are the following: (i) clusters are distributed at random over the substrate surface; (ii) deposition of the new phase occurs by nucleation and growth processes; (iii) the growth law is the same for all clusters; (iv) clusters are not mobile; (v) the size of the smallest stable nucleus is considered to be zero. The solution given by Avrami for the kinetics of the surface coverage, S, is [69–71]:

$$S(t) = 1 - e^{-S_e(t)} \tag{41}$$

where $S_e(t)$, traditionally called the 'extended surface', is an appropriate function of time characteristic of the nucleation and growth processes. Its meaning will shortly be clear.

To demonstrate equation (41) we shall not follow Avrami's original approach

[69–71], but a recent statistical method will be described [76], which has the advantage of being direct and simpler.

Consider a set of N_0 identical discs, of radius r, randomly distributed on the unit surface. The surface covered by discs can be seen as a new phase and from now on it will be addressed as the transformed portion of the surface, S. What, then, is the probability of finding a surface point not covered by the discs? In order to answer this question it is convenient to consider the following issue: given N_0 points distributed at random on the unit surface, the probability of finding n points in a sampled circle of area πr^2 and centre c is given by the Poisson distribution:

$$p_n(m) = \frac{m^n}{n!} e^{-m} \tag{42}$$

where $m = N_0 \pi r^2$ is the average number of points in πr^2. According to equation (42) the probability that no points ($n = 0$) will be found in the given circle is $p_0(m)$. In other words, the centre c lies at a distance greater than r from the nearest of the N_0 points (Fig. 6.9) and $p_0(m)$ also defines, therefore, the probability that a surface point does not belong to the transformed phase [77]:

$$p_0 = e^{-N_0 \pi r^2} = 1 - S \tag{43}$$

Let us now consider a collection of discs distributed at random with different radii

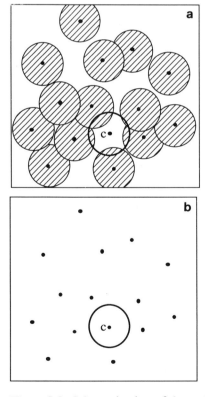

Figure 6.9. Schematic view of the grain arrangement in the case of simultaneous nucleation (a). The probability that a point does not belong to the transformed phase is found by demanding that the centre c lies at a distance greater than r from the nearest of the nucleation sites (b).

$r = r_j$ ($j = 1, 2, \ldots, M$). In this case the probability of finding a surface point not covered by the discs can be given on the basis of the following argument. Consider a collection of planes each containing n_i randomly distributed circles of radius r_i per unit surface; overlaps among circles are allowed. The planes can be stacked, for instance, in order of decreasing radius. The probability that a mobile point will pass through the set of planes without crossing any disc also gives the probability that a point on the surface will not belong to the transformed phase, and is given by:

$$P_0 = \prod_{i=1}^{M} P_{0i} = \prod_{i=1}^{M} e^{-m_i} = \exp\left(-\sum_{i=1}^{M} m_i\right) \tag{44}$$

and, therefore, we can write

$$S(M) = 1 - \exp\left(-\sum_{i=1}^{M} m_i\right) = 1 - \exp\left(-\pi \sum_{i=1}^{M} n_i r_i^2\right) = 1 - e^{-S_e(M)} \tag{45}$$

The collection of planes in Fig. 6.10 represents a configuration of the system at a given time in the kinetics. In the case which leads to equation (43) the picture would reduce to just one plane and mimic a Dirac δ nucleation, i.e. all nuclei start growing simultaneously. The continuous limit of equation (45) is expressed by:

$$S_e(t) = \pi \int_0^t r^2(t,z) \frac{dN_p}{dz} \, dz \tag{46}$$

where $r = r(t,z)$ is the radius of a cluster which starts growing at time z, t is the running time and dN_p/dz is the nucleation rate. We observe that S_e represents the contribution of all clusters to the transformed surface regardless of the possible overlaps among them. Although in this approach the nucleation rate is unambiguously defined, a brief comment on this point is necessary with references to Avrami's original work [69–71].

It is evident from Fig. 6.10 that a cluster belonging to plane j can completely hide a nucleus which lies on plane k in the case $k > j$ ($r_k < r_j$). Although these clusters are naturally included in the computation for obtaining the kinetics according to equations

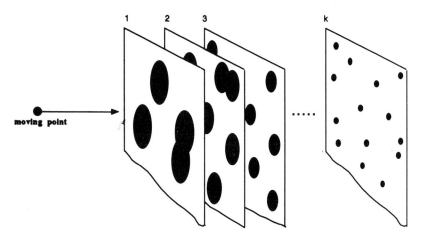

Figure 6.10. Pictorial view of the scattering problem whose solution leads to Avrami's kinetics. Clusters that lie on the same plane were born at the same time.

(41) and (46), they do not contribute to the transformed portions of the surface. Avrami called the clusters underneath the growing phase 'phantom' grains. In equation (46) the subscript p has been added just as a reminder that phantom nuclei must also be taken into account in the computation of the extended surface. Incidentally, it is possible to verify, through computer simulation, that the inclusion of phantoms is mandatory in obtaining the correct solution of the kinetics. A 2D phase transition has been simulated numerically by Sessa *et al.* on a square lattice [67,68]. A random distribution of germs which start growing following an exponential function and an a priori given growth law were considered. A perfect coincidence was found between the $S(t)$ kinetics and the analytical solution $S(t) = 1 - \exp[-S_e(t)]$ when S_e is computed taking into account the phantom contribution. On the other hand, when the random hypothesis is relaxed such an agreement is no longer obtained. In Fig. 6.11 the evolution of the transformed portion of the surface, in real space, is shown for several values of time.

Among the several examples of $S(t)$ experimental kinetics, well described by Avrami's model, it is worth mentioning the measurements recently performed by Donner *et al.* [78]. The kinetics of 5-bromocytosine film formation on a dropping mercury electrode was studied by measuring the double-layer differential capacity in an electrocapillary experiment. The experimental kinetics, $S(t)$, and the best fits obtained using Avrami's model are shown in Fig. 6.12.

It is worth noticing that the term 'coalescence', introduced in Section 3, is sometimes used to indicate the process of collision among clusters during film growth. Strictly speaking, the coalescence between two particles leads to a single particle under conservation of both mass and shape [56]. Conversely, the impingement among islands, as modelled in this subsection, implies that cluster growth ceases at the common interface without any redistribution of matter.

Equations (41) and (46) hold when nucleation occurs at pre-existing nucleation centres. Were this not the case it would be impossible to use equation (42) for the ith plane because the n_i clusters were not distributed at random on the entire ith plane. (The n_i points are, however, distributed at random in the portion of the surface left uncovered by the contributions of the $i - 1$ previous planes.)

Equation (44) can easily be generalised for any number of overlaps, i, as follows [76]:

$$P_i = \frac{[S_e]^i}{i!} e^{-S_e} \qquad (47)$$

which, for $i = 1$, gives the surface coverage of the non-overlapped portions of the grains. By differentiating equation (41) one obtains:

$$\partial_t S = [1 - S(t)]\partial_t S_e$$

or

$$\int_0^t \partial_t s(t,z) \frac{dN}{dz} dz = [1 - S(t)]\int_0^t \partial_t s_e(t,z) \frac{dN_p}{dz} dz \qquad (48)$$

where $\partial_t s (\partial_t s_e)$ is the partial derivative of the actual (extended) single nucleus. If the nucleation is promoted at specific surface sites only, randomly located at the surface, the relationship between the actual rate and those including phantom nucleation is:

$$dN = (1 - S)dN_p \qquad (49)$$

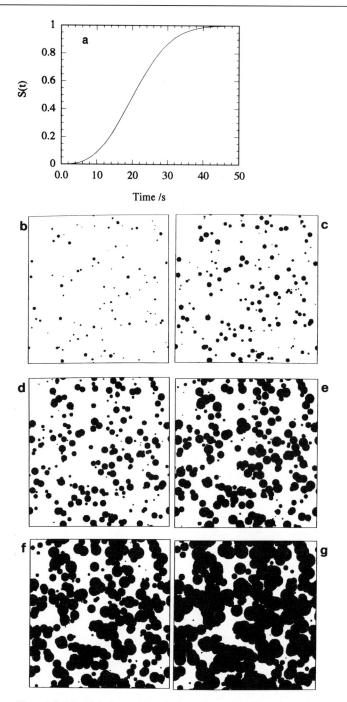

Figure 6.11. Computer simulation of a 2D phase transition on a 500×500 square lattice. (a) The $S(t)$ kinetics. (b)–(g) Graphic representations, in real space, of the phase transition for six values of time, namely 4, 8, 12, 16, 20 and 25 s. The nucleation rate and the growth law took the form $e^{-\alpha t}$ and $r = at$, respectively. Parameters for the computation are: total number of germs $N_0 = 500$, $\alpha = 0.05 \text{ s}^{-1}$ and $a = 1 \text{ s}^{-1}$.

which, after substitution in equation (48), leads to the expression:

$$\frac{\partial_t s(t,z)}{\partial_t s_e(t,z)} = \frac{1 - S(t)}{1 - S(z)} \tag{50}$$

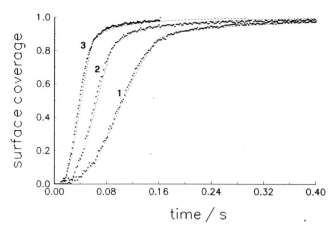

Figure 6.12. Surface coverage–time curves for the formation of a 5-bromocytosine film on a dropping mercury electrode. The electrocapillary measurements were performed at 8°C after a potential step from – 100 mV to – 220 mV (curve 1), – 229 mV (curve 2) and – 232 mV (curve 3). The points represent the experimental data, the dotted lines theoretical computations based on Avrami's theory. Reproduced with permission from [78].

Equation (50) gives a formula for the ratio between the increments of both the transformed and the extended surfaces of a single actual grain in terms of the total surface coverage. We observe that any actual grain is not randomly located in the entire surface; its position is in fact subject to the restriction that its nucleus lay somewhere outside the transformed surface $S(z)$ at the time of its origin, z. This is the physical origin of the $1 - S(z)$ term in the denominator of equation (50). We may therefore assume equation (50) to hold whether the nucleation occurs at pre-existing active sites or not, provided the distribution of clusters in the uncovered portion of the surface is random. Again, after summing both members of expression (50) over the population of actual nuclei, we obtain:

$$\ln[1 - S(t)] = -\int_0^t s_e(t,z)\,\frac{\dot{N}(z)}{1 - S(z)}\,dz \tag{51}$$

i.e. an integral equation for the unknown $S(t)$. Equation (51) has to be included in the rate equation scheme (equation (18)) in order to describe the late stage of growth. This approach has been employed for evaluating both the nucleation rate and the covered surface, up to film closure, in the case of stable dimers by solving the system of integral differential equations using Euler's method [79]. It goes without saying that if $s_e(t,z)$ is a known function and the nucleation rate can be expressed in the form:

$$\dot{N} = f(z)[1 - S(z)] \tag{52}$$

$f(z)$ being a function of z, equation (51) will reduce to Avrami's solution (equations (41) and (46)). It is interesting to show that for the atomistic model of Section 3, where special surface sites for nucleation are not present, particular conditions may also be realised for which equation (52) is satisfied. In the light of the quasi-equilibrium approximation for the subcritical islands, the high-temperature regime can lead to expression (52). In this case desorption of adatoms represents the main channel for admonomer depletion on the surface and the steady state of the adatom surface density, expressed by equation (33), is given by:

$$N_1 = J_0 N_0 / v_{-1} \tag{53}$$

Moreover, for a low value of the surface coverages of subcritical cluster conservation of the number of surface sites gives:

$$N_0 + N_1 \cong N_a (1 - S) \tag{54}$$

where N_a is the surface density of adsorption sites at the bare surface. The nucleation rate is:

$$\dot{N} = \frac{\sigma_n D N_1 N_n}{(1 - S)} \tag{55}$$

where $N_1/(1 - S)$ is the monomer surface density on the uncovered portion of the surface and σ_n is the capture factor of adatoms at the critical cluster edge. By making use of the law of mass action to express the density of critical clusters in terms of the surface density of admonomers, equations (53)–(55) lead to:

$$\dot{N} = \frac{D \sigma_n}{(1 + J_0/v_{-1})^2} \left(\frac{J_0}{v_{-1}} \right)^{n+1} e^{-\Delta G_n^0/kT} N_a^2 (1 - S) \tag{56}$$

where the contribution due to the direct impingement of gas monomers at the critical cluster surface has been neglected. If adatoms surface diffusion is the main process for cluster growth and considering the adatom diffusion length, λ, to be much lower than the cluster dimension, the island growth law is:

$$\rho_v r \frac{dr}{dt} = \frac{DN_1}{1 - S} \frac{2}{\lambda}$$

where ρ_v is the monomer density in the hemispherical cluster [76].

In the framework of Avrami's model, the kinetics of the total cluster perimeter of the film has been evaluated on the basis of equation (50) [80]:

$$P(t) = [1 - S(t)] \int_0^t 2\pi r(t,z) \frac{\dot{N}(z)}{1 - S(z)} dz = [1 - S(t)] P_e(t) \tag{57}$$

where P is the total cluster perimeter (per unit area) and P_e is the total 'extended' perimeter which, by analogy with the total extended surface, represents the sum of the perimeters of all circular clusters, independently of overlaps. Modelling the total perimeter is by no means a trivial issue, since there exists experimental evidence about the role played by atoms at the cluster perimeter, for example, in heterogeneous catalysis [81,82] and in interface physics [83,84]. This issue is further discussed in the next section.

4.2   Dirac δ nucleation

As reported in Section 3.1, in several physical cases the measured nucleation rate does not deviate substantially from the Dirac δ function [54,57]. In other words, the nucleation stage is very short compared to the time required to attain $S = 1$. The mathematical complexity of the kinetic expressions derived in the previous section is dramatically simplified when the nucleation rate reduces to the Dirac δ function. In this case and for nucleation at active sites, the nucleation rate is $\dot{N} = \dot{N}_p = \Omega_0 \delta(z)$, Ω_0 being the surface density of nucleation sites. On the other hand, in the framework of the atomistic model based on the kinetic equations, the rate for stable cluster formation is $\dot{N} = N_s \delta(z)$, where N_s is the surface density of clusters at saturation whose typical expression, in the case of stable dimers, has been reported in equation (30). Accordingly, the kinetics of the surface fraction reduces to:

$$S(t) = 1 - e^{-\pi M_0 r^2(t,0)} \tag{58}$$

where M_0 stands for either Ω_0 or N_s. By virtue of equations (57) and (58) the total cluster perimeter can be expressed in terms of the covered surface [80]:

$$P(S) = [4\pi M_0]^{1/2} (1 - S) \left[\ln \frac{1}{1-S} \right]^{1/2} \tag{59}$$

This function attains a maximum at $S_M = 1 - e^{-1/2} = 0.3935$.

Through an STM study Trafas *et al.* [57] showed that, on GaAs(110) surface, silver nucleation terminates after a few angstroms of metal deposit and $S = 1$ is attained for very long deposition times. In this case the nucleation rate is well approximated by a Dirac δ function. Arciprete *et al.* [83] studied the same interface via high-resolution electron energy-loss spectroscopy. The loss spectra show a broad feature centred at 1.2 eV assigned by the authors to transitions between occupied and empty states induced by the overlayer and situated at the interface. Since the 3D silver clusters are very thick, it is reasonable to consider the signal as just coming from the borders of the clusters. As a consequence it is expected that the intensity of the loss feature will reach its maximum at a surface coverage of about 40%. Indeed, by expressing the silver deposit in terms of surface coverage, the authors find the maximum of the signal at around 41% coverage. Recently, Polini *et al.* have verified equation (59) for the entire range of the surface coverage ($0 \leqslant S \leqslant 1$) [85]. To this purpose, diamond deposition on a mechanically deformed Si(100) substrate was investigated as a 'model' system for Dirac δ nucleation. The distribution of nuclei was deemed to be random at the surface on the basis of the Poisson statistical analysis. The time evolution of both the surface coverage S and the perimeter P were measured by the authors and very well described by Avrami's model. In particular, it was shown that equation (59) reproduces the experimental kinetics quite well and provides an estimation of the nucleation density M_0.

In the framework of Dirac δ nucleation it is interesting to tackle the problem of collision among clusters, for a random arrangement of nuclei, at a given value of S. Equation (42) was found to be quite useful for evaluating the probability, $p_k(S)$, of a nucleus being overlapped by k other nuclei. This probability is given by inserting $m = M_0 \pi [2r(t,0)]^2$ in equation (42), where $r(t,0)$ is given, as a function of S, through inversion of equation (58)

$$p_k(S) = \frac{[-4\ln(1-S)]^k}{k!} (1-S)^4 \tag{60}$$

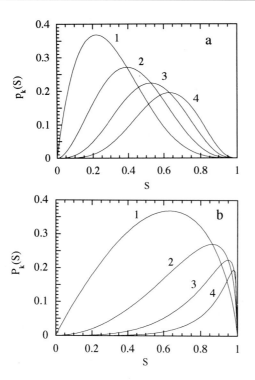

Figure 6.13. (a) The probability that a grain collides with $1,2,\ldots,k$ clusters as a function of S for Dirac δ nucleation. (b) The fraction of surface overlapped by $1,2,\ldots,k$ islands.

Similarly, equation (47) becomes

$$P_k(S) = \frac{1-S}{k!}\left(\ln\frac{1}{1-S}\right)^k \tag{61}$$

where $S_e = \pi M_0 r^2(t,0)$ has been calculated from expression (58). Figure 6.13(a) graphs equation (60) and Fig. 6.13(b) equation (61), in each case for $k = 1,2,3,4$. Notably $p_k = p_{k+1}$ is fulfilled for the maximum of p_{k+1}, i.e. when the covered surface is $S_k = 1 - e^{-(k+1)/4}$. The case $k = 1$ leads to $S_1 = S_M$. In other words, the total cluster perimeter reaches its maximum when a nucleus has the maximum probability of being overlapped by two other nuclei, this probability coincides with that for binary collisions, p_1.

5   Recent experimental results

In recent years the development of both new experimental techniques and new-generation synchrotron radiation sources has lent considerable support to the investigation of microscopic processes occurring during the formation of clusters on solid surfaces.

In our opinion STM is at present the experimental technique providing most of the fundamental insights into the nucleation process [86]. Through this technique it is in fact possible to study the morphology of the film, identify the size of critical clusters and measure the nucleation kinetics as a function of total surface coverage. Two excellent works have been published in which the authors explored the possibility of tailoring the morphology of the film, from regular-shaped islands up to fractal aggregates, by means of a correct interplay between substrate temperature and deposition flux [52,53]. In particular, the temperature determines the perimeter mobility of adatoms and the flux the growth velocity of clusters. The authors were able to identify the critical size of Ag clusters on Pt(111) as well as to measure the nucleation kinetics up to 0.1 ML (Figs 6.4 and 6.5).

The STM technique has been used for measuring the surface diffusion coefficient of mobile species. The surface diffusion coefficient is an important quantity because it is involved as an input parameter in the rate equations for nucleation, and knowledge of it is therefore essential for testing nucleation theories. We shall illustrate three different ways of measuring it, just to give the flavour of how this technique is rapidly evolving towards higher and higher degrees of sophistication.

Mo *et al.* measured the activation energy for the diffusion of Si on Si(001) by a combination of STM analysis and a computer simulation of clustering [87,88]. Nucleation density (N) against diffusion coefficient (D), at a given deposition flux, was simulated for both isotropic and anisotropic diffusion. Differences in the sticking coefficient at the sides and ends of the cigar-shaped islands were also considered. Notably, for isotropic diffusion and anisotropic bonding their simulation was found to be consistent with the curve of N against D obtained by solving the rate equations. STM gives the number density of islands as a function of reciprocal temperature which is utilised to generate, by using the simulated curves of N against D, the Arrhenius plot for D in both isotropic and anisotropic diffusion cases. The activation energy for surface self-diffusion of Si adatoms was found to be 0.67 ± 0.08 eV [87,88].

The second method, due to Bott *et al.* [89], is based on a combination of low-temperature STM and kinetic Monte Carlo simulation. Basically the investigators carried out the STM study in a temperature range in which the only dominant kinetic process is the surface diffusion of monomers. In this way the assumptions for the Monte Carlo simulation are oversimplified, the activation energy for diffusion and the attempt frequency being the two free parameters in the computation. These parameters are changed in order to achieve the best fit to the experimental data. The method, applied to the self-diffusion of Pt on Pt(111), gives an activation energy of 0.26 ± 0.01 eV, a value in excellent agreement with that obtained by field ion microscopy (FIM) [90]. Incidentally, it is worth mentioning that the FIM technique has been used for measuring adatom and cluster diffusion coefficients as well as cluster nucleation, since about the mid-1960s [91]. Cohesion energy and stability of small clusters (few adatoms) can also be determined by FIM. Until now FIM has mainly been confined to the investigation of metal on metal systems. The linear dimension of the sample being investigated is of the order of 20 Å, and this limits the possibility of studying nucleation kinetics. For a detailed review of FIM the reader is referred to the already cited excellent work by Kellogg [91].

The third STM method of studying surface diffusion, in principle but not in practice the simplest, is due to Swartzentruber [92]. It consists of following the diffusion species using a scanning tunnelling microscope equipped with feedback electronics which maintain the tip over the species, tracking its coordinates as it diffuses over the surface. In this way it is possible to report the frequency of residence of the adatoms at the adsorption site as a function of the residence time. The residence-time probability

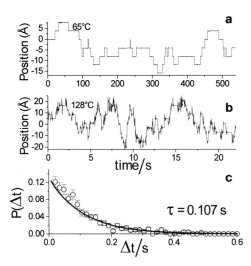

Figure 6.14 Addimer position as a function of time at (a) 65°C, (b) 128°C. (c) The measured residence time probability distribution at 128°C. Reproduced with permission from [92].

distribution is shown in Fig. 6.14 for the diffusion of Si dimers on Si(001) at 128°C. The average of the distribution is equal to the mean residence time, τ. Distributions obtained for different temperatures lead to the evaluation of the diffusion activation energy, $E_d = 0.94$ eV, being $\tau \propto e^{E_d/kT}$.

Among the most used indirect experimental techniques for studying the nucleation and growth processes, there are the electron and photoelectron spectroscopies. A number of studies have been performed, exploiting techniques such as Auger electron spectroscopy, electron energy-loss spectroscopy, X-ray photoelectron spectroscopy and ultraviolet photoelectron spectroscopy. We will not go into the details of the techniques, and the interested reader can easily find many good books and reviews on the subject [93]. Also a wide-ranging review of experiments and results can be found in [7]. What is of interest here is the fact that, apart from few rare exceptions, until now the growing film has been studied by analysing a succession of frozen situations. Conversely, a real kinetic evolution would require the changes of the spectra under examination to be followed in real time. By using these spectroscopies the possibility of studying the dynamics of surface processes is limited by the acquisition time of the spectrum, which has to be shorter than the time-scale of the process under study. Recently, by using photoelectron spectroscopy, thin-film growth has been monitored as a function of time by Tucker *et al.* [94] at SRS Daresbury Laboratories. One of their results is reported in Fig. 6.15, where the growth kinetics of gadolinium on a faceted W(110) substrate is shown. Because of the not very high photon flux the kinetics are obtained by monitoring the variation of the height of the gadolinium 4f peak. In this respect we observe that it would be more convenient to follow the time evolution of the shape of the whole line during growth. As a matter of fact this goal can be achieved by a very high photon flux which can be only provided by new-generation synchrotron radiation sources. These are characterised by high brightness and high electron beam current, not to mention the possibility of exploiting insertion devices. Consequently, the acquisition time of a single spectrum can be reduced by as much as an order of magnitude when compared to an old synchrotron source. As an example of the potentiality of the new sources we refer to

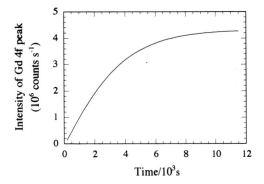

Figure 6.15. Kinetics of the Gd growth on a faceted W(110) substrate, studied by synchrotron radiation spectroscopy. The intensity of the gadolinium 4f peak is reported as a function of the growth time. The photon energy was 146 eV with an analyser resolution of about 1 eV. Reproduced with permission from [94].

the real-time measurements recently obtained by Baraldi *et al.* of reaction kinetics at solid surfaces [95,96]. Due to the high brilliance of the third-generation synchrotron radiation source ELETTRA in Trieste, the total time required to measure a single spectrum was a very impressive 6 s. These remarkable achievements open up new horizons for the study of nucleation and growth in real time. In fact, since these spectroscopies are, in principle, sensitive to the state of bonding of atoms at the surface, information related to the energetics of clustering could be extracted from the spectra. However, one should remember that in order to extract any significant information, these data should be analysed on the basis of a theoretical model.

Although this section has mainly been devoted to the nucleation and diffusion processes at surfaces, it is worth noting that experimental techniques do also exist for studying the growth law for clusters. For this purpose a widely used technique is low-energy electron diffraction (LEED), where the angular profile of the diffracted beams provides quantitative information on the island sizes and growth.

Consider, for simplicity, a one-dimensional problem. In particular, assume a random distribution of N clusters of the same size, La, with $\mathbf{a} = 2\mathbf{b}$, $\mathbf{a}$ and $\mathbf{b}$ being the primitive vectors of the overlayer and of the substrate, respectively and L is an integer. At the 'fundamental reflection' the LEED intensity is [97–99]

$$I \propto N^2 L^2 \tag{62}$$

Far from the 'fundamental reflection' the random phases from different islands sum to zero and the LEED intensity is now given by:

$$I(\mathbf{K}) \propto N \frac{\sin^2(L\mathbf{Kb})}{\sin^2(\mathbf{Kb})} \tag{63}$$

where $\mathbf{K}$ is the momentum transfer. For $\mathbf{Kb} = 2\pi h$ (with h an odd integer), equation (63) gives $I = NL^2$, namely, the 'sublattice reflections'. This simple interpretation allows one to determine L and N by measuring $I(\mathbf{K})$ under the condition that impingement among clusters is negligible. The assumption of single size clusters, which implies Dirac δ nucleation, can be relaxed and the calculation generalised to a distribution of sizes [97].

Figure 6.16(a) shows the LEED angle profile for the Pd–Al$_2$O$_3$–NiAl(110) system, where Pd was deposited on a thin Al$_2$O$_3$ layer (5 Å in thickness) grown on the NiAl(110) substrate [100]. The corresponding growth kinetics are shown in Fig. 6.6(b). The island sizes were obtained from LEED patterns by calculating profiles and comparing them with the experimental data.

Another method which has been used for studying thin-film growth is grazing-incidence small-angle X-ray scattering (GISAXS) [101]. In the experimental scattering geometry for GISAXS, the sample is aligned at the substrate critical angle for total external reflection. The incident beam is reflected and refracted by the substrate, where the wavevector of the refracted beam, $\mathbf{S}_r$, is parallel to the sample surface. The refracted beam encounters the islands which produce small-angle scattering around the refracted beam. The scattering vector, $\mathbf{h}$, is $\mathbf{h} = \mathbf{S}_r - \mathbf{S}_s$, where $\mathbf{S}_s$ is the wavevector of the scattered beam that is detected by the X-ray analyser positioned parallel to the surface. The vector $\mathbf{h}$ is resolved into two components, h_p and h_n, parallel and perpendicular to the surface. In particular, $h_p = (4\pi/\lambda) \sin \vartheta$ where λ is the wavelength of the X-rays and 2ϑ is the scattering angle. During film growth the GISAXS intensity is plotted as a function of h_p ($0 < h_p < 0.02$ Å^{-1}). When the average edge-to-edge island spacing becomes less than or equal to the average island diameter, peaks are recorded in the scattering

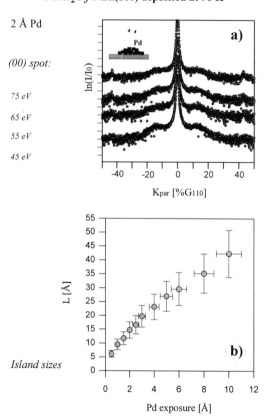

Figure 6.16. Growth kinetics of Pd clusters on Al$_2$O$_3$–NiAl(110) substrate studied by LEED. (a) The LEED spot profile for the (00) reflex at different energies. (b) Calculated island diameters. Reproduced with permission from [100].

patterns which are related to intercluster interference. The average centre-to-centre island spacing, d, is given by $d = 2\pi/h_p$, where h_p is the peak position. The average radius of islands, the Porod radius, can be obtained directly from the scattering patterns [102]. The advantages of this technique are that it can be done *in situ*, is not limited to conducting substrate and does not require synchrotron radiation.

6 Summary and conclusions

The experimental conditions usually employed during thin-film growth by vapour phase lead to critical clusters so very small in size that the classical nucleation theory becomes inapplicable. In this case, at variance with classical theory, where the nucleus is considered a thermodynamic system, models aiming at describing the nucleation kinetics necessarily have to deal with atomic-level quantities such as adsorption energy of adatoms and interaction energy among admonomers. In this context, rate equations have been used successfully since the beginning of the 1960s. Recently, this approach was compared with kinetic Monte Carlo (KMC), which is a more sophisticated method that provides a 'true' solution of the kinetics. The test led to the following conclusions: the mean field rate equations approach and KMC simulations are in excellent agreement with respect to the average quantities (the average island size, etc.).

On the other hand, the description of the entire kinetics of growth, up to the film closure, demands an accurate statistical treatment of the collision events among nuclei. The Avrami model achieves this, provided that the islands are distributed randomly and nucleation occurs at pre-existing 'active sites'. Work is in progress on including Avrami's model of cluster impingement in the rate equation schemes in which nucleation does not occur at pre-existing nucleation sites. It is hoped that in the near future it will be possible to describe the entire film growth within the framework of atomic-level approaches using rate equations.

Several experimental techniques are currently available which allow accurate measurement of many quantities related to the nucleation and the film-growth processes such as the number of stable islands, the size of stable clusters and adsorption and diffusion activation energies of adatoms. In particular, the STM technique and the spectroscopies carried out using latest-generation synchrotron radiation sources seem to show considerable promise for measuring activation energies of surface processes as well as kinetics of clustering in real time for a wide range of systems.

7 Acknowledgements

We are grateful to Professor M.W. Roberts who gave us the opportunity to write this paper and to Professor E. Molinari for his encouragement and useful advice during the draft of the manuscript. We also acknowledge Professor F. Battaglia and Dr A. La Monaca for fruitful discussions, and we would like to express our appreciation to Mr M. Brolatti for his help in realising some of the drawings.

8 References

1 Bauer E, Poppa H. *Thin Solid Films* 1972; **12**: 167.

2 Rhead GE, Barthés M-G, Argile C. *Thin Solid Films* 1981; **82**: 201.

3 King AD, Woodruff DP (eds) *The Chemical Physics of Solid Surfaces and Heterogeneous Catalysis*. Amsterdam: Elsevier, 1984.

4 Dufour LC, Perdereau M. In Dufour LC, Monty C, Petot-Ervas G (eds) *Surface and Interfaces of Ceramic Materials*. Dordrecht: Kluwer, 1989.

5 Campbell CT. *J. Chem. Soc. Faraday Trans.* 1996; **92**: 1435.

6 Spear KE, Frenklach M. In Spear KE, Dismukes JP (eds) *Synthetic Diamond*. New York: Wiley, 1994; 243.

7 Zinke-Allmang M, Feldman LC, Grabow MH. *Surf. Sci. Rep.* 1992; **16**: 377.

8 Evans JW, Bartelt MC. *J. Vac. Sci. Technol. A* 1994; **12**: 1800.

9 Kang HC, Weinberg WH. *J. Chem. Phys.* 1989; **90**: 2824.

10 Bales GS, Chrzan DC. *Phys. Rev. B* 1994; **50**: 6057.

11 Frankl DR, Venables JA. *Adv. Phys.* 1970; **19**: 409.

12 Venables JA. *Phil. Mag.* 1973; **27**: 697.

13 Stoyanov S, Kashchiev D. In Kaldis E. (ed.), *Current Topics in Materials Science*, Vol. 7. Amsterdam: North Holland, 1981; 69.

14 Venables JA, Spiller GDT, Hanbücken M. *Rep. Progr. Phys.* 1984; **47**: 399.

15 Venables JA. *Surf. Sci.* 1994; **299–300**: 798.

16 Kukushkin SA, Osipov AV. *Progr. Surf. Sci.* 1996; **51**: 1.

17 Hill T.L. *An Introduction to Statistical Mechanics*. New York: Dover, 1986.

18 Hirth JP, Pound GM. *Progress in Material Science*. Oxford: Pergamon, 1963.

19 Lifshitz IM, Slyozov VV. *J. Phys. Chem. Solids* 1961; **19**: 35.

20 Chakraverty BK. *J. Phys. Chem. Solids* 1967; **28**: 2401.

21 Chakraverty BK. *J. Phys. Chem. Solids* 1967; **28**: 2413.

22 Listvan M. *Surf. Sci.* 1986; **173**: 294.

23 Farrell JE, Valls OT. *Surf. Sci.* 1988; **199**: 586.

24 Zinke-Allmang M, Stoyanov S. *Jpn J. Appl. Phys.* 1990; **29**: L1884.

25 Zinke-Allmang M, Feldman LC, Grabow MH. *Surf. Sci.* 1988; **200**: L427.

26 Cossmann H-J, Fisanick GJ. *Scanning Microsc.* 1990; **4**: 543.

27 Frenkel J. *Kinetic Theory of Liquids*. New York: Dover, 1995.

28 Kelton KF, Greer AL, Thompson CV. *J. Chem. Phys.* 1983; **79**: 6261.

29 Walton D. *J. Chem. Phys.* 1962; **37**: 2182.

30 Zinsmeister G. *Vacuum* 1966; **16**: 529.

31 Lewis B, Campbell DS. *J. Vac. Sci. Technol.* 1967; **4**: 209.

32 Lewis B. *Surf. Sci.* 1970; **21**: 273.

33 Lewis B. *Surf. Sci.* 1970; **21**: 289.

34 Zinsmeister G. *Thin Solid Films* 1968; **2**: 497.

35 Zinsmeister G. *Thin Solid Films* 1969; **4**: 363.

36 Zinsmeister G. *Thin Solid Films* 1971; **7**: 51.

37 Logan RM. *Thin Solid Films* 1969; **3**: 59.

38 Halpern V. *J. Appl. Phys.* 1969; **40**: 4627.

39 Stowell MJ. *Phil. Mag.* 1970; **21**: 125.

40 Stowell MJ. *Phil. Mag.* 1972; **26**: 361.

41 Stowell MJ, Hutchinson TE. *Thin Solid Films* 1971; **8**: 41.

42 Stowell MJ, Hutchinson TE. *Thin Solid Films* 1971; **8**: 411.

43 Schmidt AA, Eggers H, Herwig K, Anton R. *Surf Sci.* 1996; **349**: 301.

44 Biham O, Barkema GT, Breeman M. *Surf. Sci.* 1995; **324**: 47.

45 Rakocevic Z, Strbac S, Bibic N, Nenadovic T. *Surf. Sci.* 1995; **343**: 247.

46 Wen J-M, Evans JW, Bartelt MC, Burnett JW, Thiel PA. *Phys. Rev. Lett.* 1996; **76**: 652.

47 Ruckenstein E, Pulvermacher B. *J. Catal.* 1973; **29**: 224.

48 Tomellini M. *Appl. Surf. Sci.* 1996; **99**: 67.

49 Pound GM, Simnad MT, Yang L. *J. Chem. Phys.* 1954; **22**: 1215.

50 Sigsbee RA. *J. Appl. Phys.* 1971; **42**: 3904.

51 Tomellini M. *Ber. Bunsenges. Phys. Chem.* 1995; **99**: 838.

52 Röder H, Hahn E, Brune H, Bucher JP, Kern K. *Nature* 1993; **366**: 141.

53 Brune H, Romainczyk C, Röder H, Kern K. *Nature* 1994; **369**: 469.

54 Brune H, Röder H, Boragno C, Kern K. *Phys. Rev. Lett.* 1994; **73**: 1955.

55 Venables JA. *Phys. Rev. B* 1987; **36**: 4153.

56 Vincent R. *Proc. Roy. Soc. London A* 1971; **321**: 53.

57 Trafas BM, Yang YN, Siefert RL, Weaver JH. *Phys. Rev. B* 1991; **43**: 14 107.

58 Jones GW, Marcano JM, Norskov JK, Venables JA. *Phys. Rev. B* 1990; **65**: 3317.

59 Robins JL. *Appl. Surf. Sci.* 1988; **33–34**: 379.

60 Wilkinson JH. *The Algebraic Eigenvalue Problem.* Oxford: Oxford University Press, 1965; 71.

61 Tomellini M. *Ber. Bunsenges Phys. Chem.* 1996; **100**: 1199.

62 Liu H, Dandy DS. *Diamond Chemical Vapor Deposition, Nucleation and Early Growth Stage.* Park Ridge, NJ: Noyes, 1995.

63 Molinari E, Polini R, Sessa V, Terranova ML, Tomellini M. *J. Mater. Res.* 1993; **8**: 785.

64 Polini R. *J. Appl. Phys.* 1992; **72**: 2517.

65 Jiang X, Schiffmann K, Klages C-P. *Phys. Rev. B* 1994; **50**: 8402.

66 Frenklach M, Wang H. *Phys. Rev. B* 1991; **43**: 1520.

67 Sessa V, Fanfoni M, Tomellini M. *Phys. Rev. B* 1996; **54**: 836.

68 Sessa V, Fanfoni M, Tomellini M. *Mater. Sci. Forum* 1996; **203**: 181.

69 Avrami M. *J. Chem. Phys.* 1939; **7**: 1103.

70 Avrami M. *J. Chem. Phys.* 1940; **8**: 212.

71 Avrami M. *J. Chem. Phys.* 1941; **9**: 177.

72 Holloway PH, Hudson JB. *Surf. Sci.* 1974; **43**: 123.

73 Tomellini M. *J. Appl. Phys.* 1992; **72**: 1589.

74 Stiegler J, Kaenel Y, Cans M, Blank E. *J. Mater. Res.* 1996; **11**: 716.

75 Kashchiev D. *J. Crystal. Growth* 1977; **40**: 29.

76 Fanfoni M, Tomellini M. *Phys. Rev. B* 1996; **54**: 9828.

77 Evans UR. *Trans. Faraday Soc.* 1945; **41**: 365.

78 Donner C, Pohlmann L, Baumgärtel H. *Surf. Sci.* 1996; **345**: 363.

79 Fanfoni M, Tomellini M. *J. Electr. Spectr.* 1995; **76**: 283.

80 Tomellini M, Fanfoni M. *Surf. Sci.* 1996; **349**: L191.

81 Carley AF, Davies PR, Roberts MW. *Topics Catal.* 1994; **1**: 35.

82 Spencer MS. *Catal. Today* 1992; **12**: 453.

83 Arciprete F, Colonna S, Fanfoni M, Patella F, Balzarotti A. *Phys. Rev. B* 1996; **53**: 12 948.

84 Feenstra RM, Mårtensson P. *Phys. Rev. Lett.* 1988; **61**: 447.

85 Polini R, Tomellini M, Fanfoni M, Le Normand F. *Surf. Sci.* 1997; **373**: 230.

86 Binning G, Rohrer H. *Rev. Mod. Phys.* 1987; **59**: 615.

87 Mo YW, Kleiner J, Webb MB, Lagally MG. *Phys. Rev. Lett.* 1991; **66**: 1998.

88 Mo YW, Kleiner J, Webb MB, Lagally MG. *Surf. Sci.* 1992; **268**: 275.

89 Bott M, Hohage M, Morgenstern M, Michely T, Comsa G. *Phys. Rev. Lett.* 1996; **76**: 1304.

90 Feibelman PF, Nelson JS, Kellog GL. *Phys. Rev. B* 1994; **49**: 10 548.

91 Kellog GL. *Surf. Sci. Rep.* 1994; **21**: 1.

92 Swartzentruber BS. *Phys. Rev. Lett.* 1996; **76**: 459.

93 Briggs D, Seah MP. (eds) *Practical Surface Analysis* (2nd edn.) Chichester: Wiley, 1990.

94 Tucker NP, Blyth RIR, White RG, Lee MH, Robinson AW, Barrett SD. *J. Synchrotron Rad.* 1995; **2**: 252.

95 Baraldi A, Barnaba M, Brena B *et al. J. Electr. Spectr.* 1995; **76**: 145.

96 Baraldi A, Comelli G, Lizzit S, Cocco D, Paolucci G, Rosei R. *Surf. Sci.* 1996; **367**: L67.

97 Lu TM, Wang GC, Lagally MG. *Surf. Sci.* 1981; **108**: 494.

98 Lu TM, Wang GC. *Surf. Sci.* 1981; **107**: 139.

99 Pukite PR, Lent CS, Cohen PI. *Surf. Sci.* 1985; **161**: 39.

100 Bäumer M, Libuda J, Sandell A. *et al. Ber. Bunsenges. Phys. Chem.* 1995; **99**: 1381.

101 Levine JR, Cohen JB, Chung YW, Georgopoulos P. *J. Appl. Cryst.* 1989; **22**: 528.

102 Levine JR, Cohen JB, Chung YW. *Surf. Sci.* 1991; **248**: 215.

7

Surface Reactions of Hot Electrons at Silver–UHV and Silver–Electrolyte Interfaces

A. OTTO, D. DIESING, H. JANSSEN, M. HÄNISCH, M.M. LOHRENGEL*, S. RÜßE*,
A. SCHAAK, S. SCHATTEBURG, D. KÖRWER, G. KRITZLER and H. WINKES

Lehrstuhl für Oberflächenwissenschaft (IPkM), Heinrich-Heine-Universität, Düsseldorf, D-40225 Düsseldorf, Germany

* AGEF-Institut an der Heinrich-Heine-Universität, Düsseldorf, D-40225 Düsseldorf, Germany

1 Introduction

1.1 *Definition of hot electrons*

In the following, we define hot electrons in metals as electrons in the energy range between the Fermi level and the vacuum level, more than kT above the Fermi level, of very low density, which only interact with the 'cold' electrons in the ground-state Fermi–Dirac distribution. We exclude, for instance, the excitations of the quasi-free electron gas of a metal under high-power laser irradiation with a Maxwell distribution not equilibrated with the lattice or the non-equilibrated electron distribution in femtosecond laser-induced dynamical quantum processes on solid surfaces as it is discussed, for instance, in the desorption of adsorbates induced by multiple electronic transitions (DIMET) [1–3].

In the sense of our definition, hot electrons are, for instance, the low-energy end of the secondary or cascade electron distribution below the vacuum level in photoemission, low-energy electron diffraction (LEED) and electron energy-loss spectroscopy (EELS) or a final state of the electron incident in inverse photoemission spectroscopy.

It is straightforward to extend the definition to 'hot holes' in metals. According to the creation or excitation mechanism of hot electrons, one may further differentiate between primary and secondary electrons (see Section 2.3). Secondary electrons have often led to radiation damage of adsorbates, which is especially cumbersome in LEED investigations. Nevertheless, these and similar effects may also be considered in a more positive light.

1.2 *Surface reactions of hot electrons*

1.2.1 SURFACE PHOTOCHEMISTRY

Recent years have witnessed extended research into surface photochemical effects [4]. Moreover, intra-adsorbate optical transition reactions of single hot electrons, optically excited in the metal, are believed to trigger desorption or dissociation of the adsorbates by transfer to the unoccupied or only partly occupied adsorbate orbitals.

1.2.2 SURFACE ENHANCED RAMAN SPECTROSCOPY

In this as yet not fully understood spectroscopy it is believed that part of the

163

enhancement is caused by a coherent photo-driven charge transfer between the metal and the adsorbates [5].

1.2.3   SURFACE PHOTOEFFECT AND NONLINEAR SURFACE OPTICS

This field will be covered by a forthcoming monograph [6]; we will refer to some of this work in the discussions below.

1.3   *Creation of hot electrons*

The examples above involved optical excitations of electrons into hot electron states. Hot electrons can be injected into a metal by tunnelling from the tip of a scanning tunnelling microscope, for instance in ballistic electron emission microscopy or spectroscopy.

If one wants to have free access to the metal surface for the usual surface-science spectroscopies in the case of the metal–ultrahigh vacuum interface or to electrochemical techniques or optical spectroscopies at a metal–electrolyte interface, it is of advantage to inject hot electrons from the interior of the metal towards the interface. This is possible for thin metal films whose thickness does not exceed the mean free elastic path of the hot electrons. One possibility would be given by reverse bias of a semiconductor-metal Schottky barrier. However, in this case the energy of the hot electrons will not exceed the height of the Schottky barriers, which are below 1 eV [7,8]. In order to have hot electrons also of higher energies, we used metal–insulator–metal (MIM) tunnelling contacts [9].

2   Metal–insulator–metal contacts

2.1   *Characterisation of MIM junctions*

We use Al–aluminium oxide (AlOx)–Ag tunnel junctions. These are prepared by first vacuum-depositing 99.999% Al strips on to cleaned microscope slides (Fig. 7.1). The Al deposition rate is 0.2 nm s^{-1} and the thickness of the Al film is 30 nm. During the

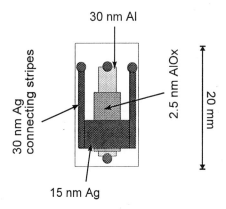

Figure 7.1. Configuration of Al–AlOx–Ag tunnel junctions and electric connecting strips, prepared on glass slides.

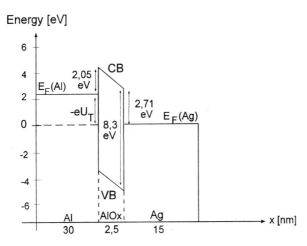

Figure 7.2. Energy levels and layer thicknesses of Al–AlOx–Ag junctions. CB is the lower edge of the conduction band and VB the upper edge of the valence band of the aluminium oxide.

deposition, the system is kept at a pressure of 1×10^{-8} mbar. The oxide film is grown by exposing the system for 2 h to an oxygen atmosphere at 1 bar pressure and room temperature. The resulting oxide thickness is 2.5 nm [10]. Then a (99.999%) silver film with a thickness of 15 nm is deposited at a rate of 0.05 nm s^{-1} and a pressure of 1×10^{-7} mbar.

Figure 7.2 shows the one-electron energy levels and the thicknesses of the three layers of an Al–AlOx–Ag junction. The silver top electrode is earthed. The applied tunnel voltage is assumed to be $U_T = -2.3$ V. The negative value of the voltage means that the electrons tunnel from the aluminium side through the oxide layer to the silver side. This tunnel current, I_T, is detected with an ammeter.

There are many different values in the literature for the barrier height between the metals and the oxide conduction band. For instance, Shepard found a value of 1.1 eV [11] for the contact voltage of aluminium and aluminium oxide, and Shu and Ma found a value of 2.9 eV [12]. Most often the contact voltages are measured by the photocurrent method. Here we apply an increasing absolute tunnel voltage $|U_T|$ and measure the tunnel current I_T. If $|U_T|$ becomes greater than the contact voltage then the thickness of the tunnel barrier decreases for the hot electrons and so-called Fowler–Nordheim tunnelling sets in [13]. This can be detected with the change of the slope in the current–voltage characteristic.

Figure 7.3 shows the $\log(|I_T|)$ versus $|U_T|$ characteristic measured at a temperature $T = 77$ K. Curve (a) shows a change in slope at a tunnel voltage of $U_T = +2.05$ V, indicating the value of the contact voltage between aluminium and aluminium oxide (see Fig. 7.2). Curve (b) shows the change in slope at -2.71 V. This is the contact voltage between silver and aluminium oxide [14] (see Fig. 7.2).

The position of the valence band of the aluminium oxide is not exactly known, but we expect electron correlation effects to be small and therefore we position it according to the optical bandgap [15] 8.3 eV below the conduction band. Since the offset voltages of the valence band with respect to the Fermi levels in the metals are larger than the contact voltages discussed above, tunnelling via the conduction band of the AlOx will prevail in any case over hole tunnelling via the valence band. However, electron

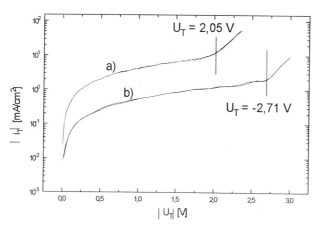

Figure 7.3. Current–voltage characteristic of an Al–AlOx–Ag junction at $T = 77$ K: (a) at positive bias; (b) at negative bias. After [14].

tunnelling in the direction from Ag to Al via the conduction band at positive bias is equivalent to hole tunnelling from Al towards Ag. This becomes observable if a suitable hole acceptor is present in the electrolyte (see [16] and Fig. 6 therein). This will not be further discussed in this paper.

When the MIM contacts are within an electrochemical cell, the tunnelling barrier is not altered by ion migration into the oxide, driven by the electrochemical potential. This is discussed in Section 3.4.

2.2 *The tunnelling process*

During tunnelling [17,18], an electron is in a coherent eigenstate extending in the aluminium, the insulating oxide and the silver top electrode. The electron in this state may become localised as a hot electron in the silver electrode only by breaking the coherence of the tunnelling state by dephasing or elastic or inelastic scattering. Eventually it will end in a state close to the Fermi level of the silver electrode, contributing to the tunnel current, measured as a 'cold electron current' in the external circuit connecting the silver and aluminium electrode (see also Section 5.2).

Inelastic scattering may be given by electromagnetic, electronic and vibronic excitations of the material in the insulating gap (in the latter case this is applied as so-called inelastic tunnelling spectroscopy [9]) or by excitations in the interior or at the outer surface of the silver electrode. In the interior, the electron–electron scattering determines the elastic mean free path of the primary hot electrons, whereas the scattering by phonons is weaker. The latter determines the d.c. resistivity of the 'cold' electrons but it may be neglected in our case [19]. As the experimental results in Section 5.2 show, for a thin silver electrode the coherence may also be broken at the outside interface of the electrode.

In the following section we neglect all aspects of tunnelling and surface effects and assume hot electrons injected with defined energy into the silver at the interface between the oxide and silver, with momentum normal to the interface. Silver is treated as a degenerate free electron gas with an electron density equal to the density of silver atoms. The energy distribution of the electrons reaching the outside surface has been calculated under the assumption that every primary or secondary hot electron reaching this surface is converted into a cold electron. Formally this corresponds to a zero reflection coefficient at the outer Ag interface.

2.3 *Monte Carlo simulation (MCS) of hot electron distribution [20]*

The transition probability $w(k_1, k_2, k'_1, k'_2)$ from the collision between the hot electron with wavevector k_1 and the Fermi electron of the metal with wavevector k_2 to the states with wavevectors k'_1 and k'_2 is given by the screened Coulomb interaction. The exchange interaction is neglected. The quantum-mechanical perturbation theory yields [21]

$$w(k_1, k_2, k'_1, k'_2) = \frac{2\pi e^4}{\hbar \varepsilon_0^2} \sum_q \frac{1}{q^4} \frac{1}{|\varepsilon(q, \Delta E)|^2} \delta_{k'_1, k_1+q} \delta_{k'_2, k_2-q}$$

$$\times \delta(E_1 + E_2 - E'_1 - E'_2)[1 - f(E'_1)][1 - f(E'_2)]f(E_2) \tag{1}$$

The one-electron energies E correspond to the wavevectors k, $f(E)$ is the Fermi–Dirac distribution and the screening by the 'cold' electrons of the metal ('in the Fermi sea') is taken into account with the Lindhard dielectric function [22] $\varepsilon(q, \Delta E)$, where q is the momentum and ΔE is the energy exchanged between the two electrons.

From equation (1) follows the probability $w(k_1, k'_1)$ of scattering the hot electron from state k_1 into state k'_1:

$$w(k_1, k'_1) = -\frac{4e^2}{\varepsilon_0} \frac{1}{|k'_1 - k_1|^2} \operatorname{Im} \frac{1}{\varepsilon(|k'_1 - k_1|, E_1 - E'_1 + i0^+)}[1 - f(E'_1)] \tag{2}$$

Due to the isotropy of the free electron gas, this transition probability is only a function of the energies E_1 and E'_1 and the angle θ between k_1 and k'_1:

$$w(E_1, E'_1, \theta) = \frac{16\pi e^2}{\varepsilon_0 \hbar^2} \frac{\sqrt{E_1}}{q^2} \frac{\varepsilon_2(q, E_1 - E'_1)}{|\varepsilon(q, E_1 - E'_1)|^2} \sin\theta \tag{3}$$

where

$$q = q(E_1, E'_1, \theta) = |k'_1 - k_1| = \frac{\sqrt{2m}}{\hbar} \sqrt{E_1^2 + E'^2_1 - 2\sqrt{E_1 E'_1} \cos\theta} \tag{4}$$

In this scattering transitions of the hot electron all the 'cold' electrons with initial wavevector k_2 are involved, which are characterised by the absolute value

$$k_2(\vartheta) = \frac{2m}{\hbar^2} \frac{\Delta E - \frac{\hbar^2}{2m} q^2}{2q \cos\vartheta} \tag{5}$$

where $\Delta E = E_1 - E'_1$ and ϑ is the angle between k_2 and $(k_1 - k'_1)$ within the limits

$$\vartheta_{max} = \arccos\left(\sqrt{\frac{2m}{\hbar^2(E_F + \Delta E)}} \frac{|\Delta E - \frac{\hbar^2}{2m} q^2|}{2q}\right) \tag{6}$$

$$\vartheta_{min} = \arccos \left(\sqrt{\frac{2m}{\hbar^2 E_F}} \; \frac{\left| \Delta E - \dfrac{\hbar^2}{2m} q^2 \right|}{2q} \right) \tag{7}$$

Monte Carlo simulations [23] with the setting of the values of E'_1 and θ by the so-called direct method [24] were performed to obtain the mean free path λ_{ee} of the primary electrons with the parameters $E_F = 5$ eV and $r_s = 3.08$. This result is compared in Fig. 7.4 to the analytic derivation by Quinn [25].

$$\lambda_{ee} = \frac{1.45 E_F^{3/2} E}{\sqrt{m^*/m} \, \arctan(\beta^{-1/2}) + \dfrac{\beta^{1/2}}{\beta + 1}} \; \frac{1}{(E - E_F)^2} \tag{8}$$

with $m^*/m = 1.1$, and $\beta = 0.166 \, r_s$; E and E_F are in electron-volts, λ_{ee} in nanometres.

Since the agreement is satisfactory, the MCS of the energy distributions at the outside interface, as described above, was performed with Quinn's approximation of the mean free path in order to keep computation time to within reasonable values.

In order to keep account of the cascade of all the secondary electrons, values for the stochastic free path (derived after the mean free path is known, delivering the point of appearance of the two secondaries with wavevectors k'_1 and k'_2), E'_1, θ and δ were obtained by the direct method [24]. The results for the primary energies $E_1 - E_F = 2.5$ eV and thicknesses of the silver layer of 10, 20 and 30 nm are given in Fig. 7.5(a). The number of electrons reaching the outside surface ballistically without energy loss diminishes with silver film thickness, reflecting the increasing ratio of thickness to the mean free path (see Fig. 7.4). The yield of secondaries increases up to a thickness of 0.1 μm. As expected, the distribution is not changed by a further introduction of elastic scattering into the MCS. Since the relative energy loss in hot electron–phonon scattering is small, we expect only a small change of the energy distributions. The appearance of primary hot electrons at the silver–vacuum interface is experimentally demonstrated by electron emission into the vacuum (see, for instance, Fig. 4 in [26]) and at the silver–electrolyte interface by the emission into the conduction band of water (see Section 3.2).

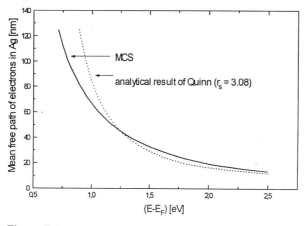

Figure 7.4. Mean free path of electrons in silver, numerical results from MCS and analytical theory. From [20].

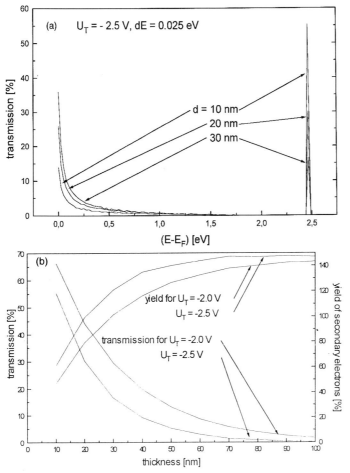

Figure 7.5. (a) Energy distribution at the surface of silver top electrodes of a tunnel junction with starting energy $E_1 - E_F = 2.5$ eV. The energy intervals ΔE for the MCS are 0.025 eV. (b) Transmission of primary electrons to the surface (*left scale*) and yield of secondary electrons (*right scale*) with energy $E > E_F + 50$ meV at the surface as a function of the thickness of the silver layer.

3 Electrochemical reactions of hot electrons

3.1 *Electrochemical set-up*

In the following set-up the silver film of the MIM will be made the working electrode in an electrochemical cell. By changing its electrochemical potential (i.e. the difference between the electrostatic potentials of the reference and the working electrode), different states of the metal–electrolyte interface, consisting of the metal surface, adsorbates and fully or partly hydrated ions, are achieved. Especially important for this paper is the shift of the electronic levels within the bulk electrolyte (electron acceptor or donor levels on solute molecules or ions, so-called redox couples) with respect to the Fermi level of the silver electrode: if the electrochemical potential is changed by x volts, then the levels are shifted by x electron-volts. In this way electron transfer processes are controlled by the setting of the electrochemical potential.

To perform these measurements two electrical circuits are necessary (see Fig. 7.6).

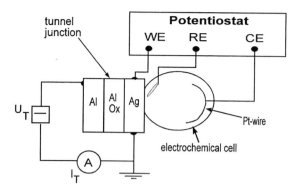

Figure 7.6. Tunnel junction and electrochemical circuitry. WE, working electrode; RE, saturated calomel reference electrode; CE, counter-electrode.

One circuit contains the voltage source U_T and the ammeter to control the current I_T through the tunnel junction. The other circuit controls the electrochemical potential E_{SCE} of the silver top electrode of the MIM contact which is made the working electrode, whereas the counter-electrode is a platinum wire. We use the saturated calomel electrode (SCE) as a reference electrode. The silver working electrode is at common mass. The active tunnel area of the samples is typically 0.12 cm^2 and in contact with the electrolyte. Outside this area the sample is covered with a protective lacquer.

Figure 7.7 shows the electron energy scheme of the tunnel junctions in the electrolyte, the threshold PE for photoemission into an aqueous electrolyte at 0V$_{SCE}$ (see section 3.2), and the threshold electrochemical potentials (versus SCE) of the reduction of Fe^{3+} and the hydrogen evolution HER observed with our MIM contacts at $U_T = 0$ V. Note that these values are not necessarily the thermodynamic redox potentials — for instance, the threshold of hydrogen evolution involves a considerable overpotential [27]. The Fermi level of the saturated calomel reference electrode E_{SCE} is given in Fig. 7.7 at the electrochemical potential of 0 V$_{SCE}$. The threshold potentials are given on the scale of the one-electron energy levels in the MIM contact, but they should not be mistaken for the levels of electronic acceptor or donor states.

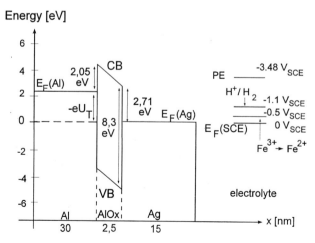

Figure 7.7. Energy scheme of the tunnel junctions and the threshold potentials of some electrochemical reactions (see text).

3.2   Electron injection into electrolytes

First we give the results of our own experiments on photoemission from polycrystalline silver electrodes in 0.1 mol KCl aqueous electrolyte, before and after roughening the electrode by an oxidation–reduction cycle [28]. For various laser frequencies ω the yield Y of the photocurrent into the electrolyte was monitored as a function of the electrochemical potential E_{SCE}. The threshold potential of emission $E_{SCE}(Y = 0)$ was evaluated from the relation [29]

$$Y \begin{cases} \propto (\hbar\omega - \hbar\omega_0 - eE)^{5/2}, & \hbar\omega - \hbar\omega_0 - eE > 0 \\ = 0, & (\hbar\omega - \hbar\omega_0 - eE) > 0 \end{cases} \tag{9}$$

and is plotted against $\hbar\omega$ in Fig. 7.8. The threshold $\hbar\omega_0 = 3.44$ eV for the non-roughened silver electrode, in good agreement with the literature [30], is given by the lower boundary of the electron conduction band in aqueous electrolytes. The photoelectron is prevented from proceeding back to the electrode by scavengers, in this case probably by the formation of OH^- ions from water [31]. A strong roughening of the silver surface by an oxidation–reduction cycle involving 37 mC cm^{-2} lowers the emission threshold only by 0.2 eV (see Fig. 7.8) and increases the photocurrent at constant laser power by a factor of 3 [28].

In the following experiments we insert an unroughened tunnel junction in a 0.9 mol NaAc buffer (pH = 5.9) with 50% water and 50% ethylene glycol. This electrolyte was chosen because the MIM contacts do not fail as easily under hydrogen evolution as they do when inserted into a pure aqueous electrolyte.

Figure 7.9 shows the electrolyte current i_{El} of the silver top electrode of the tunnel junction against applied tunnel voltage U_T for three electrochemical potentials E_{SCE}. There appears a clear threshold of U_T, shifting with E_{SCE}, below which electrons are injected into the electrolyte. These threshold values are plotted against E_{SCE} in Fig. 7.10. The data points are well approximated by the relation $E_{SCE} + U_T = -3.3$ V. This value is between the thresholds of photoemission into the electrolyte of unroughened and roughened silver electrodes (see Fig. 7.8). This demonstrates that primary tunnelling electrons have been injected into the conduction band of the electrolyte (of

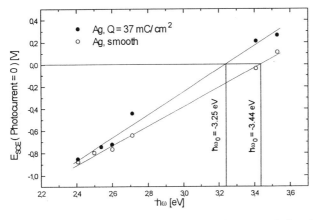

Figure 7.8. Evaluation of the threshold of photoemission from unroughened and roughened (oxidation–reduction cycle of 37 mC cm^{-2}) into a 0.1 mol KCl aqueous electrolyte.

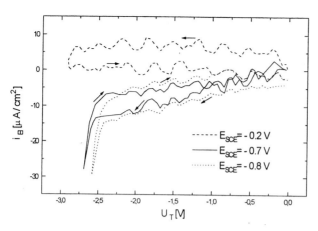

Figure 7.9. Injected electron current density i_{EI} as a function of the tunnel voltage U_T at various electrochemical potentials E_{SCE}.

the order of 1% of the tunnelling electrons, given the currents in Figs 7.3 and 7.9) but that electrons of lower energy than the position of the electronic conduction band in water are not injected into the electrolyte or scavenged in this electrolyte and return to the electrode. This behaviour changes when the silver electrode is activated (see Section 3.4).

Note that the injected currents are about three orders of magnitude higher than the photocurrents emitted from an unroughened silver electrode, though the incident laser power of some 100 mW [28] is about one order of magnitude above the electric energy dissipated in the MIM contacts.

3.3   Reduction of Fe^{3+}

In the following experiments we inserted the tunnel junctions in the glycol/water/ acetate electrolyte with an additional concentration of 2 mmol Fe^{3+} sulfate. The voltammogram taken without biasing the tunnel contact in Fig. 7.11 shows in the

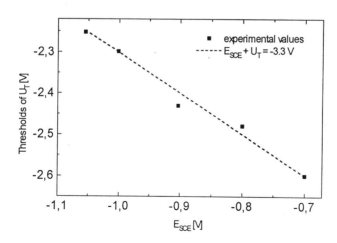

Figure 7.10. Threshold values of the tunnelling voltage for electron injection versus electrochemical potential of the silver electrode in a 0.9 mol NaAc buffer (pH = 5.9) with 50% water and 50% ethylene glycol. Below the dotted line, electrons are injected into the electrolyte.

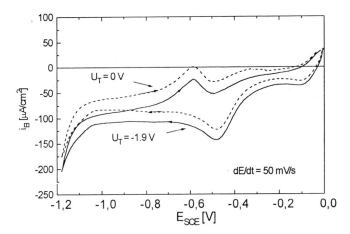

Figure 7.11. Cyclic voltammograms of a non-roughened silver top electrode of an MIM junction in a 50% water, 50% ethylene glycol, 0.9 mol NaAc, 2 mmol $Fe_2(SO_4)_3$ electrolyte, at tunnelling voltages 0 V and – 1.9 V.

cathodic scan the onset of the reaction $Fe^{3+} + e^- \rightarrow Fe^{2+}$ at $E_{SCE} = -0.5$ V and the reverse reaction at $E_{SCE} \approx -0.6$ V. These are the well known electron transfer processes between silver and the hydrated Fe ions, without hot electrons present. Applying a tunnel voltage of $U_T = -1.9$ V connected with a tunnel current $i_T = -2.5$ mA cm^{-2} changes the voltammogram (see Fig. 7.11).

In contrast with the electron injection experiments of the previous section, a well-defined onset of the cathodic reaction of the hot electrons is not detected in the difference in the electrolyte current densities between the voltammograms with $U_T = -1.9$ V and $U_T = 0$ V. The cathodic change of current with the tunnel voltage increased with negative potential and with Fe^{3+} sulfate concentration (see Fig. 7.12). It should be noted that at $U_T = -1.9$ V and $E_{SCE} > -1.2$ V (this is the range in Fig. 7.12) the primary hot electrons are below the threshold of electron injection into the electronic conduction band of the electrolyte. Therefore, one may safely exclude the possibility that Fe^{3+} in the bulk of the electrolyte acts as scavenger in this case. Rather, the hot electrons react with the same Fe ions with which the 'cold' electrons react. Our

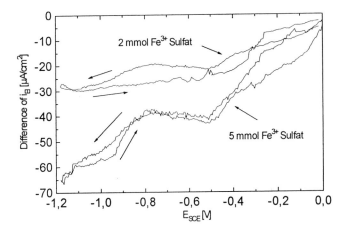

Figure 7.12. Difference in the electrolyte current densities of the two voltammograms in Fig. 7.11 and for the analogous case with 5 mmol $Fe_2(SO_4)_3$.

experiment does not allow us to assign the hot electron current into the electrolyte to only the primary or secondary hot electrons. The overall transfer rate of hot electrons, taken as the ratio R of the extra current $\Delta(I_{El})$ into the electrolyte and the tunnel current I_T, is 1.2% at $E_{SCE} = -1.2$ V, $U_T = -1.9$ V, and 2 mmol Fe^{3+}; for the same E_{SCE} and U_T and 5 mmol Fe^{3+}, it is 2.2%.

The addition of a europium salt leads to comparable results [14].

3.4   *Hydrogen evolution and hydrogen oxidation*

In the experiments with the pure 0.9 mol NaAc buffer (pH = 5.9) with 50% water and 50% ethylene glycol electrolyte the only hot electron process observed on smooth silver surfaces is the injection of the primary electrons into the electronic conduction band of the electrolyte [14]. However, when the MIM contacts are intentionally produced on a rough substrate or activated (see below) hydrogen gas evolves by a hot electron process, at electrochemical potentials, where neither the normal hydrogen evolution reaction by 'cold' electrons nor the injection of primary hot electrons into the electronic conduction band of the electrolyte (see Section 3.2) is possible.

The roughness of the silver surface is obtained by two methods.

1 Undercoating the tunnel junctions with a CaF_2 layer 100 nm thick. These junctions are mainly produced for light emission (see Sections 4.1 and 4.2). Capacity measurements of the metal–electrolyte interface which can be performed with current transients after potential steps (see Fig. 7.13) yield differential capacities of 19 μF cm^{-2} for the smooth and 21 μF cm^{-2} for the CaF_2 undercoated surface. Accordingly the accessible surface area is increased by about 10% by the undercoating.

2 So-called activation of the silver top electrode in the same electrolyte by a potential pulse to 0.5 V_{SCE}, with the oxidation and reduction charge always below 3 mC cm^{-2}. Both methods deliver equal experimental results in terms of the hydrogen evolution [14].

Figure 7.14 shows voltammograms of a tunnel junction prepared on CaF_2 at tunnel voltages $U_T = -1.8$ V, -1.9 V and -2.1 V [16]. The voltammograms depend on the

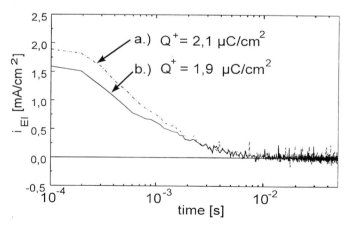

Figure 7.13. Current transients and corresponding charge Q^+ after a potential step from $E_{SCE} = -0.9$ V to -0.8 V in a 0.9 mol NaAc buffer with 50% water and 50% ethylene glycol of (a) a smooth MIM contact (b) and an MIM contact on 100 nm CaF_2.

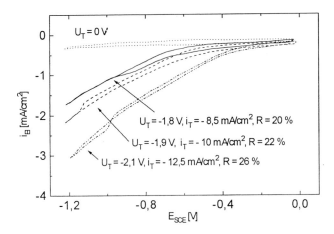

Figure 7.14. Cyclic voltammograms of a silver top electrode of an MIM junction on CaF_2 in a 0.9 mol NaAc buffer with 50% water and 50% ethylene glycol at tunnelling voltages 0 V, – 1.8 V, – 1.9 V and – 2.1 V. The corresponding tunnelling currents at E_{SCE} = – 1.2 V_{SCE} are indicated. R is the ratio of the hot current into the electrolyte and i_T at – 1.2 V_{SCE}.

applied tunnel voltage, even though the hot electrons are certainly below the threshold of injection into the electronic conduction band of the electrolyte as discussed above. The onset of the hot electron-induced cathodic current is shifted 300 mV to positive potentials when the tunnel voltage is raised by 300 mV. The ratio R, defined in Section 3.3, increases at E_{SCE} = – 1.2 V from 20% at U_T = – 1.8 V to 26% at U_T = – 2.1 V. These R values are a factor of about 10 bigger than the measurements with Fe^{3+} and 'smooth' tunnel junction surfaces.

Under stationary conditions, the hot electron-induced cathodic current leads to formation of gas bubbles. Therefore we assign the reaction of the hot electrons to a hydrogen evolution reaction. The hydrogen evolution can be characterised by the factor b, given by

$$b = - dE_{SCE}/dlog\ (i_{el}) \qquad\qquad (10)$$

derived from the Tafel plots in Fig. 7.15 of the hydrogen evolution with 'cold' electrons (U_T = 0) and with hot electrons at U_T = – 2.1 V. For the hydrogen evolution with hot electrons, b is about four times greater. The values for different tunnel voltages U_T are given in Table 7.1. The relation between b and the tunnel voltage shows clearly that the dynamics of the hydrogen evolution reaction are determined by the hot electrons. It should be noted that at a tunnel voltage U_T = – 2.1 V hot electrons reach the threshold for injection into the electrolyte only at E_{SCE} < – 1.2 V (see Figure 7.7). Therefore the hydrogen evolution by hot electrons in Fig. 7.15 is not a scavenger process.

Figure 7.16 shows a voltammogram with U_T = – 2.1 V and a second voltammogram

Table 7.1 b factor for hydrogen evolution with hot electrons.

U_T (V)	b (V)
– 1.8	1.07
– 1.9	1.01
– 2.1	0.9

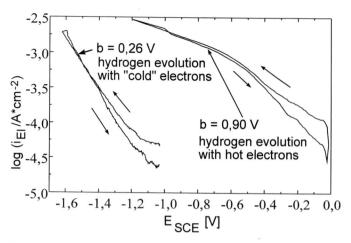

Figure 7.15. Tafel plots and *b* factors of the hydrogen evolution by 'cold' electrons and hot electrons at $U_T = -2.1$ V, in a 0.9 mol NaAc buffer with 50% water and 50% ethylene glycol. Scan velocity $dE/dt = 50$ mV s^{-1}.

with $U_T = 0$ in which a tunnel voltage $U_T = -2.1$ V is switched on at times A and B, and switched off at times C and D. One can observe the reversibility between reactions of the hot electrons and the 'cold' electrons. After switching off at C and D there are transients of the anodic current. The following experiments study these phenomena with pulsed tunnel voltages and a time resolution of 1 μs.

Figure 7.17 shows I_{EI} versus the logarithm of time after the application of three pulses of the tunnelling voltage U_T from 0 to -3 V and back to $U_T = 0$ after 10^{-2} s, 10^{-1} s and 1 s. The electrochemical potential is $E_{SCE} = -0.8$ V. The measurements demonstrate that the electrolyte current reaches a constant value after 10^{-5} s, corresponding to $R = 0.33$. This time is given by the charging of the MIM capacity and thus by the transient of I_T, as measured spearately by impedance spectroscopy and current transient spectroscopy of the MIM junction [14]. This delay is 1–10% of the charging time of the double layer capacity [14,32]. The electrolyte current driven by the hot electrons is constant after a charge transfer of less than 1.5 μC, which corresponds to less than 10^{-2} of a monolayer of ions. Both the short time and the small charge exclude

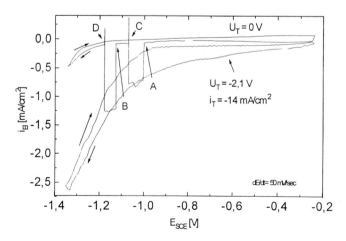

Figure 7.16. Switching between reactions of hot electrons and 'cold' electrons (see text).

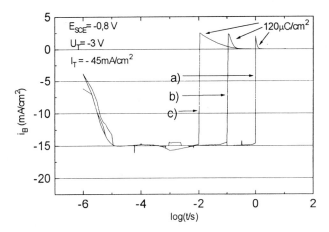

Figure 7.17. Electrolyte current during and after pulses of U_T from 0 to -3.0 V and back at constant E_{SCE}. Time t starts at the onset of the pulses. The electrolyte is 50% water, 50% ethylene glycol, 0.9 mol NaAc.

ionic effects within the oxide barrier driven by the electrochemical potential. Apparently the silver film is tight enough not to allow potential modulated migration of ions or atoms from the silver–electrolyte interface into the tunnelling barrier, influencing the tunnelling current.

After switching off the tunnel voltage there is an anodic current transient within about 10^{-1} s. The related charge density is $120 \, \mu C \, cm^{-2}$ in all three cases in Fig. 7.17, though the cathodic charges injected by the pulses are different: by pulse (a), of 1 s duration, $15 \, mC \, cm^{-2}$, but by pulse (c), of 0.01 s duration, only $150 \, \mu C \, cm^{-2}$. Further experiments with shorter pulse durations (5–100 ms) show that the anodic charge density saturates as a function of the pulse duration at a value of $115 \, \mu C \, cm^{-2}$. This is shown in Fig. 7.18 for the example of $U_T = -2.2$ V. Up to the cathodic charge of $115 \, \mu C \, cm^{-2}$ all the charge is recovered in the anodic transient.

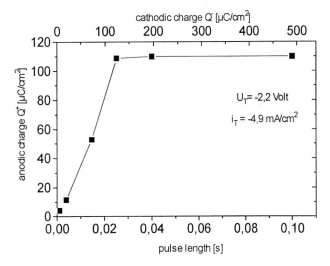

Figure 7.18. Charge density Q^+ in the anodic transient versus hot electron pulse length (below) and corresponding cathodic charge density Q^- (above), calculated from the indicated values of i_T and the pulse length.

These results are compared with the hydrogen evolution by 'cold' electrons for an MIM contact in the same electrolyte activated by an oxidation pulse as described above (see Fig. 7.19). The tunnel voltage is constant at zero. The electrochemical potential is switched at $t = 0$ from $E_{SCE} = -0.8$ V up to $E_{SCE} = -1.8$ V to evolve hydrogen. The stationary current due to hydrogen evolution reached after about 50 ms is about 2 mA cm^{-2}. Subtracting the charge density due to stationary hydrogen evolution in the 300 ms pulse from the overall charge density of 705 μC cm^{-2} leaves a charge density of about 100 μC cm^{-2} unaccounted for. Loading the differential capacity of the silver–electrolyte interface of about 20 μF cm^{-2} (see Fig. 7.13) will need only about 20 μC cm^{-2}. Therefore we do not exclude the possibility that the cathodic transient in Fig. 7.19 is caused either by a transient hydrogen evolution from protons or by unknown cathodic reactions with ethylene glycol, whereas the stationary hydrogen evolution in neutral aqueous electrolytes involves only the water molecule [27]. After 10 ms, 50 ms and 300 ms, respectively, the electrochemical potential is stepped from $E_{SCE} = -1.8$ V back to $E_{SCE} = -0.8$ V outside the range of the hydrogen evolution reaction. Again an anodic transient is observed, saturating at 105 μC cm^{-2}. Only about 20 μC cm^{-2} can be assigned to discharging of the differential capacity.

The anodic current transients in the experiments with hot and 'cold' electrons are nearly equivalent; differences may be accounted for by the different ways of roughening. The saturation value of about 120 μC cm^{-2} or 85 μC cm^{-2} corresponds to the oxidation of about half a monolayer of adsorbed neutral atoms.

In order to study further the differences in the hydrogen evolution at smooth and 'activated' silver surfaces, we used thin single-crystalline Ag(111) films grown on Si(111) [27] in a 0.1 mol KClO$_4$ aqueous electrolyte. A similar anodic transient to that shown in Fig. 7.19 after hydrogen evolution was observed only when the surface was activated by an anodic pulse increasing the surface area by 10% [27]. Any reactions with organic molecules could be excluded by using pure neutral aqueous electrolytes [27].

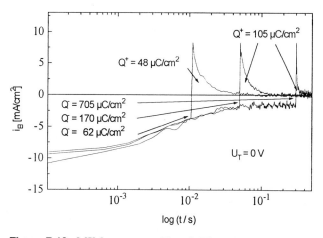

Figure 7.19. MIM contact at $U_T = 0$. Time dependence of the electrolyte current density during and after cathodic pulses of E_{SCE} from -0.8 V to -1.8 V between $t = 0$ and $t = 10$ ms, 50 ms, 300 ms. Charge densities accumulated in the cathodic pulses and the anodic current transients are indicated.

Taking all the experiments described in this section into account, we assign the anodic charge Q^+ (see Fig. 7.18) to a transient layer of adsorbed atomic hydrogen, produced during the hydrogen evolution, reoxidised to atomic hydrogen within about 0.1 s after the end of the hydrogen evolution. The coverage of about 0.5 cannot be assigned to adsorption at sites of atomic roughness only, because these sites can account for at most 10% of the silver electrode area, e.g. to the increase in area by the activation procedure. The hydrogen evolved beyond the formation of this monolayer, e.g. the molecular hydrogen gas in the bubbles formed after some time and adhering to the surface is not reoxidised.

4 Surface optical effects with hot electrons

4.1 *Light emission of tunnel junctions in UHV*

It has been known since the mid-1970s [33] that Al–AlOx–Au or Ag contacts prepared on a rough substrate emit light under an appropriate bias. Comprehensive experiments on light emission from rough CaF$_2$ substrate–AlOx–Ag contacts prepared in ultrahigh vacuum (UHV) [26] have clearly demonstrated that the light emission is due to the excitation of surface plasmon polaritons [34] at the silver–vacuum interface. The surface plasmon polaritons, which may be considered as 'photons bound to a plane surface', are converted into photons, propagating in free space ('light') by scattering at the surface roughness. In this scattering the most important spatial Fourier components are those with a correlation length of the order of the wavelength of the emitted light. These correlation lengths are provided with sufficient amplitude by the CaF$_2$ undercoating (see, for instance, [35] and references therein).

The surface plasmon polaritons at the silver–UHV interface are excited by hot electrons, which is indicated by a 30 times higher light emission yield Y (photons per tunnelling electron) under negative tunnel voltage $-|U_T|$ than at positive tunnel voltage $+|U_T|$. The interaction between the hot electron and the surface plasmon polariton seems to take place right at the surface, as is indicated by the quenching of Y by a submonolayer coverage of atomic oxygen [26] and the increase in Y by a factor of 2.5 after covering the clean silver surface with about 0.6 nm of potassium (see Fig. 7.20).

The same effect is observed by deposition of about 0.5 nm of caesium [26]. The deposition is accompanied by a rise of the tunnel current from 0.17 mA up to 0.55 mA. This will be discussed in Section 5.2.

The comparison with reflection measurements of exposed and unexposed silver top electrodes shows that the enhancement of light emission cannot be explained due to the variation of the macroscopic optical properties [26]. The adsorption of oxygen and the deposition of potassium alter the electronic properties of the surface. The quenching or enhancement of light emission is caused by a variation of the coupling of hot electrons with the SPP at the silver–vacuum surface.

We explain our results with the help of Sommerfeld's surface photoeffect at a free electron gas–vacuum interface as 'inverse surface photoemission with hot electrons'. We use the modern version of internal surface photoemission from the comprehensive work of Liebsch [6] on surface-electron–photon coupling at the jellium edge with density functional theory, especially the theoretical results in Fig. 7.21.

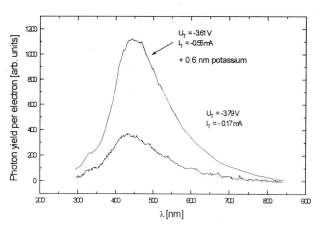

Figure 7.20. Light intensity normalised by the tunnelling current i_T of a tunnel junction with a silver top electrode before and after the deposition of about 0.6 nm K.

In Fig. 7.21(a) the electron density n_d in the jellium is equal to the atomic density of metallic silver. Formally, the caesium coverage is modelled by a jellium background layer of the thickness t of one monolayer of Cs; the positive charge density in this layer

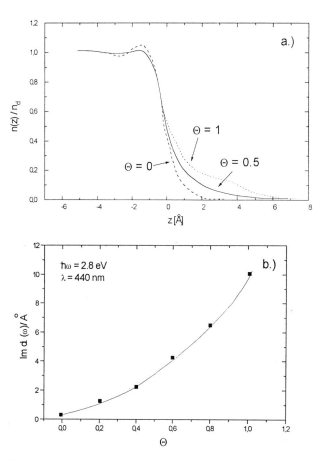

Figure 7.21. (a) Ground-state electron distribution $n(z)$ for Cs overlayers on Ag at different coverages Θ (adapted from Figure 2 of [36]); z is normal to the surface, the edge of the positive background of 'Ag jellium' is at $z = 0$; $n_d = n(z)$ for z towards minus infinity. (b) Calculation of the internal photoemission yield for different Θ (adapted from Figs 1 and 2 of [37]).

is taken as $\Theta e n_d(\text{Cs})$, where n_d is the bulk atomic density in metallic caesium. This corresponds to an 'extra surface charge' of 65 $\mu\text{C cm}^{-2}$, equivalent to 0.29 electrons per surface atom at an Ag(111) surface. We will come back to this point in Section 4.2. The photoemission yield Y_p is given for a flat surface and p-polarised light of frequency ω incident at 45° to the surface normal of the jellium by

$$Y_p(\omega) = \sqrt{8}\alpha\omega[1 - (\omega^2/\omega_p^2)] \text{ Im } d_\perp(\omega) \tag{11}$$

$d_\perp(\omega = 0)$ is the distance of the centre of gravity of the static screening charge (caused by the application of an external static field) from the edge of the positive background (at $z = 0$ in Fig. 7.21(a)). When an electric field normal to the surface is applied at optical frequencies, ω, the screening charge has a phase lack with respect to the electric field, taken into account by a complex value of $d_\perp(\omega)$. As in all phenomena of time retardation between action and reaction, the imaginary part of the response function describes the energy dissipation. In our case the dissipation channel is the optical excitation of the 'cold' electrons, which is observed as electron emission into the vacuum ('photoemission') and into the interior of the jellium ('internal photoemission'). When $\hbar\omega$ is smaller than the work function, which applies to the Ag–Cs jellium model at $\hbar\omega = 2.8$ eV, there is only internal photoemission. So Fig. 7.21(b) shows the internal photoyield in arbitrary units as a function of the Cs coverage Θ, which is formally equivalent to an extra surface charge. One should note that the increase in the internal photoelectron yield is connected with the increase in the width of the tail of the electron density distribution $n(z)$ at the jellium surface.

In Section 5.3 we describe our first attempts to observe the internal surface photoemission. The time-inverted process of internal surface photoemission would be 'inverse internal surface photoemission' — in other words, light emission into the vacuum by hot electrons propagating from the bulk of the silver into the inhomogeneous electron gas at the jellium surface, connected with a de-excitation of the electrons right there.

As yet the theory has only given the yield of internal photoemission, not the detailed energy distribution of the hot photoelectrons, whereas in principle both the energy of the primary electrons and the energy of the emitted photons are measurable. Nevertheless, we explain the increase in the yield after covering the MIM contact with K (Fig. 7.20) qualitatively with the results of the jellium calculations in Fig. 7.21. We are confident that, in contrast to the important and little-understood role of surface roughness in the hydrogen evolution reaction (Section 3.4) and the influence of the electrochemical potential on the tunnelling current (Section 5.2), the surface roughness achieved by undercoating the MIM contacts by CaF_2 is only necessary to convert the surface plasmon polaritons into light, but that they are anyway excited at a smooth silver surface, and just not converted into light. This will be proved in Section 4.3.

4.2 *Enhancement of photon yield by strongly charging the silver surface*

In the previous subsection it was pointed out that covering a silver jellium with a jellium layer representing a caesium coverage corresponds formally to an extra surface charge of 65 $\mu\text{C cm}^{-2}$. Surface charges of this order of magnitude can be induced at silver–electrolyte interfaces.

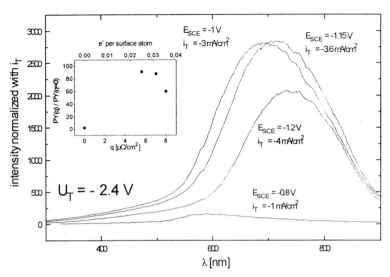

Figure 7.22. Light emission spectra of a rough tunnel junction in an ethylene glycol electrolyte with 0.9 mol NaAc at $U_T = -2.4$ V for different parameters E_{SCE} normalised by the tunnel current density i_T which depends on E_{SCE} (see Section 5.2). Inset: normalised integrated photoyield PY(q) as a function of the negative surface charge density q on a scale with PY($q = 0$) set equal to 1.

In the following experiments Al–AlOx–Ag junctions undercoated with 100 nm CaF$_2$ are inserted into an ethylene glycol electrolyte with 0.9 mol NaAc [38]. The applied tunnelling voltage is $U_T = -2.4$ V. We do not use pure aqueous electrolyte because the heavy hydrogen evolution destroys the thin silver layer. The light emission is detected for various values of the electrochemical potential E_{SCE}. Figure 7.22 shows the increase in light emission by shifting the potential E_{SCE} from -0.8 V up to -1.15 V. Typical integral intensities at $U_T = -2.4$ V and $E_{SCE} = -1.15$ V are 5×10^{-5} photons per tunnelling electron. No light is detected for $E_{SCE} > -0.8$ V. The decrease at $E_{SCE} = -1.2$ V is probably caused by the diffusion of light by the microbubbles of hydrogen gas which is evolved under these conditions (see Section 3.4). The tunnel current I_T depends on the electrochemical potential as discussed in Section 5.2.

The surface charge of the silver top electrode, dependent on the electrochemical potential, can be calculated if the potential of zero charge E_{PZC} and the double layer capacity are known. Therefore we measured the double layer capacity of the silver top electrode (see Fig. 7.23). The structure at $E_{SCE} = -0.8$ V indicates the so-called potential of zero charge E_{PZC} at which the electrostatic potential in the bulk of the silver and the electrolyte are equal. At $E_{SCE} < E_{PZC}$ the silver electrode is charged negatively. A shift of 0.4 V to more cathodic potential enhances the surface charge up to 8 μC cm^{-2} or 0.04 e^- per surface atom of Ag(111). This negative surface charge should be compared with the formal surface charge in deposition of alkali metals as discussed above. We envisage the negative surface charge at the silver electrolyte to be distributed in a tail like that depicted in Fig. 7.21(a), extending towards the hydrated K$^+$ ions.

The inset of Fig. 7.22 shows an increase in the integrated intensity of the emitted light as a function of the surface charge of about two orders of magnitude. We explain this increase by analogy with the case of the potassium coverage of the MIM contact in UHV (Section 4.1).

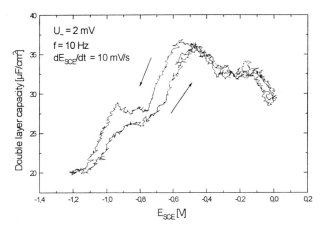

Figure 7.23. Differential double-layer capacity as a function of the electrochemical potential.

The results of Fig. 7.22 demonstrate dramatically the advantage of the photon in–out, electron up–down methods over photon in–photon out methods, for instance electroreflectance, which yields only changes in reflectivity of the order of 10^{-2}. Apparently, with the photon in–out, electron up–down methods one suppresses heavily the bulk response and reaches a surface sensitivity otherwise only given by nonlinear optical methods (see also Section 5.3).

No light is observed for positive tunnelling voltages, when hot holes reach the interface, corroborating the results with the tunnel junctions in UHV. Probably hot holes do not interact with the photons, because their wave functions do not extend beyond the centre of gravity of the optical screening charge, which is mainly composed of electronic states at the Fermi energy. In contrast, hot electron wave functions reach beyond the screening charge, approximately reaching their reversal point only when their energy equals the image potential in front of the so-called image plane which is at $z = \mathrm{Re}\{d\perp(\omega)\}$. Loosely speaking, the hot electrons 'see the light', whereas the hot holes do not.

4.3 Light emission of smooth contacts
employing attenuated total reflection configurations

In the optical experiment described thus far, the SPP were converted into light by surface roughness. Using a so-called Weierstrass prism [39] (made of rutile, with refraction index $n = 2.8$, or ZrO_2, with $n = 2.2$) in an attenuated total reflection (ATR) configuration [34] is an alternative method to convert SPPs into light without the need to roughen the surface of the tunnel junctions. The prism enlarges the momentum $\hbar\omega/c$ of photons within the prism by the factor n. The SPP can be converted into a photon in the prism with an angle of emission α given by $(n\omega/c)\sin\alpha = k_p(\omega)$, where k_p is the two-dimensional wavevector of the SPP parallel to the surface.

We prepared tunnel junctions without CaF_2 undercoating and not intentionally roughened, in the Otto [40] and the Kretschmann [41] configuration. In the Kretschmann configuration the tunnel junction is deposited on the prism. In the Otto configuration the tunnel junction is deposited on a glass substrate and the surface of the silver top electrode is separated by an electrolyte gap of the order of the wavelength of the observed light from the prism surface.

The light emission spectra in the Kretschmann configuration are similar to those in Fig. 7.22. The integrated photoyields, normalised by the tunnelling current, are shown in Fig. 7.24. The electrolyte is ethylene glycol with 0.9 mol NaAc, and U_T is -2.4 V. With more cathodic electrochemical potential the light emission increases. The photon yield per tunnelling electron is 10^{-4} for $E_{SCE} = -1.3$ V, which is twice as much as in the experiments with rough tunnel junctions described in Section 4.2. The emitted light power is 6.5×10^{-8} W cm^{-2}. Activating the silver electrode by oxidation–reduction cycles in 0.01 mol KCl with a charge transfer of 240 μC cm^{-2} shows up by a stronger influence of E_{SCE} on i_T (this will be discussed in Section 5.2), but the normalised integrated photoyield as a function of E_{SCE} is not changed by the activation.

The advantage of the Otto configuration is the ability to optimise the coupling between SPP and photon with distance between the top electrode and the prism; the disadvantage is the relatively slow electrochemical response, due to the narrow gap. The optimal gap width d in our experiment is 200 nm. Figure 7.25 shows the dependence of light emission on this width.

If the gap exceeds 1 μm in thickness, the light emission is only caused by the scattering of the SPPs by the residual surface roughness of the silver surface of the tunnel junction [32]. Under optimum outcoupling by ATR, the integrated normalised photon yield is increased by a factor of 15. Based on the results of this section we come to the following conclusions.

1 The excitation rate of SPPs is only a function of the tunnelling current density i_T and the charge density q on the surface.

2 The modification of the surface by the activation affects only the tunnelling current density i_T, but not the conversion rate of hot electron energy into SPPs.

3 Because of **1** and **2**, SPPs are also excited at ideally smooth surfaces by hot electrons.

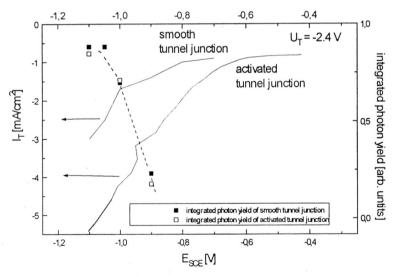

Figure 7.24. Normalised integrated photoyield as a function of E_{SCE} from a 'smooth' and subsequently 'activated' tunnel junction in the Kretschmann configuration, at constant U_T and simultaneously measured tunnelling current densities i_T.

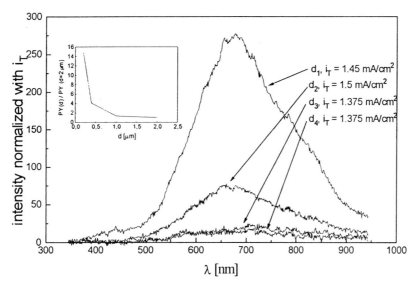

Figure 7.25. Normalised light emission spectra of 'smooth' tunnel junction in Otto configuration at constant tunnel voltage and electrochemical potential. The parameter d is the width of the electrolyte gap, d_1 = 200 nm, d_2 = 400 nm, d_3 = 1 μm, d_4 = 2 μm. Inset: integration normalised emission as a function of d.

Therefore we can apply qualitatively the theory based on smooth jellium surfaces, as we have already done in Section 4.1.

5 Preliminary results and outlook

5.1 *Non-adiabatic oxidation of adsorbed atomic hydrogen, with the MIM contact acting as a retarding field spectrometer*

McIntyre and Sass have injected hot electrons from tetrapropylammonia-tetra-fluoroborate into Au(111) electrodes by periodic modulation of the electrochemical potential. The de-excitation of the injected hot electrons was observed as emitted light [42].

Here we register hot electrons which have tunnelled into the aluminium electrode, after they were released during the oxidation of atomic hydrogen, which is adsorbed at the silver–electrolyte interface (see Section 3.4). Since the electrons are not transferred when the electronic level of adsorbed hydrogen passes the Fermi level of silver, this is a non-adiabatic electronic effect.

First we apply a pulse of the electrochemical potential of 100 ms length from E_{SCE} = − 0.9 V up to E_{SCE} = − 1.5 V in order to evolve hydrogen. We measure simultaneously the densities of the electrolyte current i_{El} (Fig. 7.26(a)) and the tunnelling current i_T (Fig. 7.26(b)) at a tunnel voltage U_T = + 0.3 V. The cathodic current i_{El} does not influence the tunnel current. In other words, holes are injected into the silver, but they do not noticeably tunnel towards the aluminium.

However, the anodic current i_{El} (see Fig. 7.27(a)) — measured as always by our potentiostat as a current at the counter electrode (see Fig. 7.6) — during the oxidation of the layer of hydrogen adsorbed at the silver–electrolyte interface at the electrochemical

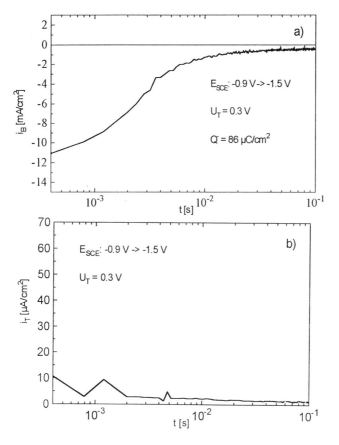

Figure 7.26. Simultaneously recorded current densities i_{El} (a) and i_T (b) during a 100 ms electrochemical pulse from -0.9 V_{SCE} to -1.5 V_{SCE} starting at $t = 0$. The charge Q^- transferred into the electrolyte by the pulse is 86 µC cm^{-2}.

potential $E_{SCE} = -0.9$ V, is split into a current into the silver top electrode and a hot electron current i_T into the aluminium (see Fig. 7.27(b)). Of an anodic charge of 81 µC cm^{-2} which flows from the electrolyte into the silver electrode, 0.56 µC cm^{-2} reaches the aluminium electrode. This corresponds to a transmission rate of $T = 7 \times 10^{-3}$, which is a surprisingly high value. The decay time of i_{El} is 4.5 ms, that of i_T only 2 ms.

In order to demonstrate the hot electron character of the electrons released in the experiment described in Fig. 7.27, the tunnel voltage U_T is varied in steps from 'accelerating' at 0.3 V to 'retarding' at -0.4 V. The constant tunnel currents induced by the application of U_T are seen at about 0.1 s after the end of the anodic transient. At $U_T = -0.2$ V the transient is not yet fully suppressed, demonstrating that we do not observe a thermal effect.

5.2   *Modulation of the tunnel current by the electrochemical potential*

With MIM contacts prepared on rough CaF$_2$ substrates or activated (see Section 3.4), in an electrolyte consisting of 50% water, 50% ethylene-glycol and 0.9 mol NaAc, we observed (see Fig. 7.29(a) that the tunnelling characteristic depends reversibly on the electrochemical potential E_{SCE} in the cathodic range, where the silver–electrolyte interface becomes negatively charged and acetate desorbs (see Section 4.2). Fig. 7.29(b) shows the

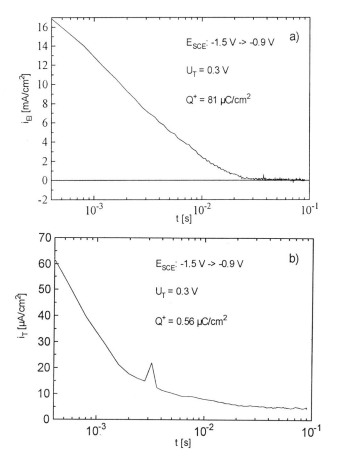

Figure 7.27. Simultaneously recorded current densities i_{EI} (a) and i_T (b) directly after the end ($t = 0$) of the pulse described in the caption to Fig. 7.26. The charge densities of the anodic transients are indicated.

same reversible behaviour at a fixed tunnel voltage and with a cyclic variation of E_{SCE}.

At a constant tunnel voltage $U_T = -2.5$ V the tunnel current rises by about a factor of 5 when the electrochemical potential is varied from $E_{SCE} = -0.2$ V to $E_{SCE} = -1.2$ V. Figure 7.30 shows how the tunnel current follows a change in the electrochemical

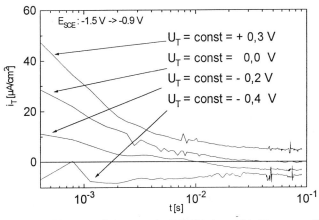

Figure 7.28. Experiments as in 7.27(b), but with U_T set to different values, from 'accelerating' to 'retarding'.

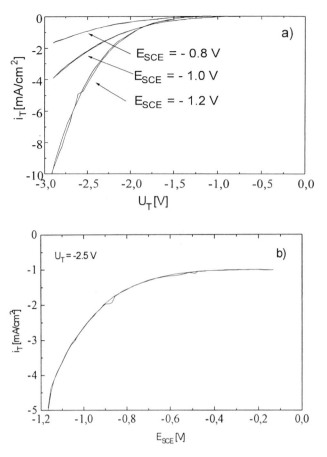

Figure 7.29. (a) Tunnel characteristic $i_T(U_T)$ of an activated MIM contact in a 0.9 mol NaAc buffer with 50% water and 50% ethylene glycol at various indicated constant cathodic potentials E_{SCE}. Scan velocity 50 m V s^{-1}. (b) Tunnelling current i_T at constant $U_T = -2.5$ V during the E_{SCE} cycle. Scan velocity 50 mV s^{-1}.

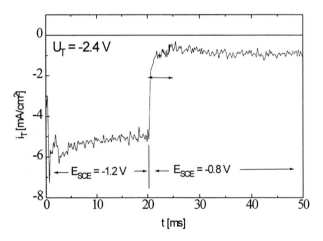

Figure 7.30. Time dependence of the tunnelling current density at constant $U_T = -2.4$ V during a step of E_{SCE} from -1.2 V to -0.8 V. The horizontal arrow following the step characterises the time of reordering the electrode double layer. From [14].

potential from $E_{SCE} = -1.2$ V to $E_{SCE} = -0.8$ V. The tunnel current reaches within 5 ms the new stationary value. This is the time needed for the decharging of the double-layer capacity. Apparently the tunnelling current reacts instantaneously on the electronic structure of the silver–electrolyte interface.

The enhancement of the tunnel current at rough MIM contacts seems to be related to the enhancement of hydrogen evolution (Section 3.4) and the light emission (Section 4.2). On the other hand, the hydrogen evolution is not a precondition of the enhancement of the tunnelling current [43]. A change in the tunnelling current was observed first with MIM contacts on CaF_2 substrates in UHV [26]. A submonolayer coverage of atomic oxygen achieved by an exposure at a sample temperature of 40 K diminished I_T by a factor of 2, a coverage of about 0.6 nm of potassium increased I_T by a factor of 3.

The low temperature in the UHV case and the reversibility and the fast response in the electrolyte case preclude explanations of the effect on the basis of ions migrating into the oxide barrier. It is not possible to understand the strong variation of i_T with the results of the Monte Carlo simulation in Section 2.3. Let us assume a starting rate J_0/e of electrons at the oxide–silver interface, as in the MCS. Let us further assume a ballistic (elastic) current J_{el} at the silver–electrolyte interface at first encounter, as obtained by the MCS, an elastic reflection coefficient p_{el} at the silver–electrolyte interface and a coefficient t_{el} of tunnelling into Al after elastic return to the silver–oxide interface. One easily derives, for the tunnel current J_T,

$$J_T = J_0 - J_{el}p_{el}(J_{el}/J_0)t_{el} \tag{12}$$

If one wishes to explain the observed increase by a factor of 5 in terms of a variation of p_{el} from 1 to 0, even when assuming $t_{el} = 1$ one needs to assume $J_{el}/J_0 > 0.9$ at a silver film thickness of 15 nm. This is equivalent to a mean free path of 142 nm at 2.5 eV above E_F, which is very unlikely, since the mean free path λ_{ee} is about 20 nm according to Quinn and the MCS (see Fig. 7.4). We therefore dispense with the notion of a current J_0 defined above and discuss tentatively a qualitative quantum-mechanical picture, already sketched in Section 2.2.

The transfer of an electron from Al to Ag as a real event with probability 1 is only given after the first scattering event on the silver side, breaking the coherent electronic wave function, expanded both on the Al and the Ag sides. Before this event, the 'primary electron wave' incident from the Al side on to the barrier has amplitude 1 in Al but only a small amplitude t_B (barrier transmission coefficient) in Ag. Only after scattering does it have amplitude 1 in the silver film. Thus, in contrast to the assumption of electron–electron interaction within the random phase model, which leads to Lindhardt's dielectric constant, the amplitudes of the wave functions of the 'tunnelling electrons' above E_F and the electrons within the Fermi sea, with which they interact, are not equal. A very simple approximation is:

$$J_T \propto t_B^2((1 - \exp(-d/(t_B^{-2}\lambda_{ee}))) + \exp(-d/(t_B^{-2}\lambda_{ee})) W_S \tag{13}$$

where d is the thickness of the silver film, λ_{ee} is the elastic mean free path obtained from the Monte Carlo simulation and W_s is the probability of de-excitation or breaking of coherence at the surface. This equation yields the usual expectation for J_T for $d \rightarrow \infty$. For tunnel barriers with small t_B, and relatively small d, as in our case, it follows that

$$J_T \propto t_B^2 W_S \tag{14}$$

This means that the tunnel current is mainly controlled by the conditions at the surface.

We offer some speculations on W_S. Elastic scattering by atomic roughness may break the coherence. That alone cannot explain the dependence of J_T on the electrochemical potential. Inelastic scattering of the electrons out of the coherent state may predominantly happen within the inhomogeneous electron gas at the surface, whose extension is increased by the application of a cathodic potential or by a coverage of one monolayer of an alkali metal. We resort to the jellium model of silver. The primary hot electrons are reflected by the image potential barrier, which is further out than the exchange correlation potential barrier which keeps the 'cold electrons' bound to the positive background. In this way the primary hot electrons reach the surface layer of low electron density. Here electron–electron scattering has a higher probability than in the bulk, because more phase space is available to accommodate the scattered electrons. By modulating the width of the tail of the inhomogeneous electron gas, one controls the inelastic scattering rate and thus the tunnel current J_T.

An alternative idea is the breaking of the orthogonality between surface and bulk electronic states by atomic roughness. In this way the hot electrons may penetrate the surface states, reside there for some time and lose coherence.

5.3 *Optical surface absorption*

MIM junctions have been prepared on an optical grating, which allows for a high optical absorption of p-polarised light under a special angle by the resonant excitation of surface plasmon polaritons (see [44] and references therein). The grating was inscribed into an SiO_2 layer on a Si substrate with a grating parameter of 500 nm and a modulation height of 20 nm. The MIM contact was prepared as described above, the thicknesses being: 40 nm Al; 3 nm AlOx; and 40 nm Ag. A 1 mmol $KClO_4$ aqueous electrolyte was chosen, because the perchlorate ion adsorbs only at potentials more positive than $-0.6\,V_{SCE}$, as demonstrated by the increase of the electrical d.c. resistance of a thin epitaxial Ag(100) film in a 0.01 mol aqueous $KClO_4$ electrolyte (see Fig. 7.31) [45].

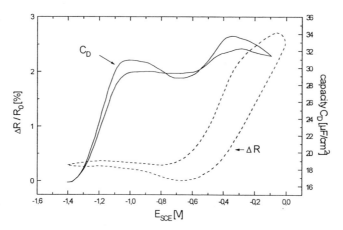

Figure 7.31. Surface resistance in comparision to the interface capacity at weak adsorption.

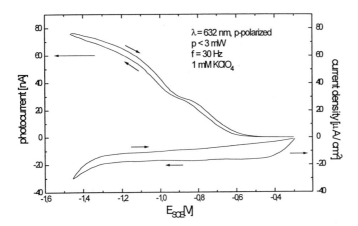

Figure 7.32. Dependence of internal photocurrent on the potential of the illuminated silver electrode.

The light beam of a HeNe laser incident at about 12° was chopped at a frequency of 30 Hz and the internal photocurrent picked up by the aluminium electrode was measured with a lock-in amplifier at $U_T = 0$. Without phase detection Fig. 7.31 compares the cyclic voltammogram with the simultaneously recorded internal photocurrent. This current sets in at cathodic potentials of $-0.6 \, V_{SCE}$ when the surface is not covered by an adsorbate and the silver surface becomes negatively charged. (In this case the double layer capacity consists of the K^+ cations, held at a distance from the negatively charged surface by their solvation shell.) By applying a retarding negative U_T we found, with the sample in air, that the chopped photocurrent persisted up to $U_T = -0.5 \, V$, thus excluding thermally excited electrons tunnelling into the aluminium electrode. The internal photoemission starts at about zero and grows with increasing negative surface charge. At present, it is not yet clear whether this can be explained by the surface photoeffect or an optical effect in the double layer. The latter effect is indicated by phase delays between illumination and internal photocurrent.

Notwithstanding this optical surface absorption, the conspicuous absence of a hot electron signal from the bulk of the silver film, as apparent at a potential of $-0.4 \, V_{SCE}$, may indicate that the bulk absorption of silver in the spectral range far below the frequencies of the bulk d-band excitations is not caused by phonon-assisted intra-sp band-transitions but by an analogue of energy dissipation by a.c. resistivity. A compensation effect of the photocurrents from Al into Ag and vice versa is very unlikely, according to preliminary experiments.

5.4 *Outlook*

The future application of MIMs may include hot electron and hot hole induced processes in adsorbed molecules, supplementing research in surface photochemistry. MIM contacts also appear appropriate for the study of non-adiabatic electronic effects in energy dissipation processes at surfaces during sticking, Langmuir–Hinshelwood surface reactions, crystallisation and tribology. The effects described in Sections 5.2 and 5.3 may have a potential for chemical sensors.

6 Acknowledgements

We thank Dr D. Schumacher for discussions and his help in the first preparation of MIM contacts for use in UHV and for the early designs of the tunnel junction circuitry, and Professor J.W. Schultze for discussions. This research would not have been possible without the support of the Ministerium für Wissenschaft und Forschung des Landes Nordrhein-Westfalen.

7 References

1 Misewich JA, Heinz TF, Weigand P, Kalamarides A. In Dai H-L, Ho W (eds) *Laser Spectroscopy and Photochemistry on Metal Surfaces.* Singapore: World Scientific, 1995; 764.
2 Ho W. *J. Phys. Chem.* 1996; **100**: 13050.
3 Busch DG, Ho W. *Phys. Rev. Lett.* 1996; **77**: 1338.
4 Ho W. *Surf. Sci.* 1994; **299–300**: 996.
5 Otto A, Mrozek I, Grabhorn H, Akemann W. *J. Phys.: Condensed Matter* 1992; **4**: 1143.
6 Liebsch A. *Electronic Excitations at Metal Surfaces: Application of Local Density Theory.* New York: Plenum Press (in preparation).
7 Mönch W (ed.). *Electronic Structure of Metal–Semiconductor Contacts.* Kluwer Academic Publishers, 1990.
8 Mönch W. *Semiconductor Surfaces and Interfaces.* Berlin: Springer-Verlag, 1993.
9 Hansma PK (ed.). *Tunnelling Spectroscopy.* New York: Plenum Press, 1982.
10 Diesing D, Rüsse S, Lohrengel MM. *Proc. Electroceramics IV*, Vol. 2. Aachen: Verlag der Augustinus Buchhandlung Aachen, 1994; 1295.
11 Shepard KW. *J. Appl. Phys.* 1965; **36**: 796.
12 Shu QQ, Ma WG. *Appl. Phys. Lett.* 1992 **61**: 2542.
13 Fowler RH, Nordheim L. *Proc. Roy. Soc. A.* 1928; **119**: 173.
14 Diesing D. Doctoral thesis, Universität Düsseldorf, 1996.
15 American Institute of Physics Handbook, 3 edn, chap. 9, p. 20, McGraw-Hill 1972.
16 Diesing D, Janssen H, Otto A. *Surf. Sci.* 1995; **331**: 289.
17 Duke CB. *Tunnelling in Solids.* Place: Academic Press, 1969.
18 Wolf EL. *Principles of Electron Tunnelling Spectroscopy.* Oxford: Oxford University Press, 1989.
19 Kanter H. *Phys. Rev. B* 1970; **1**: 522.
20 Schaak A. Diploma thesis, Universität Düsseldorf, 1995.
21 Rösler M, Brauer W. Springer Tracts in Modern Physics 122 (1991) 1.
22 Lindhard J. Dan K. *Vidensk. Selsk. Mat.-Fys.Medd.* **18**, 1.
23 Reklaitis A. *Phys. Lett.* 1982; **7**: 7367.
24 Jacoboni C, Reggiani L. *Rev. Mod. Phys.* 1983; **55**: 647.
25 Quinn JJ. *Phys. Rev.* 1962; **126**: 1453.
26 Hänisch M, Otto A. *J. Phys.: Condensed Matter* 1994; **6**: 9659.
27 Diesing D, Winkes H, Otto A. *Phys. Stat. Sol. B* 1997; **159**: 243.
28 Körwer D. Doctoral thesis, Universität Düsseldorf, 1993.
29 Brodsky AM, Gurevich YY, Levich VG. *Electrokhimiya* 1967; **3**: 1302.
30 Furtak TE, Kliewer KL. *Comments Solid State Phys.* 1982; **4**: 103.
31 Gerischer H, Meyer E, Sass JK. *Ber. Bunsenges. Phys. Chem.* 6 (1972) 1191.
32 Kritzler G. Diploma thesis, Universität Düsseldorf 1996.
33 Lambe J, McCarthy SL. *Phys. Rev. Lett.* 1976; **37**: 923.
34 Otto A. *Adv. Solid State Phys.* 1974; **15**: 1.
35 Arnold M, Otto A. *Optics Commun.* 1996; **125**: 122.
36 Liebsch A, Hinzelin G, Lopez-Rios T. *Phys. Rev. B* 1990; **41**: 10463.

37 Liebsch A, Benemanskaya GV, Lapuschkin MN. *Surf. Sci.* 1994; **302**: 303.
38 Janssen H, Diesing D, Otto A. *Surf. Sci.* 1995; **331**: 1267.
39 Wittke W, Hatta A, Otto A. *Appl. Phys. A* 1989; **48**: 289.
40 Otto A. *Z. Phys.* 1968; **216**: 398.
41 Kretschmann E, Raether H. *Z. Naturforsch.* 1968; **23a**: 2135.
42 McIntyre R, Sass JK. *Phys. Rev. Lett.* 1986; **56**: 651
43 Janssen H. Doctoral thesis, Universität Düsseldorf, 1997.
44 Arnold M, Bussemer P, Hehl K, Grabhorn H, Otto A. *J. Modern Optics* 1992; **39**: 2329.
45 Winkes H. Doctoral thesis, Universität Düsseldorf, 1996.

8   Computer Modelling of Surfaces and Interfaces

C.R.A. CATLOW, J.D. GALE*, D.H. GAY, M.A. NYGREN and D.C. SAYLE

The Royal Institution of Great Britain, 21 Albemarle Street, London W1X 4BS, UK

*Department of Chemistry, Imperial College of Science, Technology and Medicine, Exhibition Road, London SW7 2AZ, UK

1   Introduction

Atomistic computer modelling techniques have, since the 1980s, become increasingly powerful and general tools in both the physical and biological sciences. Using these methods we are able, with growing accuracy and reliability, to model the structures, properties and reactivities of complex molecules — e.g. proteins [1] and polymers [2] — and condensed phases, whether liquid [3], crystalline [4] and amorphous [5]. The incentive to develop and apply the techniques to surface and interface chemistry is especially strong in view of the well-known difficulties in obtaining experimentally unambiguous models for surface structures and for the structures, dynamics and reactivities of sorbed molecules.

In this paper we therefore review the current status of computational methods in constructing models of the surface and interfacial chemistry of complex inorganic materials. We show how with these techniques we can develop detailed models for surface and interface structures, calculate surface and attachment energies which we may then relate to the macroscopic crystal morphologies, model surface and interfacial impurity and defect states, and simulate the structures and reactivities of sorbed molecules.

Surface modelling makes use of the increasingly powerful range of methods available to the computational chemist and physicist. As in other fields of atomistic computer modelling, these can be usefully divided into two broad categories: (i) *electronic structure* methods, which attempt to solve the Schrödinger equation for the system modelled, at some level of approximation; and (ii) *interatomic potential* (or forcefield) based methods in which all knowledge of electronic structure is subsumed into an interatomic potential which writes the energy of the system simulated as a function of its nuclear coordinates. Explicit electronic structure methods based on both Hartree–Fock and local density functional (LDF) methods are being applied increasingly to surface problems; nevertheless, despite the continuing growth in computer power, their scope remains limited and there is a continuing role for the interatomic potential based techniques — including minimisation, Monte Carlo and molecular dynamics methods — for modelling structures and dynamics of the surfaces of complex materials.

In this paper we describe how this range of methods has been implemented and adapted to the modelling of surfaces. We also review recent applications, especially to the surface chemistry of oxide systems.

2   Modelling surfaces using interatomic potentials

Modelling of the bulk properties of crystals is now well developed [4]. Lattice energy minimisation techniques may be used to model the structures and energies of crystals, using standard summation procedures (including the well-known Ewald method for the electrostatic term) to calculate lattice energies, coupled with minimisation methods to calculate the lowest-energy structures of the unit cell. Such calculations employ three-dimensional periodic boundary conditions using a specific unit cell.

The methods used for modelling surfaces are quite similar to those used for modelling bulk systems. The main difference is how the simulation cell is chosen and what it represents. For bulk systems the asymmetric unit or the P1 lattice cell is generally used. There is usually no information gained by including more cells than the fundamental unit. Likewise, for surfaces and interfaces, all the information is contained in a cell that is repeated in two dimensions, instead of three. The question of cell termination enters into the bulk system only as an adjustment to the Ewald sum, and the lattice is treated as infinite. For the two-dimensional (2D) system, the lattice is infinite in two directions, but finite in the third. The question of how to treat the third direction makes surface simulations more demanding than those for bulk systems. The description of surface modelling that follows describes the techniques used by the recently developed program MARVIN [6], but could be used for any surface simulation technique.

The fundamental unit of a surface is a 2D cell, which has planar periodic boundary conditions parallel to the surface, as shown in Fig. 8.1. The energy is computed with a 2D lattice summation using only the atoms contained within the cell. The cell consists of one or more blocks divided into two regions. The first region, which is near the surface or interface, is designated as region 1. Atoms in region 1 are allowed to relax during energy minimisation. The other region is designated region 2, and these atoms are held in fixed locations. The purpose of region 2 is to reproduce the effect of the bulk crystal on the region 1 atoms.

The main types of calculation than can be handled in this manner are surfaces, adsorbed molecules and interfaces. For modelling a free surface, one block is used. This block consists of a region 1 and a region 2, the sizes of which are determined by the energetics. Adsorption (or docking) of molecules in this formalism can be described with two blocks, one containing the surface and the other the adsorbate, while an interface is two blocks with the region 1s adjacent. The use of blocks and fixed regions is important in reducing the problem to a size that can be computed in a reasonable time.

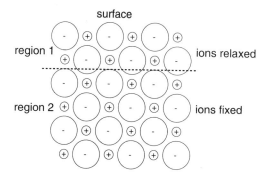

Figure 8.1. Schematic diagram of the MARVIN simulation model. The simulation cell is repeated infinitely in two dimensions. The region 1 atoms are allowed to relax, while those in region 2 are kept fixed.

Like the bulk solid, it is important to treat all the unique atoms individually and usually counterproductive to include symmetry-replicated atoms. For surfaces, it is crucial that sufficient layers of atoms be included in region 1, so that the lowest layer of atoms does not move during minimisation. The depth of region 2 is important as well, since the lowest layer of atoms should not 'see' the vacuum beyond the bottom of region 2. In practice this can be done by increasing the size of region 2 until the total energy does not change. Then the depth of region 1 can be increased until the surface energy does not change with increasing region 1 depth. The relaxed surface energy cannot be calculated a priori, hence further calculations are necessary to determine if the depth is sufficient.

The region sizes depend upon the convergence properties of the lattice summations. Of the long-range interactions, it is the electrostatic type that is the hardest to deal with. In 3D systems, as noted, the Ewald method is used to get a converged value for the electrostatic energy [7]. Parry produced a derivation appropriate for 2D slabs [8,9]. A more complete derivation is given by Heyes $et\ al.$ [10]. For systems with dipoles normal to the surface, the total energy is a function of the system size and requires either defects or an adlayer to compensate the net surface dipole.

Our discussion now continues with an account of the forcefields (or potentials) employed in describing the interaction between atoms in these calculations. As noted in Section 1, interatomic potentials express the total potential energy of the system simulated as a numerical or more commonly analytical function of the nuclear coordinates. This many-body function is broken down into two-, three-body and higher-order terms. In practice many potential models for non-metallic systems are restricted to two-body terms, although the inclusion of three-body terms is commonly necessary; while the only higher-order terms that are routinely included in non-metallics are intramolecular torsional terms.

There are two general categories of interatomic potential for insulating condensed phases. The first, the Born model, rests on the ionic description, that is, it includes Coulomb and short-range interactions between ions, writing the total potential, V, as:

$$V = \frac{1}{2} \sum_{ij}' \left[\frac{q_i q_j}{r_{ij}} + V_2(r_{ij}) \right] + \Phi_{MB}, \tag{1}$$

where the prime on the summation indicates that we exclude terms in which $i = j$; where q_i and q_j are the charges on the ions i and j, r_{ij} is the distance between them; $V_2(r_{ij})$ is the short-range two-body interatomic potential which depends solely on the r_{ij}; Φ_{MB} are the many-body terms. For ionic systems the most commonly used function is the Buckingham or 6-exp potential:

$$V_2(r_{ij}) = A_{ij} \exp(-r_{ij}/\rho_{ij}) - C_{ij} r_{ij}^{-6}, \tag{2}$$

where A, ρ and C are variable parameters, but other functional forms, for example the Lennard–Jones and Morse function, may be employed. The commonest form of many-body term is the 'three-body', harmonic bond-bending function:

$$V_{TB} = \frac{1}{2} k_{ijk} (\theta_{ijk} - \theta_0)^2, \tag{3}$$

where θ_{ijk} is the angle subtended at atom i by atoms j and k, θ_0 is the equilibrium value of this bond angle and k_{ijk} is the force constant. Additionally, Born model potentials generally include a representation of ionic polarisation, most commonly by the shell model [11] which is based on a simple mechanical model of a polarisable atom in which a massless shell is coupled to the core by a harmonic spring, the development of a dipole moment being described in terms of the displacement of the shell relative to the core. This simple model has proved to be most effective in modelling lattice dynamical and dielectric properties of solids.

The second general class of potential models is based on the 'molecular mechanics' concept which views the solid as a covalently bonded network whose energy depends on the deviation of bond lengths and angles from equilibrium values. The former terms are commonly described using simple bond-harmonic functions:

$$V_B(r_{ij}) = \frac{1}{2} k_{ij}(r_{ij} - r_0)^2, \tag{4}$$

where k_{ij} is the force constant and r_0 is the equilibrium bond length. Bond angle terms of the type described in equation (3) are normally used. Torsional terms depending on the position of four atoms may be added, as may cross-terms between bond-stretch and bond-bending interactions. Non-bonded interactions including Coulomb and short-range terms (the latter being most commonly described by Lennard–Jones functions) must also be included.

Both Born model and molecular mechanics potentials include variable parameters for which values are derived either by fitting to empirical data, i.e. known structures and properties of crystals and molecules, or, increasingly, by the use of electronic structure techniques which may calculate the energy of a cluster or periodic array of atoms for a range of geometries; the resulting variation in energy is then fitted to an interatomic potential function. A good example of this lattter approach is given by the work of Gale et al. [12] on α-Al_2O_3.

In the systems discussed later in this paper, the solid and its surface have generally been described using a Born model potential, and sorbed molecules by the molecular mechanics approach. Good parameter sets are now available for a wide range of molecules and solids. The greatest uncertainty arises in describing the molecule surface interactions. Standard non-bonded potentials are normally employed, but the derivation of suitable parameters is difficult as relevant empirical data are very limited. Parameterisations via direct quantum-mechanical calculations are probably the most promising way forward, although such calculations raise a number of difficulties. The development of improved parameter sets for molecule–surface interactions remains a high priority for the field.

3   Applications of interatomic potential based methods

3.1   Morphology

Morphological prediction, based on the internal structure of a compound, has enjoyed many decades of active study. Initial approaches were based solely on the interplanar spacings, d_{hkl}, of the different crystal faces, with the lowest growth rates occurring at the faces with the greatest d_{hkl} [13,14]. Donnay and Harker improved this simple idea when they took account of the symmetry of a crystal by reducing the d_{hkl} growth slices at the faces containing a screw axis or glide plane, since the slice at these faces can be reduced into identical subslices [15]. These arguments took no account of the detailed structure or energetics of the material.

Hartman and Perdok [16] revolutionised this field by considering the role of intermolecular forces in crystallisation. They classified crystal faces into three groups depending on the number of periodic bond chains (PBCs) in the growth slice, where a PBC is an uninterrupted chain of nearest-neighbour particles. A slice with two or more PBCs is an F face, those with one PBC are S faces and the rest are K faces. They proposed that only F faces can grow according to a layer mechanism, with the other two having a continuous mode of growth. Thus the growth of F faces will be slow and therefore they will be the most important morphologically. They proposed that the attachment energy (the energy per molecule released when a new slice of depth d_{hkl} is attached to the crystal face) be proportional to the growth rate and hence inversely proportional to its morphological importance [17]. Again the depth of the slice is reduced at the faces containing a screw axis or glide plane. A large slice energy in absolute value implies large interactions between ions in the slice, which in turn means that there are at least two PBCs in the slice. This quantity has been used to predict the morphologies of a wide variety of materials [18–21]. A similar quantity has been defined by Dowty [22]. The morphology derived by this methodology is known as the growth morphology, since it is based on the idea of layers attaching themselves to a growing crystal.

Another approach to crystal morphology uses the surface energy, i.e. the difference in energy between the surface ions and those in the bulk normalised to unit surface area. The surface energy must be a minimum for a given volume [23]. Hence the morphological importance of a face is inversely proportional to the surface energy. The morphology derived in this way is termed the *equilibrium morphology*. There has been some debate over which morphology is more likely to be observed [22,24]. The equilibrium morphology has been found to agree well with small experimentally grown crystals less than a micrometre in diameter [25,26]. For larger crystals, and especially geological specimens, the growth morphology has been found to reproduce best the experimental results. Results obtained using these techniques are illustrated in Fig. 8.2 for the case of zircon ($ZrSiO_4$), a widely studied mineral. The results show the sensitivity of the predicted morphology to the details of the methodology used in its calculation. Further details are given by Gay and Rohl [6].

In concluding this subsection, we note that both methods of calculating the crystal morphology have traditionally ignored the effects of surface relaxation, i.e. the tendency of surface ions to move away from their ideal lattice sites. Although this seems a reasonable approximation for non-polar organic molecules, for inorganic systems it is known to be unrealistic.

3.2 *Docking*

Molecular recognition at the interface between inorganic materials and organic templates is known to be a powerful force in the control of crystallisation and mineralisation in both synthetic and biological systems. Perhaps the simplest and best-known example of this recognition process occurs during crystal growth in the presence of morphology-modifying additives. In the past, morphological and structural data have

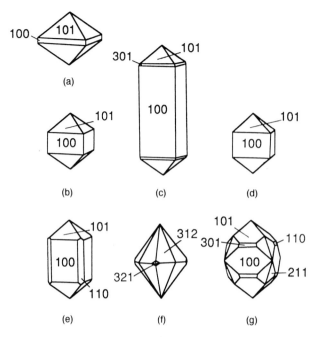

Figure 8.2. Calculated zircon morphologies: (a) Donnay–Harker; (b) unrelaxed growth morphology; (c) unrelaxed growth morphology assuming (202) growth slices; (d) relaxed growth morphology; (e) unrelaxed equilibrium morphology; (f) relaxed equilibrium morphology; (g) relaxed equilibrium morphology with faces with invalid attachment energies removed.

been used to infer the essential elements of the interactions involved [27,28]. The control and inhibition of crystal growth is a major problem in several contemporary technologies. In many inorganic systems, the most efficacious growth modifiers are found to be polyanions, in which several charged groups are joined by a backbone. A particularly important case is provided by $BaSO_4$ [29], whose precipitation in oil pipes is a significant problem in oil extraction and transport [30]. The most commonly used inhibitors for the growth of crystals of this compound are polyphosphonates, of which the most widely studied are diphosphonates [31,32]. Growth inhibition of baryte by these additives is thought to involve the incorporation of the inhibitor into a growing terrace, where it occupies the sites of two sulfate groups. Experimental evidence shows that diphosphonates are actually incorporated into the surface rather than just adsorbed on to the surface. In particular, it is found that similar dicarboxylates and disulfonates have no effect on crystallisation [32]. The most obvious explanation of the latter is the mismatch between the single negative charge of the terminal groups and the double negative charge on the sulfate group. Moreover, scanning force microscope experiments [33] have shown that if commercially available polyphosphonate growth inhibitors are added to growing calcite crystals, they appear on the images as step edges. Once the diphosphonate is in place, it blocks the crystal growth when a step edge reaches it. Indeed, Monte Carlo simulations [34] have shown that the presence of a single ion impurity is enough to block step growth. However, a large impurity that binds simultaneously at two surface sites will be much more efficient at blocking step growth than a single ion impurity.

The docking of the diphosphonate involves the replacement of surface sulfates. Thus to determine which crystal faces will be affected by the additive, we must calculate the energetics of the process in which two surface sulfate ions are replaced by the additive. First, however, the surface structure of each growing face must be investigated. In general, faces corresponding to a given Miller index may have several different surface structures in which the crystal is cleaved in different places [35]. In baryte, only the (010) face has a unique surface structure. For all the other faces considered, we calculated for each possible structure both the attachment energy and the surface energy. If the crystal growth is controlled by thermodynamic factors, then the cut with the lowest surface energy will be dominant in the growing surface. If, on the other hand, the growth is kinetically controlled, then the cut with the lowest absolute attachment energy will be most favoured.

For each surface, there are many combinations of pairs of sulfates which can be replaced by a diphosphonate. Since the P–P distance in the solution conformation of diphosphonate as determined by MOPAC calculations is $6.17 \, \text{Å}$, it seems reasonable to exclude those pairs of sulfates with S–S distances greater than $7.5 \, \text{Å}$, although the list becomes more extensive after relaxation. For each pair of sulfates removed, a diphosphonate was inserted into the vacated area and both the surface ions and diphosphonate were allowed to relax to minimise the energy.

The most physically significant term is the replacement energy. This involves computing the total energy of the system with and without the inhibitor docked on the surface, and the solvation energies of sulfate ions and the inhibitor. Despite the simplicity of this approach, it may be expected to yield a reasonable estimate of the replacement energy.

The two most stable configurations calculated in this way for the incorporation of an alkyldiphosphonate are shown in Fig. 8.3. These demonstrate clearly how the alkyldiphosphonate is included in the surface structure of the baryte crystal. The value calculated for the energy to replace two surface sulfates by a docked diphosphonate is $-303 \, \text{kJ mol}^{-1}$ for the (100) surface and $-298 \, \text{kJ mol}^{-1}$ for the (011). The inclusion of the alkyldiphosphonate at these sites is not entirely expected, based upon simply comparing S–S ($5.47 \, \text{Å}$ and $5.66 \, \text{Å}$) and P–P ($6.17 \, \text{Å}$) distances. There are sites that should provide a better fit based simply on distance measurements; the value of the detailed calculation of energetics is therefore clearly shown. Moreover, the results show how the alkyldiphosphonate docking controls the growth of the baryte crystal.

3.3 *Hydroxylated surfaces*

In many cases the ions and molecules in the surface undergo chemical reactions. Not only are they exposed to the environment, but a clean surface will expose either dangling bonds or ions with incomplete coordination that may be stabilised chemically. It might also be that the chemically most stable cut of the crystal is for a surface which would otherwise be unstable due to a permanent dipole moment, but for which the chemical reaction cancels the dipole.

In this subsection we will discuss two cases where reaction with water stabilises the surface. The first is α-quartz in which protonation of the oxygens saturates the dangling bonds and adsorption of hydroxide ions fills the coordination shell for silica ions. The second is α-Al$_2$O$_3$, where the oxygen-terminated basal-plane surface is a polar cut and halving the surface layer charge by hydroxylation cancels the dipole.

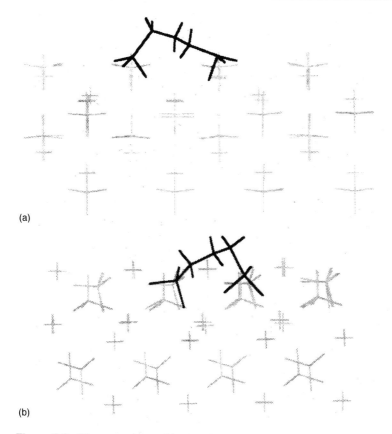

(a)

(b)

Figure 8.3. The two most stable docking sites for an alkyldiphosphonate on the surfaces of baryte: (a) on the 5.47 Å site on the (100) surface; (b) on the 5.66 Å site on the (011) surface. The diphosphonate is coloured black and the surface baryte ions grey.

When cleaving quartz, the silicon exposed will always lack one of its four oxygens in the coordination sphere. For each silicon ion exposed there is one exposed oxygen. This means that by adsorbing protons and hydroxides in pairs, with the hydroxide terminating the silicon and protonating the oxygen, the surface is saturated.

The basal plane surface of α-Al_2O_3 possesses one aluminium-terminated electrostatically stable cut. For this cut, the surface aluminium ions lack three out of the six oxygens in its coordination sphere; it is thus highly coordinatively unsaturated. When relaxing the clean aluminium-terminated surface, the outermost planes move inwards in a dramatic fashion, in calculations using both atomistic potentials [6,36,37] and periodic Hartree–Fock methods [38]. Large relaxations and high coordinative unsaturation are usually signs of instability, and it is likely that a more stable situation may be found.

The only possible oxygen-terminated cut results in a net dipole across the crystal. Moving half of the oxygens from the top to the bottom cancels the dipole, giving an electrostatically stable situation. Reaction of the resulting surface with water results in a fully hydroxylated surface without any dangling bonds and coordinative unsaturation. This type of water surface reaction differs from the type occurring on quartz in that the reaction not only saturates the surface but also cancels the dipole, removing the electrostatic instability.

The surface energy is defined from the energy difference between the surface simulation cell and a part of the bulk of corresponding size. This is the work needed to cleave the crystal; however, when the surface undergoes a chemical reaction the reaction enthalpy must be included in the surface energy. The reaction giving rise to the resulting surface is for both cases:

$$E_{surf} = \frac{E_{region1} - N.E_{bulk}}{A_{unit}} \qquad (5)$$

To calculate reaction enthalpies involved in the formation of defects and surfaces, one usually introduces a hypothetical gas phase reaction, via a Born–Haber cycle, for which the reaction enthalpy may be estimated.

$$H_2O_g \rightarrow OH_g^- + H_g^+ \qquad (6)$$

$$O_{surf}^{2-} \rightarrow O_g^{2-} \qquad (7)$$

$$O_g^{2-} + H_g^+ \rightarrow OH_g^- \qquad (8)$$

$$2OH_g^- \rightarrow 2OH_{surf}^- \qquad (9)$$

It should be noted that the Born–Haber cycle includes O^{2-} gas and therefore the second electron affinity of oxygen enters into the calculated energy. Since O^{2-} gas is an unstable species the second electron affinity of oxygen cannot be defined in a general manner. Indeed, this quantity is only meaningful when defined in connection with the formation of the bulk material, and is thus structure-dependent. Nevertheless, using this procedure, it is possible to calculate the reactive surface energy of α-Al_2O_3.

For a clean surface the surface energy must be a positive quantity since it would otherwise be favourable for the crystal to maximise its surface area, i.e. disintegrate. When the surface undergoes a chemical reaction this is not necessarily true. A negative surface energy is possible if further reaction is either unfeasible on kinetic grounds or sterically inhibited. The best-known example of such a situation is the oxide coating covering aluminium metal. If we work out the corrections to the surface energy, using standard thermochemical data to estimate the reaction enthalpies [39], the surface energies stay positive for α-quartz but the α-Al_2O_3 basal plane surface energy is negative. It must be remembered that our estimate of the surface energy may show considerable uncertainty owing to the uncertainty in the definition of the oxygen second electron affinity. All aspects of the results clearly point to the stability of the basal plane of α-Al_2O_3.

3.4   *Surfaces of zeolites*

The structure of zeolitic surfaces will clearly control the access of the molecules to the interior of the crystal. Moreover, surface energies are known to influence strongly the resulting crystal morphology. Very little is, however, known experimentally concerning the surface chemistry of zeolites. Computer modelling can therefore have an important predictive role.

The techniques for modelling surfaces can be applied to zeolitic systems as easily as to any other inorganic surface. Recently, simulations have been carried out on sodalite [40] and faujasite [41]. The surfaces of these zeolites were terminated with hydroxides to saturate the broken bonds and create a more natural surface, as shown in Fig. 8.4. During energy minimisation, the surface structure changes little; for example, the pore diameter for siliceous faujasite changes from 10.97 Å to 10.89 Å during minimisation. Further work needs to be done with different compositions of the framework to see if this has an additional effect. The relaxed equilibrium morphology for sodalite agrees well with experiment.

The calculations discussed above show the value of interatomic potential based studies. Detailed modelling of both adsorption and reactivity require the use of quantum-mechanical techniques, which are discussed in the next section.

4   *Ab initio* calculations using cluster models of surfaces

Adsorption on surfaces and chemical reactions on surfaces are localised processes. It is therefore reasonable to describe them in terms of a small cluster, surrounded by the surface, upon which the adsorption or reaction takes place. The cluster and the adsorbents may then be treated by the same quantum-chemical methods as used in ordinary molecular calculations. In this section we will discuss how cluster models may be used to study reactions and adsorption on the surfaces of ionic materials. We will pay special attention to the need to embed the clusters properly, with some approximate representation of the surrounding lattice.

The evaluation of the accuracy of cluster modelling is to a large extent based on comparisons with experiments on well-defined surfaces under ultrahigh vacuum (UHV) conditions. Molecular adsorption on oxide films has been studied extensively during the last decade, notably by Freund *et al.* [42] and Goodman [43]. We now illustrate the current state of the art by two case studies relating to adsorption and reaction.

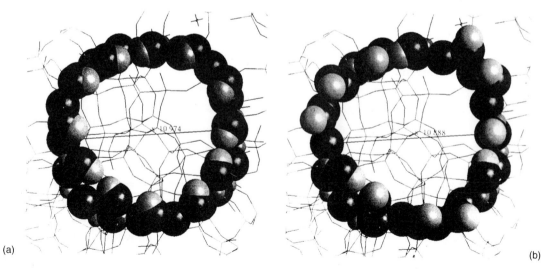

(a)   (b)

Figure 8.4. Calculated surface structure for a siliceous faujasite pore (a) before and (b) after relaxation.

4.1 *Molecular adsorption*

The system that has been most widely studied is CO adsorbed on MgO(001), and we will use this example to illustrate especially the need for proper embedding to describe the surface–adsorbate interaction. The latest experimental value for the binding energy is 0.43 eV [44], whereas the most accurate calculations performed give a binding energy of 0.08 eV [45]. This large discrepancy between the calculated and the experimental binding energy shows that there is still much to be learned from this simple system about the interactions between surfaces and molecules.

Calculations using embedding methods show that the most stable adsorption configuration is with the carbon end downwards on top of a magnesium cation. The bonding is of an electrostatic nature, without significant covalent contributions. The surface–adsorbate distance is determined from the balance between the attractive electrostatic forces and the Pauli repulsion. How these different effects contribute to the net binding energy may be analysed by relaxing the cluster adsorbate wave function in several steps. Starting from the wave function of the non-interacting system, each step allows the system to relax with respect to one interaction. Such an analysis has been performed by Pacchioni *et al.* [46] showing that the Pauli repulsion cancels the first-order electrostatic interaction and that the binding energy is built up largely from the polarisation terms.

Both the computed electrostatic potential and the strength of the Pauli repulsion depend critically on the embedding used. The long-range electrostatic interaction is normally included via a point charge array. If only a point charge array is used, the Pauli repulsion between the ions in the quantum-mechanically treated cluster and the surrounding ions is completely neglected. This results in a spurious polarisation of the cluster ions towards these naked point charges. For an anion-terminated cluster, such unrealistic polarisations have a large effect on the adsorbate–surface interaction. Such polarisations may be avoided by surrounding the cluster by model potentials. The most sophisticated choice available is the so-called *ab initio* model potential (AIMP) which, for a saturated cluster basis set, corresponds to freezing the orbitals of the entire ion. For technical details of the AIMP we refer to work by Barandiaran and co-workers [47–49], but note that an ordinary effective core potential (ECP) [50] or even just a simple projection operator description of the surrounding ions [51] is much better than using only point charges.

As noted above, the two most important [46] contributions to the interaction between CO and MgO(001) are the electrostatic attraction and the Pauli repulsion. Both of these forces are affected if the terminal cluster ions are allowed to polarise towards the surrounding point charges. Using an MgO_5^{8-} cluster embedded in point charges to describe the surface, the oxygen ions will polarise away from the central magnesium ion, and with the consequent reduced Pauli repulsion the CO molecule may approach closer to the surface. This artificial lowering of the Pauli repulsion is removed if the cluster is extended with the next layer of magnesium ions; however, the computationally cheap AIMP description gives the same result. With the CO closer to the surface, the electrostatic field strengths at the molecule are increased and the binding energy increases from 0.02 eV to 0.17 eV [52] at the restricted Hartree–Fock

(RHF) level of theory employing the Boys–Bernardi [53] correction to estimate the basis set superposition error (BSSE).

Extending the cluster to $Mg_{14}O_5^{18+}$, the Pauli repulsion will be correctly described [52] but there will still be problems with the electrostatic potential. Ferrari and Pacchioni [89] have investigated the convergence of the electrostatic potential above the surface with cluster size. The electrostatic potential shows oscillations with a size similar to those for the CO binding energy. The oscillations of the electrostatic potential depend to a large extent on the polarisation of the cluster border ions. Removing these spurious polarisations by embedding the cluster in AIMPs results in the oscillations in the electrostatic potential disappearing, as demonstrated by Pascual and Pettersson [54].

We can thus see that the usage of AIMPs to embed the cluster allows one to use much smaller cluster sizes. As noted above, a simple projection operator description may be sufficient, but if only a point charge expansion is used, care must be taken to have a cation-terminated cluster.

The discrepancy between experiment and theory regarding the CO binding energy on MgO(001) remains problematic. Interestingly enough, it increases with time as the calculations become more accurate and experiments more elaborate [45]. The experiments have been carried out at liquid nitrogen temperatures, and since calculations predict that CO bound to regular five-coordinated sites desorbs at lower temperatures, it is tempting to speculate that the molecules observed on the surface at these temperatures are bound to defect sites, whereas the cluster calculations model perfect lattice sites. Further work is clearly needed to solve this tantalising problem.

4.2   *Modelling reactivity*

When atoms are adsorbed, covalent bonds between the surface ions and the adsorbed atom may be formed and, as is usual when forming covalent bonds, atomic excitations may occur before the bound state can be formed. If it is necessary to excite the surface ion before the bond is formed, the surface influence on the excitation energy will enter directly into the reaction enthalpy. The strength of the bond will also be affected by the flexibility of the surface charge density.

A good example is provided by our recent study of the adsorption of hydrogen atoms on a copper-terminated $Cu_2O(001)$ surface where the hydrogen atom binds to a copper ion. The bound state is a doublet with the unpaired electron in the copper d-shell. The ground state for the copper ion in the surface is a filled d-shell, so one electron has to be excited from the d-shell to the 4s-shell. This excitation energy is, of course, highly dependent on the surface structure, and on relaxing the surface the binding energy decreases from 2.27 eV to 2.06 eV. This decrease in binding energy corresponds to an increase in excitation energy from 1.89 eV to 2.06 eV using a CuO cluster embedded in AIMPs [55].

Oxygen adsorption is often an important step in redox processes. Adsorbing oxygen atoms on to CaO(001) and MgO(001), the incoming oxygen binds to a surface oxygen ion, forming a peroxy bond. The bond is formed by the p_Z orbitals and is of a σ type. The p_Z orbital of the surface oxygen ion is doubly occupied, so the p_Z orbital of the incoming atom must be emptied before the bond can be formed. This involves an excitation from the ^3P ground state to the lowest lying singlet, ^1D. In this case the excitation takes place on the incoming atom so there is no dependence of the excitation energy on the surface structure. Nevertheless, the binding energy shows a large dependence on the lattice parameter. The lattice larger parameter of CaO (9.09 Å as compared to 7.96 Å for MgO) gives a lower Madelung potential, which reduces the ionisation energy of the oxygen anion. The difference in cation size also reduces the ionisation energy [56]. These effects allow for a larger charge transfer to the incoming oxygen, and the bond is strengthened from 0.45 eV for MgO to 1.68 eV for CaO, the calculations being at the modified coupled pair functional (MCPF) [57] level of theory using a single oxygen ion surrounded by projection operators as a cluster model [56].

Using solid catalysts in redox processes such as reduction of NO_x and oxidation of CO, surface oxidation and reduction are key reaction steps. A good example of how cluster calculations can be used to study these kinds of reaction is provided by a series of papers by Snis *et al.* concerning CaO(001). The molecular adsorption of N_2O is found to give only weakly bound states (less than 0.1 eV) for the regular surface [58]. The decomposition of N_2O into N_2 and an oxygen adsorbed on the surface is predicted to be energetically favourable for CaO(001) but not for MgO(001) [56]. The transition state for the N_2O decomposition is a linear configuration on top of the oxygen anion where the peroxy species is formed [58] and the barrier is computed to be 1.1 eV [58]. This value is lower than the experimentally found activation energy of 1.5 eV [59]. For the reduction of the oxidised surface, two reactions have been studied: oxidation of CO to CO_2 and desorption of molecular oxygen. The transition state in the CO oxidation reaction is also a linear geometry with a computed barrier of 1.4 eV [60]. The lowest computed barrier for forming molecular oxygen from two neighbouring oxygen species was computed to be 1.6 eV [61], but here it should be noted that the surface ions were kept fixed. The surface ions may move appreciably in the temperature regime (600–900 K) where the oxygen recombination occurs. From temperature-programmed desorption (TPD) experiments, the activation energy for desorption of molecular oxygen has been estimated to be 1.3 eV [59].

Having explored the applications of cluster models in electronic structure calculations, we now consider a different approach that is increasingly being applied.

5   Modelling surface adsorption using 2D periodic *ab initio* techniques

Cluster models, as discussed above, are by far the most widely used approach to modelling the adsorption and reactivity of molecules at surfaces, at least for insulators. This is largely due to the greater availability of well-developed packages for the study of molecular quantum-mechanical problems and the lower computational cost. However, as discussed in the previous section, great care must be taken when generating and embedding the finite cluster model so as not to introduce spurious interactions between the cluster and its embedding matrix. Furthermore, for partially covalent materials there is the problem of saturating dangling bonds without introducing artefacts.

Given these difficulties, it is not surprising that there has been increasing interest in the use of periodic boundary condition methods to study surface chemistry, as generating a 2D slab model for a surface is more natural than a cluster and eliminates many boundary effects. The principal remaining difficulty is to ensure that images of defects and sorbed molecules are sufficiently far apart that their interaction has minimal effect, which can be readily checked in principle by the use of increasingly large supercells.

The majority of supercell calculations of surface adsorption performed to date, fall into one of two categories. Within the context of localised basis functions, the program CRYSTAL [62] has made it possible to perform, originally, Hartree–Fock and, more recently, density functional calculations on genuinely two-dimensionally periodic slabs. Other density functional codes with a true surface capability are also now available, for example the program BAND [63] which offers a variety of basis sets. There have also been a large number of plane-wave pseudopotential density functional calculations for surfaces. By necessity these are 3D periodic, but by introducing a region of vacuum between the slabs in one direction it is possible to approximate a surface in the absence of a dipole across the slab. The benefits of the second approach are the computational simplicity of plane waves, making forces readily calculable, and the systematic convergence of the basis set [64]. A major disadvantage is the computational cost in treating the vacuum region, making it harder to work with truly large gaps between the slabs.

Periodic studies of surfaces have been applied to a wide range of systems, including metals (where such an approach is essential), semiconductors [65] and insulators [66]. As this review cannot cover all aspects, we shall concentrate on oxide surfaces for illustrative purposes. Again a range of oxides have been considered, including the adsorption of hydrogen fluoride at the (0001) surface of corundum [67] and the interaction of water and carbon monoxide with the (100) facet of α-MoO_3 [68]. However, by far the most studied surface has been the (001) plane of MgO due to its simplicity and relevance to experimental measurements, in addition to the ability to contrast the results with cluster calculations [69].

Recently there have been a number of illustrative studies of the adsorption of water and the question of whether it can dissociate so as to hydroxylate the surface. This problem has been investigated by Scamehorn *et al.* [70], using periodic Hartree–Fock methods with density functional correlations, and also by Langel and Parrinello, using plane wave techniques [71].

Both of the above approaches demonstrate that the adsorption of water at the (001) surface is exothermic and that hydroxylation is unfavourable. However, there is disagreement as to the precise binding energy and geometry for the water molecule. The periodic Hartree–Fock-based study suggests that the favoured orientation for water is with both hydrogens bridging across two surface oxygens with a binding energy of about $50 \, kJ \, mol^{-1}$ depending on which correlation functional is chosen. The plane wave results have only one hydrogen interacting with the surface and the oxygen of water above magnesium with a lower binding energy of $20 \, kJ \, mol^{-1}$. Examination of the relative energies of different configurations of neighbouring water molecules in the former study suggests that surface wetting will not take place as the water–water interactions are stronger than those with the surface.

For defective surfaces containing steps or corners, both studies conclude that hydroxylation is possible. Scamehorm *et al.* [72] find that while the energies of physisorbed and chemisorbed water are close at an edge, hydroxylation is strongly exothermic and preferred at a corner. In the absence of analytic gradients it is difficult to characterise fully the energy surface for defective systems due to the lower symmetry. This is the principal limitation of the above investigation. However, the adsorption of water at a step has been examined using *ab initio* molecular dynamics by Langel and Parrinello [71]. This dynamical study shows that the dissociation of water to form two hydroxyl groups at the surface step defect has little or no activation energy. An added advantage of the latter approach is that the anharmonic vibrational spectrum can be readily obtained for comparison with experimental data. Support for the exothermic dissociation of water at stepped surfaces of MgO also comes from interatomic potential calculations [73].

Although the above work has begun to explain the chemistry of water adsorption and dissociation at oxide surfaces, much further work is needed. In particular, the differences between the results of the various computational approaches must be understood, as must the effect of lower surface coverages. As the availability of programs for the *ab initio* study of periodic surface models increases, this approach is likely to increase in popularity.

6 Complex systems: supported films and surface defects — the example of surface oxygen vacancies in TiO_2-supported V_2O_5 partial oxidation catalysts

In the final section, we aim to give an impression of the current state of the art of surface modelling, by reviewing one of the most complex systems studied to date. The methodologies employed are largely based on interatomic potentials, although the simulations suggest many future studies requiring electronic structure techniques.

TiO_2-supported V_2O_5 thin films have been shown to exhibit catalytic activity which cannot be attributed to either the unsupported V_2O_5 or the TiO_2 substrate [74]. It has been proposed that the TiO_2 support facilitates the selective exposure of certain crystallographic faces [75,76], coupled with a modification of either or both components of the catalysts which then exhibit the desired activity and selectivity [77–79]. Previously we have shown, using atomistic computer simulation techniques, that significant structural differences exist between the surface of a supported V_2O_5 thin film on TiO_2 and the unsupported V_2O_5 surface [80]. Indeed, such a modification in the surface structure of the V_2O_5 monolayer influences the sorption of small hydrocarbon molecules on the surface which may facilitate their participation in the catalysis [81].

The activity of molecular oxygen in oxidation reactions is generally low and therefore the oxygen atoms from the surface of the V_2O_5 almost certainly participate in the chemical reaction; these oxygens are then replaced in a further step by gas-phase oxygen, and consequently the modification of the thin film will directly influence the catalysis. Experimental work suggests that the abstraction of a surface lattice oxygen results in reduction of vanadium to a lower valence state [87]. However, there is still some debate over the relative importance of the oxygen vacancy site, in particular whether the vacancies are vanadyl or bridging oxygen species [83]. In this present study we address the formation of oxygen vacancies at the surface of both the supported and unsupported V_2O_5 thin film, including both bridging and vanadyl oxygen sites.

To investigate this problem we have used the standard static lattice simulation techniques, available in the MARVIN program [6] discussed earlier in this paper.

Standard Born model potentials (with the shell model [11] representation of polarisability) were employed. The structure of V_2O_5, obtained from X-ray diffraction techniques, has been given by Byström *et al.* [84] with a refinement to the structure proposed in 1986 by Enjalbert and Galy [85]; it has a layered structure, built up from VO_5 square pyramids sharing edges and corners, with V_2O_5 sheets held together via weak vanadium–oxygen interactions (see Fig. 8.5). The potential model for V_2O_5, proposed by Dietrich *et al.* [86], accurately reproduces the crystal structure of the material and is employed in this study. These potential parameters, together with those for the TiO_2, are reported by Sayle *et al.* [80].

6.1   *TiO_2 (anatase) supported V_2O_5 thin films*

Our earlier work showed that supported V_2O_5 thin films undergo substantial structural modifications compared with the unsupported V_2O_5, and we therefore employ this system [80] as a framework for our defect calculations. Specifically, we consider the $V_2O_5(001)$–$TiO_2(001)$ interface where the area of the primitive interfacial unit cell is 43.98 Å.

6.2   *Defect energies*

One oxygen species per interfacial unit cell is removed from the V_2O_5 surface which gives an oxygen vacancy concentration of 5%, and charge compensation is facilitated by the reduction of two adjacent V^{5+} species to V^{4+}. The defect energies for oxygen vacancies in the supported and unsupported V_2O_5 are given in Table 8.1 with the oxygen ion site notation illustrated in Fig. 8.6. Stable defect structures could not be obtained for some configurations. Such behaviour is perhaps not surprising in light of the highly complex nature of the potential energy hypersurfaces corresponding to these systems.

Table 8.1 also gives the difference in defect energies, ΔE, between the supported and unsupported V_2O_5, which can be as high as 4 eV. Such differences can be attributed, in part, to the change in the local environment of the oxygen vacancy. For example, in the

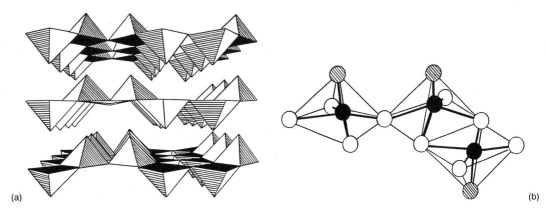

(a)   (b)

Figure 8.5. Diagrammatic representation of the V_2O_5 crystal structure, illustrating the corner- and edge-sharing VO_5 square pyramids. (b) Represents an enlargement of three square pyramids indicating the individual vanadium and oxygen species: vanadium ions are represented by filled circles, vanadyl oxygens are hatched and bridging oxygens are hollow.

Table 8.1 Oxygen vacancy formation energies in the monolayer $V_2O_5(001)$–$TiO_2(001)$ interface with oxygen ion site notation given in Figure 8.6. V and B represent vanadyl and bridging oxygens, respectively. E_u is the energy of the defect in the unsupported $V_2O_5(001)$ surface and E_s the oxygen vacancy in the TiO_2(anatase)-supported $V_2O_5(001)$ surface. ΔE is the difference between E_u and E_s. Negative values of E suggest the vacancy is stabilised with respect to the unsupported V_2O_5.

Oxygen site	E_u (eV)	E_s (eV)	ΔE (eV)	Oxygen site	E_u (eV)	E_s (eV)	ΔE (eV)
1V	139.77	—	—	11B	128.27	127.03	– 1.2
2V	139.77	139.47	– 0.3	12B	126.78	123.00	– 3.8
3V	139.77	—	—	13B	128.27	—	—
4V	139.77	139.47	– 0.3	14B	131.23	—	—
5V	140.91	136.88	– 4.0	15B	131.23	—	—
6V	140.91	141.25	+ 0.3	16B	126.78	129.11	+ 2.3
7V	140.91	136.86	– 4.0	17B	128.27	126.21	– 2.1
8V	140.91	141.74	+ 0.8	18B	126.78	123.00	– 3.8
9B	131.23	—	—	19B	131.27	—	—
10B	126.78	—	—	20B	131.23	—	—

supported V_2O_5, the oxygen vacancy is exposed to the potential generated by both the V_2O_5 and the TiO_2 substrate, and clearly the vacancy formation energy will reflect this change in the local environment. Furthermore, a large component of the interfacial energy arises from the matching of counter-ions across the interfacial plane. The removal of an oxygen species which is bound to an underlying titanium counter-ion will destabilise the interface and the defect energy is expected to be high. Similarly, if the ion–ion interaction across the interface is repulsive — for example, the oxygen is in close proximity to oxygen species from the TiO_2 support — creating an oxygen vacancy at this position will remove such deleterious interactions, enhancing the interfacial stability; consequently, the defect energy may be lower than the corresponding oxygen vacancy in the unsupported V_2O_5. In addition to the effect of the local environment, one also needs to consider other factors such as the modification to the V_2O_5 thin film to accommodate the misfit between the two lattices. All such components of the defect

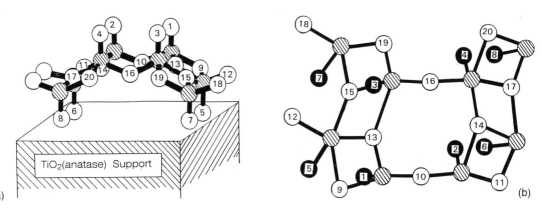

(a)   (b)

Figure 8.6. Oxygen site notation in the monolayer $V_2O_5(001)$–$TiO_2(001)$ interface. (a) Schematic diagram to aid the interpretation of the oxygen sites. (b) View looking down on the surface of the V_2O_5 thin film with relaxed ion positions. In (b) the bridging oxygens are hollow and vanadyl oxygens are filled. Vanadium species are hatched.

energy will, of course, be modelled by the simulation. We now consider the effect of the misfit on the defect energy.

For the unsupported V_2O_5, the O5 and O8 defect energies are identical as these sites are equivalent, yet they differ by about 5 eV in the supported V_2O_5. In essence, the effect of the interface is to resolve previously equivalent atomic positions as it reduces the symmetry of the system. For example, epitaxial matching of the incommensurate lattices, V_2O_5 and TiO_2, ensures that the O5 and O8 oxygen positions lie at slightly different positions above the TiO_2 lattice. In view of this small change in local environment resulting in a slight change in the potential energy of the oxygen species, we would indeed expect the vacancy energies to differ, although such effects are unlikely to account for such a large (5 eV) difference. However, when coupled with the substantial relaxation of these systems, to alleviate the strain introduced into the V_2O_5 (to accommodate the misfit), the resulting 'local environments' of O5 and O8 are substantially changed. Specifically, in this case, the O5 oxygen has nearest-neighbour oxygens of 2.85, 3.12 and 3.16 Å, yet the nearest oxygen to the O8 is 3.63 Å. Creation of an O5 vacancy therefore removes the deleterious repulsion between the nearest-neighbour oxygens, leading to an enhanced interfacial stability. Consequently, the vacancy at the O5 site is notably more stable than that at the O8 site.

Creating vacancies at any of the oxygen positions O1–O8 requires the breaking of a V=O bond. Table 8.1 suggests that this effect results in an unfavourable energy for oxygen vacancies at bridging oxygen sites in all cases. Sites O12 and O18 result in the most stable oxygen vacancies sites, both with cluster energies of 123.0 eV. Furthermore, this energy is lower than that of any oxygen vacancy on the unsupported V_2O_5 with implications for the catalysis effected by such systems. From the defect energies we can formulate an equation for the reaction, the components of which (and the resulting total energy) are given in Table 8.2.

A key factor in this analysis is that the energy of the first process corresponds to the ionisation of vanadium (IV) to vanadium (V). Experimental values for this quantity based on free ion, ionisation potentials are inappropriate as the environmental effects on the energy are considerable. We have therefore used *ab initio* calculations, based on a small core model potential description of vanadium to investigate the magnitude of

Table 8.2 Defect energies for the oxygen vacancy formation energies on the TiO_2-supported V_2O_5 surface.

	Defect process	Energy (eV)	Reference
(a)	$2V_\infty^{5+} + 2e_\infty^- \rightarrow 2V_\infty^{4+}$	-109.5 ± 5	This work
(b)	$O_\infty^{2-} \rightarrow O_\infty$	-7.3	[88]
(c)	$O_\infty \rightarrow \frac{1}{2}O_2$	-2.6	[87]
(d)	$2V_L^{5+} + O_L^{2-} \rightarrow [2V_L^{4+}, V_o^{\bullet\bullet}]$	123.0	This work
(e)	$2V_v^o + O_o^o \xrightarrow{\Delta E} [2V_v', V_o^{\bullet\bullet}] + \frac{1}{2}O_2$	3.6 ± 5	

(a) Fifth ionisation potential for vanadium
(b) First and second electron affinities
(c) Bond dissociation energy of oxygen
(d) Calculated defect energy
(e) Overall reaction

the effect of the crystal on the ionisation potential. As a first calculation, an isolated vanadium ion was embedded in a point array consisting of 559 charges designed to reproduce the bulk electrostatic site potential. The shift in the ionisation potential, relative to the gas-phase value, was then determined. We find that this shift is comparatively small at 0.5 eV in the limit of formal ionic charges. For realistic charge distributions, particularly at the crystal surface, this value will be significantly reduced. However, such calculations do not include the effect of covalent interactions, and to investigate this matter further, we have compared the ionisation potential of vanadium as part of the molecular species $(V=O)^{2+}$ with the isolated ion. In this case we find shifts in the range -8 to -13 eV for the fifth ionisation potential, which results in a value of 3.6 ± 5 eV for the overall reaction as detailed in Table 8.2. Obtaining a precise value for this quantity proves to be difficult as it is very sensitive to the nature of the oxygen basis set and degree of contraction for the vanadium valence orbitals, in addition to the level of correlation correction applied. Further work is needed to refine our prediction of the energetics of these key processes. It is clear, however, that surface vacancy formation is an energetically feasible process on the surface of TiO_2-supported V_2O_5 films. The oxygen atoms from the supported V_2O_5 almost certainly participate in the catalysis, and the presence of such labile oxygen ions on the surface of the supported material will directly influence the catalysis.

7   Summary and conclusions

The examples presented in this paper illustrate the range and limitations of current computer modelling techniques in the science of inorganic surface chemistry. These methods are increasingly able to provide a useful description of the atomic structure of the surfaces of complex materials. They are making inroads into our understanding of reactivity, although many challenges remain in the latter field.

8   Acknowledgements

We are grateful to the EPSRC and Rhône Poulenc for financial support. We are also grateful to Dr Lars Pettersson for useful discussions.

9   References

1 Van Gunsteren WF, Berendsen HJC, *Angew. Chem. Int. Ed. Engl.* 1990; **29**: 992.
2 Kremer K. In Allen MP, Tildesley DJ (eds) *Computer Simulation in Chemical Physics* Dordrecht: Kluwer, 1993.
3 Allen MP, Tildesley DJ. *Computer Simulation of Liquids*. London: Oxford University Press, 1987.
4 Catlow CRA, Thomas JM, Freemen CM, Wright PA, Bell RG. *Proc. R. Soc. London A* 1993; **442**: 85.
5 Vessal B. In Catlow CRA (ed.) *Computer Simulation in Inorganic Crystallography*. London: Academic Press, 1996.
6 Gay DH, Rohl AL. *J. Chem. Soc. Faraday Trans.* 1995; **91**: 925.
7 Deem MW, Newsam JM, Sinha SK. *J. Phys. Chem.* 1990; **94**: 8356.
8 Parry DE, *Surf. Sci.* 1975; **49**: 433.
9 Parry DE. *Surf. Sci.* 1976; **54**: 195.

10 Heyes DM, Barber M, Clarke JHR. *J. Chem. Soc. Faraday Trans. 2* 1977; **73**: 1485.

11 Dick BG, Overhauser AW. *Phys. Rev.* 1958; **112**: 90.

12 Gale JD, Catlow CRA, Mackrodt WC. *Modelling Simul. Mater. Sci. Eng.* 1992; **1**: 73.

13 Bravis A. *Études Crystallographiques.* Paris: Académie des Sciences, 1913.

14 Friedel G. *Bull. Soc. Fr. Minér.* 1907; **30**: 326.

15 Donnay JDH, Harker D. *Am. Mineral.* 1937; **22**: 446.

16 Hartman P, Perdok WG. *Acta Crystallogr.* 1995; **8**: 49.

17 Hartman P, Bennema P. *J. Crystal Growth* 1980; **49**: 145.

18 Hartman P, Perdok WG. *Acta Crystallogr.* 1995; **8**: 525.

19 Hartman P, *J. Crystal Growth.* 1980; **49**: 157.

20 Berkovitch-Yellin Z. *J. Am. Chem. Soc.* 1985; **107**: 8239.

21 Docherty R, Roberts KJ. *J. Crystal Growth* 1988; **88**: 159.

22 Dowty E. *Am. Mineral.* 1976; **61**: 448.

23 Gibbs JW. *Collected Works.* New York: Longman, 1928.

24 Mackrodt WC, Davey RJ, Black SN, Docherty R. *J. Crystal Growth* 1987; **80**: 441.

25 Hartman P. *Acta Crystallogr.* 1958; **11**: 459.

26 Lawrence PJ, Parker SC. In Catlow CRA, Parker SC, Allen MP (eds) *Computer Modelling of Fluids, Polymers and Solids.* Amsterdam: Kluwer, 1990; 219.

27 Addadi L, Berkovitch-Yellin Z, Weissbuch I, Lahav M. *Topics in Stereochem.* 1986; **16**: 1.

28 Davey RJ, Polywka LA, Maginn SJ. In Garside J, Davey RJ, Jones AG (eds) *Advances in Industrial Crystallisation.* Oxford: Butterworth-Heinemann, 1991; 150.

29 Rohl AL, Gay DH, Davey RJ, Catlow CRA. *J. Am. Chem. Soc.* 1996; **118**: 642.

30 Benton WJ, Collins IR, Cooper SD, Grimsey IM, Parkinson GM, Rodger SA. *Faraday Discuss.* 1993; **95**: 281.

31 Black SN, Bromley LA. *J. Chem. Soc. Faraday Trans.* 1991; **87**: 3409.

32 Davey RJ, Black SN, Bromley LA, Cottier D, Rout JE. *Nature* 1991; **353**: 549.

33 Hijner PE, Manne S, Hansma PK, Gratz AJ. *Faraday Discuss.* 1993; **95**: 191.

34 Van Enckevort WJP. In van der Eerden JP, Bruinsma OSL (eds) *Science and Technology of Crystal Growth.* Dordrecht: Kluwer, 1995; 355.

35 Hartman P, Strom CS. *J. Cryst. Growth* 1989; **114**: 502.

36 Mackrodt WC, Davey RJ, Black SN, Docherty R. *J. Cryst. Growth* 1987; **80**: 441.

37 Hartman P. *J. Cryst. Growth* 1987; **9**: 721.

38 Mackrodt WC. *Phil. Trans. R. Soc. London* 1992; **341**: 301.

39 Catlow CRA. *J. Phys. Chem. Solids* 1997; **38**: 1131.

40 Loades SD, Carr SW, Gay DH, Rohl AL. *J. Chem. Soc., Chem. Commun.* 1994; 1369.

41 Catlow CRA, Gale JD, Gay DH, Lewis DW. In Pinnavaia TJ, Thorpe MF (eds) *Access in Nanoporous Materials, Proceedings of the Lansing Conference, June 1995*; New York: Plenum Press, 1995.

42 Freund H-J, Kuhlenbeck H, Staemmler V. *Rep. Progr. Phys., Oxide Surfaces* 1986; **59**: 283.

43 Goodman DW. *J. Phys. Chem.* 1996; **100**: 13090.

44 He J-W, Estrada CA, Corneille JS, Wu M-C, Goodman DW. *Surf. Sci.* 1992; **261**: 164.

45 Nygren MA, Pettersson LGM. *J. Chem. Phys.* 1996; **105**: 9339.

46 Pacchioni G, Minerva T, Bagus PS. *Surf. Sci.* 1992; **275**: 450.

47 Barandiaran Z, Seijo L. *J. Chem. Phys.* 1988; **89**: 5739.

48 Barandiaran Z, Seijo L. In Fraga S (ed.) *Computational Chemistry: Structure, Interactions and Reactivity*, Studies in Physical and Theoretical Chemistry, Vol. 77B. Amsterdam: Elsevier, 1992; 435.

49 Huzinaga S, Seijo L, Barandiaran Z, Klobukowski M. *J. Chem. Phys.* 1987; **86**: 2132.

50 Polchen M, Staemmler V. *J. Chem. Phys.* 1993; **97**: 2583.

51 Stromberg D. *Surf. Sci.* 1992; **275**: 473.

52 Nygren MA, Pettersson LGM, Barandiaran Z, Seijo L. *J. Chem. Phys.* 1994; **100**: 2010.

53 Boys SF, Bernardi F. *Mol. Phys.* 1970; **19**: 553.

54 Pascual J-L, Pettersson LGM. *Chem. Phys. Lett.* (in press).

55 Nygren MA, Pettersson LGM, Freitag A, Staemmler V, Gay DH, Rohl AL. *J. Phys. Chem.* 1996; **100**: 294.

56 Nygren MA, Pettersson LGM. *Chem. Phys. Lett.* 1994; **230**: 456.

57 Chong DP, Langhoff SR. *J. Chem. Phys.* 1986; **84**: 5606.

58 Snis A, Stromberg D, Panas I. *Surf. Sci.* 1993; **292**: 317.

59 Nakamura M, Mitsuhashi H, Takezawa N. *J. Catal.* 1992; **138**: 686.

60 Snis A, Panas I, Stromberg D. *Surf. Sci.* 1994; **310**: L579.

61 Snis A, Panas I. *J. Chem. Phys.* 1995; **103**: 7626.

62 Dovesi R, Saunders VR, Roetti C, Cavsà M, Harrison NM, Orlando R, Aprà E. *CRYSTAL 95 Users Manual. Turin: University of Torino, 1996.*

63 Payne MC, Teter MP, Allan DC, Arias TA, Joannopoulos JD. *Rev. Mod. Phys.* 1992; **64**: 1045.

64 Te Velde G, Baerends EJ. *Phys. Rev. B* 1991; **44**: 7888.

65 Stich I, Payne MC, King-Smith RD, Lin JS, Clarke LJ. *Phys. Rev. Lett.* 1992; **68**: 1351.

66 Pugh S, Gillan MJ. *Surf. Sci.* 1994; **320**: 331.

67 Catlow CRA, Bell RG, Gale JD. *J. Mater. Chem.* 1994; **4**: 781.

68 Papakondylis A, Sautet P. *J. Phys. Chem.* 1996; **100**: 10681.

69 Dovesi R, Orlando R, Ricca F, Roetti C. *Surf. Sci.* 1987; **186**: 267.

70 Scamehorn CA, Hess AC, McCarthy MI. *J. Chem. Phys.* 1993; **99**: 2786.

71 Langel W, Parrinello M. *J. Chem. Phys.* 1995; **103**: 3240.

72 Scamehorn CA, Harrison NM, McCarthy MI. *J. Chem. Phys.* 1994; **101**: 1547.

73 De Leeuw NH, Watson GW, Parker SC. *J. Phys. Chem.* 1995; **99**: 17219.

74 Centi G, Perathoner S, Trifiro F. *Res. Chem. Intermediates* 1991; **15**: 49.

75 Vejux A, Courtine P. *J. Solid State Chem.* 1978; **23**: 93.

76 Vejux A, Courtine P. *J. Solid State Chem.* 1986; **63**: 179.

77 Centi G, Giamello E, Pinelli D, Trifiro F. *J. Catal.* 1991; **130**: 220.

78 Vedrine JC. *Catal. Today* 1994; **20**.

79 Kozlowski R, Pettifer RF, Thomas JM. *J. Phys. Chem.* 1983; **87**: 5176.

80 Sayle DC, Catlow CRA, Perrin M-A, Nortier P. *J. Phys. Chem.* 1996; **100**: 8940.

81 Sayle DC, Catlow CRA, Perrin M-A, Nortier P. *Catal. Lett.* 1996; **38**: 203.

82 Zhang Z, Henrich VE. *Surf. Sci.* 1994; **321**: 133.

83 Went GT, Leu LJ, Bell AT. *J. Catal.* 1992; **134**: 479.

84 Byström A, Wilhelmi K-A, Brotzen O. *Acta Chemica Scandinavica* 1950; **4**: 1119.

85 Enjalbert R, Galy J. *Acta Cryst. C* 1986; **42**: 1467.

86 Dietrich A, Catlow CRA, Maigret B. *Mol. Simul.* 1993; **11**: 251.

87 Weast RW. *Handbook of Chemistry and Physics.* Chemical Rubber Publishing Company, 1971.

88 Kaye GWC, Laby TH. *Tables of Physical and Chemical Constants.* 1996.

89 Ferrai AM, Pacchioni G. *Int. J. Quantum Chemistry* 1996; **58**: 241.

9 Surface Dynamics Studied Using Reflectance Anisotropy

M.E. PEMBLE

Division of Chemical Sciences, Science Research Institute, University of Salford, Salford M5 4WT, UK

1 Introduction

In recent years the use of optical methods to study surfaces has become increasingly popular, since very often such techniques may be applied to surfaces under conditions in which conventional surface science methods may not operate [1–4]. In particular, studies of semiconductor surfaces have concentrated upon the use of optical methods, primarily because often the processing of such surfaces — growth, etching, metallisation etc. — occurs under conditions of atmospheric pressure or high temperature. However, very often the final application of the semiconductor material involves the use of photons — light-emitting diodes, lasers, interference reflectors, etc. — and thus, via the use of light of the appropriate wavelength, the 'quality' of the material may be monitored continuously during processing.

Optical methods such as adsorption or luminescence techniques have long been used to study semiconductors, but the drive towards understanding surface processes has led to the evolution of the family of so-called '*epioptic*' techniques [2,3]. Here the prefix 'epi' simply means 'upon', implying that the techniques must provide surface information. Thus although the technique may employ photons which penetrate several micrometres into the material, the information obtained is still surface-specific. The family of epioptic techniques includes the methods of reflectance anisotropy (RA), reflectance anisotropy spectroscopy (RAS) and reflectance difference (spectroscopy) (RD(S)) [5–8], surface photoabsorption [9], spectroscopic ellipsometry [10], Raman scattering [11], infrared adsorption methods [12] and second harmonic [13,14] or sum-frequency generation [15]. Here the emphasis will be on the use of RA, since this technique has emerged as perhaps one of the simplest yet most powerful of the epioptic techniques for the study of surface dynamics.

2 Theory and experimental design

Reflectance anisotropy owes its popularity to the work of Aspnes and co-workers, who demonstrated the basic principles and instrumentation necessary for the RA experiment in 1985 [5]. The technique is essentially based upon the use of ellipsometric methods under conditions in which the light is incident upon the surface in a direction parallel to the normal. Under such conditions in which the plane of incidence is reduced to a line, the definition of s and p polarisations breaks down. For plane polarised light, the only useful points of reference become crystallographic axes present in the substrate. For materials based upon the cubic system, it is often possible to select a surface plane upon which there are two convenient orthogonal axes. Examples of this would be [110] and [1$\bar{1}$0] axes for (001) surfaces and [1$\bar{1}$0] and [100] axes for (110) surfaces. For face-centred cubic systems, which here will include the zinc-blende family of compound semiconductors, surface restructuring will often remove the isotropic nature of (001) surfaces in particular such that the [110] and [1$\bar{1}$0] directions are structurally (and hence optically) inequivalent. If (001)GaAs is taken as an example, this surface would be terminated (under ideal conditions) by a layer of either Ga or As atoms. Both species possess dangling bonds, such that Ga–Ga dimers or As–As dimers may form via rehybridisation of the dangling bond states, generating surface reconstructions. The anisotropy arises since the Ga–Ga dimers align (by convention) along [110] while necessarily the next layer of As atoms would generate As–As dimers aligned along [1$\bar{1}$0]. Thus, at its simplest, RA relies upon the difference in the surface dielectric response of the sample between the two structurally anisotropic directions. Note that the bulk material will in contrast be isotropic and thus the bulk signals for light polarised parallel to [110] and also parallel to [1$\bar{1}$0] will be equivalent. The interpretation of the spectroscopic RA response, particularly from (001) GaAs, has been made possible by combined measurements of RA (as a function of incident energy) and electron diffraction such that for (001)GaAs in particular it has been possible to assign particular RA spectra to particular surface reconstructions and further identify features within the spectra which relate to the presence of particular surface species [16,17]. From these data, particular wavelengths may be selected in order to be able to follow transient processes in real time. This is the approach that has been employed in the experiments described here.

In practice, the surface dielectric response along both directions of interest is sampled in reflection mode, using either a piezoelastic modulator (PEM) system such as first employed for this purpose by Aspnes [5,6], or via a novel system of very rapid (faster than *c.* 6 MHz) polarisation modulation using Pockels cells [18]. This latter system was developed in our laboratories. Figure 9.1 depicts schematically how the two discrete RA systems function. In the case of the PEM system it is usual to select a polarisation which has components along both axes to be sampled and to evaluate the magnitude of each of these components following reflection by applying a wavelength-dependent

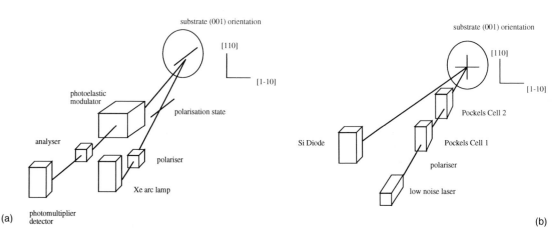

(a)

(b)

Figure 9.1. A schematic representation of the configuration and operation of (a) a photoelastic modulator-based RA system, and (b) a Pockels cell based RA system. The design of the photoelastic modulator system is after Aspnes [5], while the Pockels cell system was designed in our laboratories [18].

retardation factor using the PEM coupled to an output analyser. The specific wavelength selected is then detected using the monochromator–photomultiplier system. The Pockels cell system uses a different approach to sample the axes on the surface, in that it physically rotates the plane of polarisation through 90° such that orthogonal axes are sampled sequentially. At present the Pockels cell systems are set up to utilise a single wavelength.

Before discussing the background theory common to both experimental approaches it is appropriate to compare and contrast the two methods. PEM-based systems possess the advantage of simultaneous sampling of both axes, coupled with the ability to respond simply to a wide range of wavelengths. Thus, such systems are currently the preferred choice for the spectroscopic RA experiments in which the incident photon energy may be varied within limits such as 0.8–4.0 eV, typically using a xenon arc. In terms of the time resolution possible, since a typical PEM currently operates at a frequency of 50 kHz and given the low light levels associated with the use of broadband arc lamps and the detection system, it is possible to obtain meaningful data at a rate of up to $c.$ 10 Hz. However the true experimental limitation arises from the time taken to 'scan' a complete spectrum. Typically this varies in the range 10–50 s, dependent upon the energy range required. Thus it is currently difficult to record RA spectra in real time from surfaces which are undergoing rapid change. Richter and co-workers have demonstrated a novel means of partially overcoming this difficulty using a series of identical experiments in which time-dependent RA response is recorded at a series of incident energies [19]. Spectra corresponding to various time markers are then constructed from composites of the transient data. The drawback of this approach is that, particularly with semiconductor growth experiments, it is rather difficult to ensure that the exact surface conditions are reproduced for every experiment. Thus at present one goal is to reduce the time required to acquire a complete RA spectrum. In contrast, the Pockels cell systems currently operate at fixed wavelength — indeed, it is usual to employ a small laser as the light source rather than the arc lamp used with the PEM systems. Although losing the spectroscopic dimension, these systems have particular advantages in terms of time resolution — typically 1 ms resolution is possible for a system operating at 6 MHz switching frequency, i.e. meaningful data can be acquired at a rate of 1 kHz. Also, the increased time resolution, coupled with the use of low-noise laser systems, has been found to improve the signal-to-noise capability by a factor of 10^2 as compared to the PEM-based systems. In addition, the use of lasers brings the advantages of the use of simple Si-diode based detector systems (as compared to the use of photomultipliers with the PEM-based systems) and, because of the intensity advantage, the ability to sample through windows or reactor walls that have become partially opaque due to the deposition of thin films. In terms of the application of RA to atmospheric-pressure metal–organic chemical vapour deposition (MOCVD), this latter factor is often overlooked, but can be critical in terms of the likely success of the RA experiment.

Reflectance anisotropy is a dimensionless quantity defined in terms of the anisotropy as measured in reflection between two usually orthogonal axes, generally labelled a and b, so that the anisotropy is given by $R_a - R_b$. In order to normalise this quantity with respect to changes in bulk reflectivity which may arise due to the difference in refractive index between the substrate and the deposited overlayer or, more simply, due to surface roughening, the anisotropy is ratioed against a measure of the total reflectivity, which in turn is conveniently modelled as $\frac{1}{2}(R_a + R_b)$:

$$\Delta R/R = (R_a - R_b)/[1/2(R_a + R_b)] \tag{1}$$

It is convenient to express equation (1) in terms of the dielectric responses of the system such that

$$\Delta R/R = (8\pi/\lambda)\,\mathbf{Im}\,\{(\varepsilon_a - \varepsilon_b)d/[\varepsilon_s - 1]\} \tag{2}$$

where λ is the wavelength, d is the thickness of the anisotropic surface layer having complex dielectric function ε_a and ε_b and ε_s is the complex bulk dielectric function. Under conditions in which $d \ll \lambda$ and when ε_s is for the most part real, the reflectance anisotropy measured using the direct power reflectance terms R_a, R_b, $\frac{1}{2}(R_a + R_b)$ may be expressed as

$$\Delta R/R = (8\pi/\lambda)\,\mathbf{Im}\,[(\varepsilon_a - \varepsilon_b)d] \tag{3}$$

The imaginary term in equation (3) is often referred to as the surface-induced optical anisotropy (SIOA) [20],

$$SIOA = \mathbf{Im}\,(\varepsilon_a - \varepsilon_b)d \tag{4}$$

Since ε_a and ε_b are very similar and also small in relation to ε_s, typical values of the reflectance anisotropy, e.g. from (001)GaAs, are of the order of 10^{-3}. It is often assumed that ε_s is invariant throughout an experiment. This is an oversimplification which only applies in practice where the surface modification is subtle or, to take a specific example, where reaction results in the epitaxial growth of a material upon a substrate of the same material, e.g. the growth of GaAs on (001) GaAs by molecular beam epitaxy (MBE) or MOCVD. More usually, where the growing layer possesses a different dielectric response to that of the substrate, as soon as the layer thickness approaches $\lambda/10$, changes in the bulk reflectivity will begin to influence the measured RA. Taken further, oscillations in the bulk reflectivity will be observed due to interference effects arising from the relationship between the layer thickness and the wavelength of the light. Indeed, such oscillations are the basis of the technique known as reflectometry, which has been used in studies of semiconductor growth to provide online measurements of growth rate and adsorption coefficient [21]. In this respect it is noteworthy that an RA system is obviously also suitable for reflectometry measurements if required.

3   Applications of reflectance anisotropy to the study of surface dynamics

3.1   The Growth of AlAs on (001)GaAs by MBE

MBE is well established as a powerful means of producing precisely defined structures. In the case of semiconductor growth by MBE, the technique is usually used in conjunction with reflection high-energy electron diffraction (RHEED) which not only permits the grower to determine that the substrate has been prepared effectively but also, under conditions where growth occurs by two-dimensional (2D) nucleation on the terraces, allows the direct monitoring of layer-by-layer growth dynamics. This is achieved by monitoring the intensity of one of the reflected electron beams, since under the grazing incidence conditions employed, the reflected intensity is extremely sensitive to layer roughness. Thus it is found that, by monitoring the intensity of a particular reflected beam, oscillations in the intensity are observed which possess a period equivalent to the rate of formation of one new surface layer. In our early work on the use of RA to study semiconductor growth dynamics we found it particularly instructive to use both RA and RHEED simultaneously.

To illustrate the power of this approach, consider the data shown in Fig. 9.2. This figure depicts both RHEED and RA transients recorded as a function of time during the MBE growth of AlAs on (001)GaAs [22]. Prior to commencing the experiment, the (001)GaAs sample, mounted in a conventional MBE system, was cleaned via annealing under a flux of As_4 to remove surface contamination, while preserving surface As levels which would otherwise be depleted via preferential evaporation. This standard procedure results in the formation of a (2×4) reconstruction of the GaAs surface (as

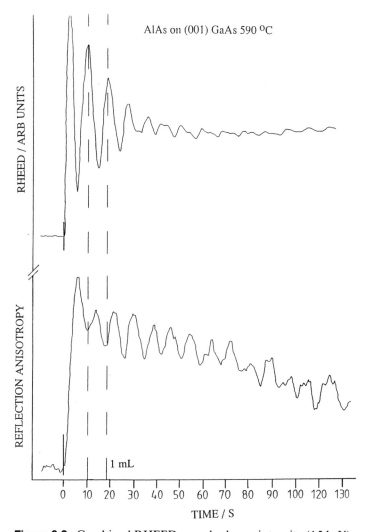

Figure 9.2. Combined RHEED specular beam intensity (15 keV) and RA data (1.96 eV) recorded as a function of time from a (001)GaAs surface at 590°C, during the growth of AlAs by molecular beam epitaxy [22].

determined by RHEED) which is quite free from contamination and is terminated in
an As-rich overlayer in which the As coverage is nominally 0.75 [23]. Growth of AlAs
was initiated following the growth of a GaAs buffer layer designed to prevent surface
effects arising from the use of a particular substrate employed influencing the experiment. While full experimental details may be obtained from the references provided, it
is noteworthy that both Al and As sources remained open throughout the AlAs growth
process such that although the AlAs layer consists of alternating Al and As layers, this is
achieved not by sequential Al and As introduction but as a result of surface diffusion
and reorganisation to produce the most energetically favourable surface. Note also that
strain does not result from this growth process since AlAs and GaAs have virtually
identical lattice parameters, although it is known that the AlAs layer grows in such a
way as to appear rougher than a corresponding layer of GaAs.

From Fig. 9.2 it is apparent that, as expected from the foregoing argument, the
RHEED intensity data reveal oscillations which may be correlated with layer-by-layer
growth. Interestingly, the RA data recorded simultaneously also depict oscillations and
it is obvious that the period of the oscillations is the same as that observed in the
RHEED data. Thus the RA is apparently sensitive to the layer-by-layer growth process.
The RA system employed was based upon the Pockels cell type of instrument and could
be easily interfaced to the MBE system via the viewport normally used for pyrometry —
this port permitted the laser to impinge upon the surface at near-normal incidence. The
energy of the incident photons, 1.96 eV (HeNe laser), corresponds to a broad resonance
observed in RA spectra from (001) surfaces of both Ga and As [24], which is believed to
originate from the metal (Al or Ga) dimer states. Indeed, calculations for (001)GaAs
support this proposal [25], but lack of supporting data, such as might be obtained by
photoemission or inverse photoemission (see data for Cu(110) described later) means
that there is still some considerable debate as to the precise nature, i.e. pure surface or
surface-to-bulk states, involved [26]. Very recent work by Bottomley has suggested that,
for non-centrosymmetric materials, such as the III–V materials, there will be a bulk
quadrupolar contribution to the measured RAS data which may be comparable in
magnitude to the surface dipolar contribution that is usually discussed exclusively [26].
Since via the use of relatively long-wavelength radiation and the use of normal
incidence it is highly unlikely that the RA method could respond to the atomic-scale
variations in surface roughness that the RHEED is sensitive to, it is generally believed
that the RA responds to the time-dependent concentrations of particular chemical
species as determined by the appropriate selection of incident wavevector, both energy
and dispersion through the surface Brillouin zone. In the case of the data shown in
Fig. 9.2 it is the Al–Al dimers which are believed to be the active species. Thus, despite
the fact that the under growth conditions the surface is likely to be a highly complex
entity in which Al and As atoms are mixed almost randomly, a further reorganisation
process leading to the formation of Al–Al dimers still occurs. Furthermore, these
dimers form despite the relatively high temperatures employed and the constant flux of
other Al atoms and As atoms. This latter point is particularly significant because to
assume a place in the growing lattice the Al atoms need to form two bonds to underlying
As atoms and thus the remaining electron must be used to form the dimer states. This
is despite the fact that the dimers must then be broken in order to accommodate the
next layer of As atoms. Thus, a complex dynamic process such as MBE growth may in

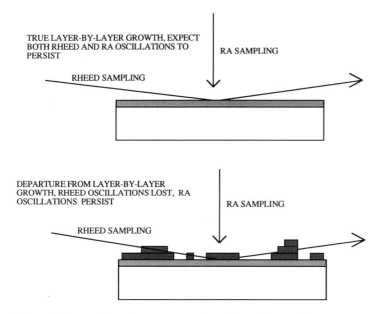

Figure 9.3. A schematic representation of the effects of incomplete layer-by-layer growth on both the RHEED and RA experiments, revealing that the presence of roughness arising from incomplete terraces does not alter the instantaneous concentration of surface dimers as sampled in the RA experiment.

fact be reduced to certain pseudo-equilibrium conditions which result principally from the time domain sampled, the area sampled ($3–5\ mm^2$) and the driving force to accommodate dangling bonds rapidly. In simplistic terms the RA resonance experiences a series of on–off cycles as the surface alternates between Al-rich and As-rich. In this respect similar responses may be observed using photons of higher energy (c. 2.7 eV) but with a 90° phase shift which arises since 2.7 eV photons are believed to excite the surface states due to the As–As dimers [24]. Figure 9.2 also reveals that the RHEED transient is damped such that, after a relatively short time, measurement of the growth rate from the RHEED intensity oscillations becomes unreliable. This is not the case for the corresponding RA oscillations which continue essentially undamped throughout the experiment. For RHEED, the damping phenomenon is well known and arises due to the fact that departure from perfect layer-by-layer growth results in a progressive roughening on a scale comparable to the surface coherence length of the RHEED experiment [27]. We have explained the absence of damping in the RA response by again citing the model of excitation of a particular chemical species: Fig. 9.3 reveals that, providing 2D nucleation and growth is the appropriate growth mode, for a method sampling directly in the surface plane (as RA does via the choice of normal incidence), the appearance of steps via deviation from perfect layer-by-layer growth does not itself influence the instantaneous surface concentration of active species.

3.2   *RA studies of the chemical beam epitaxy (CBE)*
       of Si and Si_xGe_{1-x} on (001)Si

The important observations described in Section 3.1 are also highly pertinent to the next example, which describes the use of RA to monitor the growth of Si and Si_xGe_{1-x}

on (001)Si surfaces [28–32]. This system was first examined in our laboratories since it had been previously believed that the RA experiment was inherently not possible from this surface. The reason for this belief was the knowledge that clean, nominally on-axis (001)Si surfaces adopt a double domain (1×2)–(2×1) reconstruction arising from the presence of Si–Si dimers on terraces separated in height by one atomic distance (even the most perfect of single-crystal surfaces possesses atomic height steps). It was believed that even if a suitable wavelength could be found which was sensitive to the Si–Si dimers, the fact that they exist instantaneously on the surface in two orthogonal domains would negate the RA response. Figure 9.4 depicts a combination of RHEED and RA data obtained from a clean (001)Si surface during chemical beam epitaxial growth of Si from disilane. Since again under the conditions employed the growth mode occurs via 2D nucleation, the RHEED data show the characteristic oscillations associated with the process of layer roughening and smoothing. However, in stark contrast to the data presented in the previous example, the RA data for this system again show oscillations but the rate of oscillation is half that depicted by the RHEED data.

The interpretation of the data for this system is based upon the observation, first, that the RA responds at the bilayer growth rate and, second, that the growth process involves only one type of atom and therefore the argument regarding the on–off type of response to a particular species cited for the previous example requires modification. Essentially the RA oscillation follows an on-positive to on-negative response, in that the active species, proposed to be the Si–Si dimers, are present throughout the growth cycle. The bilayer growth rate corresponds to the rate at which an initial surface, having a particular proportion of the (1×2) and (2×1) domains, returns to its starting condition. Note that it is a necessary requirement that the statistical distributions of each domain within the region sampled are not equivalent. This is easier to appreciate for vicinal surfaces where a high degree of off-cut may lead to the presence of double-height steps which effectively lock in one domain only. For nominally on-axis samples we have concluded that since for such surfaces the average size of each terrace

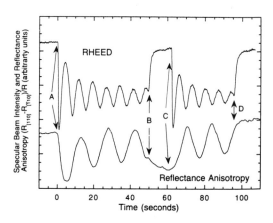

Figure 9.4. Combined RHEED specular beam intensity (15 keV) and RA data (1.96 eV) recorded as a function of time during the gas-source MBE growth of Si and $Si_{1-x}Ge_x$ on (001)Si at 600°C [28,29]. The disilane flow was initiated at points A and C, and interrupted at points B and D. Note that during the interrupt the RHEED intensity increases, indicating smoothing of the surface, yet, following reinstatement of the disilane flow, the RA continues from the level achieved prior to growth interruption.

will be larger than for off-cut surfaces (and hence also fewer in number for a given nominal 'area' of surface), for the area sampled by the RA experiment, $c.$ 2–3 mm^2, there is indeed a real statistical difference between domain sizes. This rather surprising result implies that for these surfaces the domains may be very large indeed. The nature of the surface states that are responsible for the sensitivity of the RA experiment from (001)Si are still the subject of some debate. It would appear that there is some surface-to-bulk contribution to the states, implying that the influence of dimerisation in practice extends well beyond the surface region into the bulk. Recent experiments by Zhang *et al.* have revealed that the presence of hydrogen quenches the excitation process, such that the RA experiment is no longer able to monitor layer-by-layer growth, despite the fact that it still occurs readily, as evidenced by RHEED oscillations [33]. This is an interesting observation, since the accommodation of a monolayer of hydrogen atoms on the (1×2)–(2×1) (001)Si surface occurs without a dramatic structural change, i.e. the (1×2)–(2×1) double domain reconstruction is preserved. Thus the RA is able to sense an adsorbate-induced change in surface or surface-to-bulk electronic states that may not be observed using electron diffraction. Ultimately, addition of further hydrogen breaks the Si–Si dimer structure and lifts the double domain reconstruction to produce a (1×1) surface. However, not too surprisingly, this process does not restore the RA response from the surface.

Further insight into the dynamic nature of this surface is also apparent from Figure 9.4, which also depicts the result of an interruption in growth at the growth temperature. In the sudden absence of disilane flux the surface smooths, as indicated by the increase in the RHEED intensity. However, following reintroduction of the disilane, growth restarts from a surface which is now smoother as indicated by RHEED but the relative population of the two domains is unchanged as indicated by the fact that the RA response measured depends upon the relative domain populations. Figure 9.5 describes the same type of growth interruption but concentrates on the RA data only. Figure 9.6 shows schematically what may be happening. Essentially it is believed that while atoms may move around freely upon a given terrace such that the terrace may become smoother, the same atoms do not cross the step edges such that the relative

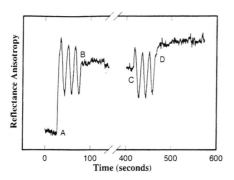

Figure 9.5. RA data (1.96 eV) depicting the results of growth interruption during the gas-source MBE growth of Si on (001)Si at 600°C [28,29] in an experiment similar to that described by Fig. 9.4. The disilane flow was initiated at points A and C, and interrupted at points B and D. Note that, following reinstatement of the disilane flow, the RA continues from the level achieved prior to growth interruption.

GROWTH, 2D NUCLEATION ON TERRACES,
DIMER FORMATION WITH SURFACE
ROUGHENING

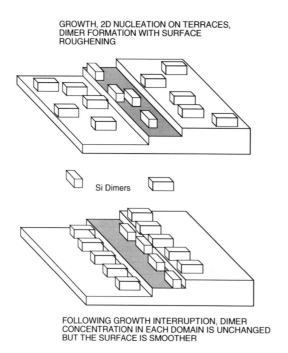

FOLLOWING GROWTH INTERRUPTION, DIMER
CONCENTRATION IN EACH DOMAIN IS UNCHANGED
BUT THE SURFACE IS SMOOTHER

Figure 9.6. A schematic representation of the processes occurring on the atomic scale during the low-temperature growth of Si on (001) by gas-source MBE using disilane. The diagram illustrates that, although the deposited Si atoms are free to move around rapidly on the terraces, which results in an overall smoothing of the surface, the atoms do not cross the step edges such that the relative proportions of the (1×2) and (2×1) domains do not alter.

proportions of the two domains do not change. The next feature of this system of interest concerns a related point which arises due to enhanced mobility of surface atoms at elevated growth temperatures. Under these circumstances it is known from RHEED and other techniques that the atoms possess too much energy to allow a layer to grow by 2D nucleation on the terraces and, upon release at the surface (from disilane), move rapidly towards the energetically more favourable higher coordination sites present at the step edges. Thus the layer grows by virtue of the steps moving (flowing) rapidly across the terraces under conditions known as *step-flow growth* [34]. Under these conditions it may be seen that the oscillatory variation in surface roughness observed at lower temperatures is efffectively lost. Thus, as Fig. 9.7 reveals, oscillations in the RHEED intensity disappear. Interestingly, the RA oscillations also disappear. This is to be expected because under step-flow growth conditions the relative proportions of the two domains do not alter appreciably. Of interest is the fact that the loss of oscillatory behaviour does not occur at the same temperatures for both types of measurement.

We have also studied the CBE growth of the direct bandgap alloy $Si_{1-x}Ge_x$ from disilane and germane [28–32]. Typically we have monitored growth oscillations in experiments in which conventional CBE growth of Si on (001)Si has been modified via the introduction of GeH_4 to the input gas flow. Figure 9.8 reveals that upon addition of the germane the bilayer growth oscillations in the RA data and the RHEED intensity oscillations increase in frequency as a result of an increase in growth rate attributed to the higher sticking coefficient of germane as compared to that of disilane at this temperature. Interestingly, there is also a change in the amplitude of the RA oscilla-

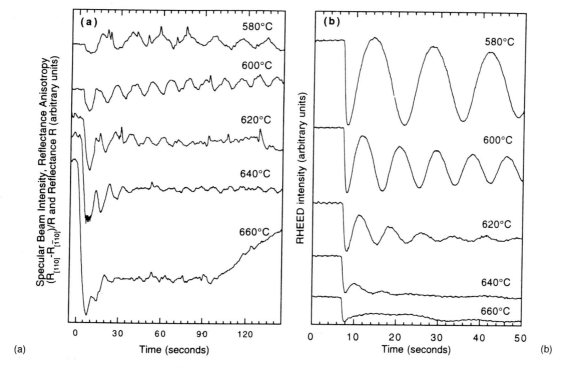

Figure 9.7. Combined RHEED specular beam intensity (15 keV) and RA data (1.96 eV) recorded as a function of time during the gas-source MBE growth of Si on (001)Si, for a series of growth temperatures [32]. These data illustrate the effect of the transition from 2D nucleation and growth to a step-flow growth mode.

tions, which we have tentatively assigned to a change in the cross-section for excitation of the surface resonance associated with dimers on terraces which may result either from the formation of some mixed Si–Ge surface dimers (as would be expected for a statistically random process) or from the formation of Ge–Ge dimers via some process in which such species were energetically more favourable. Further experiments of this nature have suggested that the change in the amplitude of the RA response may be interpreted in a semi-quantitative manner in relation to the magnitude of x, the proportion of Ge in the alloy. However, since the addition of germane also results in roughening and segregation effects [35], more work is required in order properly to establish any correlation that may exist.

Undoubtedly RA will continue to be employed in all areas of Si growth, including further studies of the growth of $Si_{1-x}Ge_x$. In addition, it is suggested that RA may be used in other areas of Si technology such as oxide desorption or growth and the use of (001)Si substrates for the growth of high-quality diamond films for high-temperature semiconductor applications.

3.3 Application of RA to atmospheric pressure InP MOCVD growth

The next example serves to illustrate the fact that as it is photon-based RA may be used in a non-vacuum environment. Figure 9.9 plots RA against time transients recorded from an (001)InP surface mounted in a metal-organic vapour phase epitaxy (MOVPE)

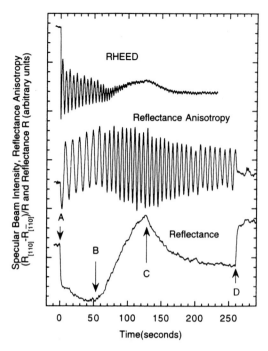

Figure 9.8. RA data (1.96 eV) recorded as a function of time during the gas-source MBE growth of Si and $Si_{1-x}Ge_x$ on (001)Si, showing the increase in bilayer oscillation that accompanies the introduction of germane into the gas flow [30]. The disilane was switched in at point A and out at point C. The germane was introduced at point B and switched out at point C. This figure also illustrates the change in amplitude of the RA oscillations that occurs under these conditions.

reactor [36]. The total pressure of the gas — mostly dihydrogen — in the reactor was 1000 mbar. Prior to commencing the RA experiment, the surface was prepared for growth using a procedure in which it was heated to 700°C in an atmosphere of dihydrogen and phosphine. This procedure is known to remove surface oxides and other possible contaminants while preserving the surface phosphorus levels which would otherwise be severely depleted by preferential phosphorus evaporation. Following cooling under H_2 and PH_3, at the times indicated the PH_3 flow was switched from the reactor and the In precursor, trimethylindium (TMIn), introduced. After some 100 s this procedure was reversed, completing the cycle.

While the primary aim of such experiments is to study the interaction of the growth precursors with the surface, for other systems this approach is often a useful means of growth. Known as atomic layer epitaxy (ALE) [37], this mode of growth has been used in the past for the deposition of ZnS–ZnSe, where the particular precursors employed were known to react readily in the gas phase above the surface, generating dust which severely degraded the quality of the growing layer. Clearly under ALE conditions a gas-phase reaction of this type is not possible. Figure 9.9 reveals that at low temperatures, although the gas switching results in an initial change in the RA, the surface does not respond to further changes in the composition of the reactor gases. In contrast, once the experiment is repeated at 325°C, a cycle is observed with one half responding to the TMIn introduction and the other half representing recovery under PH_3. Following this

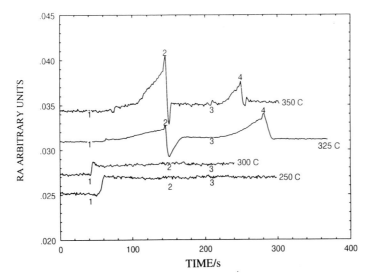

Figure 9.9. RA data (1.96 eV) recorded as a function of time from a (001)InP surface mounted in an MOVPE reactor during pseudo-atomic layer epitaxy using trimethylindium and phosphine [36]. The experiment was performed at temperatures in the range 250–350°C. The total pressure of the gas in the reactor, consisting mostly of dihydrogen, was 1000 mbar. At points 1 and 3 the phosphine flow was interrupted and the trimethylindium flow initiated. At points 2 and 4 the trimethylindium flow was interrupted and the phosphine flow reinstated.

cycle the RA response from the surface returns to the starting level. This then represents the use of RA to follow reaction dynamics at a clean (001) InP surface in real time and at atmospheric pressures. We have proposed the following explanation for the observations depicted in Fig. 9.9 which, although being consistent with established precursor chemistry, remain to be confirmed by the use of a more specific structural surface probe. At temperatures less than 325°C the TMIn interacts with the P-rich surface, disrupting the P–P dimer structure that is believed to exist (by analogy with other III–Vs) and generating the transient shown. However, the resulting surface is then unreactive towards PH$_3$, since the reaction of TMIn with the surface has resulted in a chemically unreactive layer. This somewhat obvious remark may be justified using a model in which the TMIn reacts with the P-rich surface, forming either one or two bonds with the surface, with the consequent loss of either one or two methyl fragments. Whichever of these two possibilities occurs, the In atom remains bound to at least one carbon atom. This effectively prevents formation of reactive In–In dimers and thus the surface becomes passivated towards further reaction. However, at 325°C the reaction of TMIn with the surface is able to proceed to completion and In–In dimers form which are in turn highly reactive towards the reintroduced PH$_3$ molecules. Thus the cycle is completed. Note that some caution needs to be employed in translating these data to a real continuous growth process, since it is known that although the RA may have revealed the temperature and other conditions (flow rates, reactor geometry, etc.) in which the chemical reactions leading to InP growth may occur, the material produced is quite unsuitable for device fabrication, due to the poor surface morphology which results. Thus here is an example of a process which may truly be termed heterogeneous catalysis: however, without the enhanced surface diffusion resulting from growth at much higher temperatures (typically 600°C or higher), the process is completely unsuitable for InP growth!

It is of interest to note that under normal MOVPE growth conditions, step-flow growth is again the predominant mechanism. As a result, it is not anticipated that RA will be able to monitor individual layer growth via the oscillatory responses discussed previously. However, there is enormous current interest in developing RA along with other techniques such as reflectometry, surface photoabsorption, laser light scattering and ellipsometry, in order to develop an on-line surface monitor capable of providing intelligent feedback control for MOVPE growth systems. Thus the use of RA may be envisaged in order to monitor the formation of new interfaces between materials, or simply to ensure that the growing surface maintains its required state. For example, we are currently performing measurements in which we initiate the growth of GaAs on (001)GaAs from trimethylgallium (TMGa) and tertiarybutylarsine (TBAs) and use the RA response at 1.96 eV to ensure that the growing surface maintains the required As-rich structure which is known to result in high-quality growth. Since we are confident that at this energy the RA is particularly sensitive to the presence of Ga–Ga dimers, we are using the RA signal to modify the flow rate of the TBAs. To see how this system may operate we can envisage a typical growth problem: during growth we begin to deplete the TBAs source to the point where the flow of TBAs to the reactor begins to fall. While such a change may also be monitored away from point of use by in-line sensors of various types, the RA response would tell us the precise instant at which the As/Ga surface stoichiometry begins to fall. This response could either interrupt growth to minimise layer damage or, via suitable calibration, be used to increase the flow rate of carrier gas through the TBAs container in order to compensate for the As depletion. Although these two courses of action may seem trivial, it should be noted that the cost to the semiconductor industry of failed growth processes is quite phenomenal.

3.4   *RA as a dynamic probe of adsorption and reconstruction on Cu(110)*

The final example of the use of RA to study surface dynamics concerns our work with the Cu(110) surface [32,38] which arguably is better understood in surface science terms than the semiconductor surfaces described in the previous examples. For this surface the natural choice of axes for the RA experiment to sample are the [100] and [1$\bar{1}$0] axes. However, by necessity these axes are already structurally and optically anisotropic, and thus an RA response would be anticipated even under the highly unrealistic ideal situation in which the optical components and windows along the path of the laser beam were completely free of birefringence. However, this essentially bulk anisotropy would not be expected to change during the course of a subtle surface modification and thus can be discriminated against by simply defining the 'baseline' for the RA measurement appropriately. Before commencing experimentation the Cu (110) surface was cleaned in ultrahigh vacuum using ion bombardment and annealing until the surface was 'Auger-clean' and gave rise to a sharp (1×1) LEED pattern. Figure 9.10(a) shows the result of exposing this surface to O_2 at 300 K in terms of the RA response recorded using the Pockels cell-based system operating at an incident energy of 1.96 eV (HeNe laser). Also shown on this figure is the O_2 exposure in langmuirs (where $1 L = 10^{-6}$ torr seconds). That a change in the nature of the surface

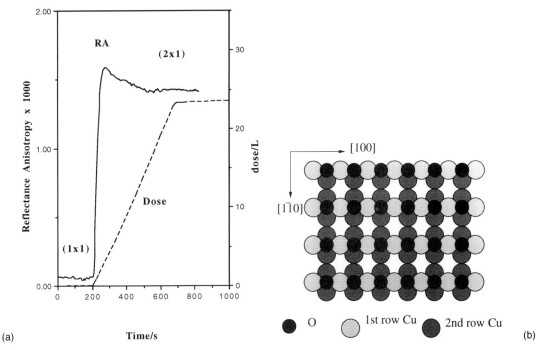

Figure 9.10. (a) RA data (1.96 eV) recorded from a clean, initially unreconstructed Cu(110) surface mounted in ultrahigh vacuum as a function of time during exposure to O_2 at 300 K. Also shown on the figure is the O_2 exposure in langmuirs [38]. (b) Schematic representation of the (1×2) reconstructed Cu(110) surface that formed following exposure of the initially unreconstructed surface to O_2 at 300 K [38].

electronic states occurs is typified in the first instance by the formation of a (1×2) reconstruction (Figure 9.10(b)). This surface was then further modified by exposure to formic acid at 300 K, which undergoes reactive adsorption to form a formate-type species on the surface together with surface oxygen atoms, while hydrogen and some oxygen atoms are lost in the form of water. Figure 9.11(a) again depicts RA/exposure data, while a schematic diagram of the resulting $p(3 \times 1)$ surface is shown in Figure 9.11(b). Finally, in a separate experiment the clean Cu(110) surface was exposed to benzoic acid at 300 K. Figure 9.12(a) depicts the variation in RA response as a function of exposure, while Fig. 9.12(b) depicts schematically a possible structure for the $c(8 \times 2)$ adlayer observed by LEED, which is consistent with the findings from other surface science techniques [39].

    Although the optical anisotropy of metal single crystals has been studied extensively by exponents of ellipsometry and related methods [40,41], the data presented here were the first examples of the application of the RA technique, as devised for the study of semiconductor surfaces, to a metal single-crystal surface. Very shortly after these measurements were made, other workers performed spectroscopic RA measurements on the Cu(110) surface, which revealed the presence of a sharp, optically active surface resonance near 2.0 eV, which could be radically altered by adsorption and desorption processes [42]. Since it appeared in an RA spectrum, it was clear that the resonance possessed an anisotropic dispersion. Examination of the literature revealed the origin of this transition via analysis of the dispersion of filled states by photoemission [43] and empty states by inverse photoemission [44]. Mapping of the Brillouin zone using these

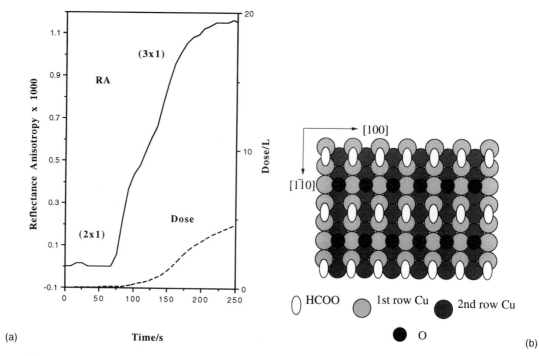

Figure 9.11. (a) RA data (1.96 eV) recorded from the initially (1 × 2) O–Cu(110) surface formed following the experiment described in Figure 9.10, as a function of time during exposure to HCOOH at 300 K. Also shown on the figure is the HCOOH exposure in langmuirs. [38]. (b) Schematic representation of the p(3 × 1) O–HCOO–Cu(110) surface that formed following exposure of the (1 × 2) O–Cu(110) surface described in Figure 9.10(b) to HCOOH at 300 K [38].

data revealed that a filled state-to-empty state direct ($\Delta k = 0$) transition existed close to the $\bar{Y}$ point along [100] but not along [1$\bar{1}$0], with the separation between these states being *c.* 2.0 eV. The anisotropy arose from the nature of the states concerned with the filled state possessing s-like symmetry and the empty state possessing p-like symmetry. This was then unequivocal evidence that the RA experiment was probing the relationship between surface states aligned in orthogonal directions. Thus we explain the RA responses observed in terms of a model in which the allowed transition measured for the clean Cu(110) surface is quenched to varying degrees by the various adlayers concerned and the reconstructions that result. However, there is some evidence in favour of the appearance of new states following the formation of these overlayers, and thus these may also contribute to the overall response.

In general it may be seen that for many metal single-crystal surfaces the data for the most part already exist from which the experimentalist can predict whether the RA experiment is viable and, if so, what axes to choose to sample and at what energy. For example, data for Ag(110) reveal the presence of a similar resonance to that observed for Cu(110) at approximately the same energy and dispersion within the surface Brillouin zone [45]. Thus it is predicted that RA will become increasingly applied to the study of metal single-crystal surfaces, particularly under conditions of high overpressures (e.g., for the study of certain catalytic reactions) and also for such surfaces immersed in liquids.

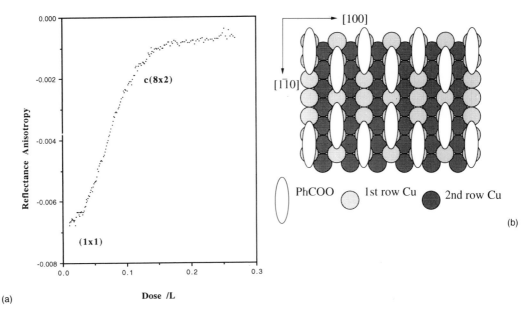

Figure 9.12. (a) RA data (1.96 eV) recorded from an initially unreconstructed, clean Cu(110) surface as a function of exposure in langmuirs to C_6H_5COOH at 300 K [38]. (b) Schematic representation of the c(8 × 2) C_6H_5COO–Cu (110) surface that is believed to form following the experiment described in Figure 9.11(a) [38].

4 Conclusions

The examples presented here reveal the power of RA for the direct monitoring of dynamic surface processes, often under conditions in which conventional surface science methods simply would not work. While for Cu(110) the origin of the RA response may be explained in terms of adsorbate-induced quenching of a well-defined optical transition, the lack of comparable data regarding the dispersion of filled and empty states on semiconductor surfaces that exhibit a number of complex, often transient, reconstructions is such that the direct interpretation of RA data is not straightforward. However, as has been shown here, it is not an absolute requirement to understand completely the true origins of the RA response when it may be readily correlated with other better defined parameters such as the appearance of broad resonances in the spectroscopic RA response that correlate well with the coverages of particular species (e.g. metal dimers on (001)GaAs or AlAs) or the proportions of orthogonal domains (e.g. (1 × 2)–(2 × 1) (001)Si). It is suggested that, by correlating spectroscopic RA data more thoroughly with local surface structure, perhaps via the combined use of RA and scanning probe microscopies, a better understanding of the nature of the optical anistropy of surfaces will be revealed, which will in turn result in a greater understanding of the data produced by RA systems.

5 Acknowledgements

The author gratefully acknowledges the opportunity to collaborate with the groups of Professor B.A. Joyce at the Semiconductor IRC, Imperial College of Science, Technology and Medicine, London, and Professor N.V. Richardson at the Surface Science IRC,

Liverpool. In addition, the particular contributions of Dr A.G. Taylor, Dr J. Zhang, Dr B. Frederick and Ms D. Raisbeck are gratefully acknowledged, along with financial support from EPSRC, the CEC via ESPRIT Project No. 6878 'EASI', Epioptics Applied to Semiconductor Interfaces, and Dr C. Pickering, DRA Malvern. Finally, the author would like to thank all members of the UMIST/Salford Solid State Chemistry Group for their invaluable contributions to the work described here.

6   References

1 McGilp JF. *J. Phys. Condensed Matter* 1990; **2**: 7985.
2 McGilp JF. *Progr. Surf. Sci.* 1995; **49**: 1.
3 Pemble ME, Armstrong SR, Curry SM *et al. Faraday Discuss.* 1993; **95**: 199.
4 Weightman P. *Phys. World* 1991; **4**: 39.
5 Aspnes DE. *J. Vac. Sci. Technol. B* 1985; **3**: 1498.
6 Aspnes DE. *IEEE J. Quantum Electronics* 1989; **25**: 1056.
7 Jonsson J, Deppert K, Paulson G, Samuelson L, Schmidt P. *Appl. Phys. Lett.* 1993; **63**: 3206.
8 Zettler JT, Rumberg J, Ploska K *et al. Physica Status Solidi* 1995; **152**: 35.
9 Kobayashi N, Horikoshi Y. *Jpn. J. Appl. Phys. 2* 1990; **29**: L702.
10 Pickering CE, Carline RT, Hope DAO, Robbins DJ. *Physica Status Solidi* 1995; **152**: 95.
11 Zahn DRT. *Physica Status Solidi* 1995; **152**: 179.
12 Patel H, Pemble ME. *J. Crystal Growth* 1992; **116**: 511.
13 Shen YR. *Nature* 1989; **337**: 519.
14 McGilp JF. *J. Phys. D, Appl. Phys.* 1996; **29**: 1812.
15 Bain CD. *Biosensors and Bioelectronics* 1995; **10**: 917.
16 Kamiya I, Aspnes DE, Tanaka H *et al. Phil. Trans. Roy. Soc. London A* 1993; **344**: 443.
17 Kamiya I, Tanaka H, Aspnes DE, Koza M, Bhat R. *Appl. Phys. Lett.* 1993; **63**: 3206.
18 Armstrong SR, Taylor AG, Pemble ME. *J. Phys., Condensed Matter* 1991; **3**: S85.
19 Richter W, *Phil. Trans. Roy. Soc. London A* 1993; **344**: 453.
20 Rossow U, Mantese L, Aspnes DE. *J. Vac. Sci. Technol. B* 1996; **14**: 3070.
21 Irvine SJC, Bajaj J, Sankur HO. *J. Crystal Growth* 1992; **124**: 654.
22 Armstrong SR, Hoare RD, Pemble ME *et al. Surf. Sci.* 1992; **274**: 263.
23 Larsen PK, Neave JH, Joyce BA. *J. Phys. C* 1981; **14**: 167.
24 Wassermeier M, Kamiya I, Aspnes DE, Florez LT, Harbison JP, Petroff PM. *J. Vac. Sci. Technol. B* 1991; **9**: 2263.
25 Chadi DJ. *J. Vac. Sci. Technol. A* 1987; **5**: 834.
26 Bottomley D. *Optics Lett.* 1996; **21**: 749.
27 Joyce BA, Zhang J, Neave JH *et al. Scanning Microscopy* 1994; **8**: 913.
28 Turner AR, Pemble ME, Fernandez JM, Joyce BA, Zhang J, Taylor AG. *Phys. Rev. Lett.* 1995; **74**: 3215.
29 Turner AR, Pemble ME, Fernandez JM, Joyce BA, Zhang J, Taylor AG. *J. Crystal Growth* 1995; **150**: 1015.
30 Zhang J, Lees AK, Taylor AG *et al. J. Crystal Growth* 1996; **164**: 40.
31 Pemble ME, Shukla N, Turner AR *et al. Physica Status Solidi* 1995; **152**: 61.
32 Zhang J, Taylor AG, Lees AK *et al. Phys. Rev. B, Condensed Matter* 1996; **53**: 10107.
33 Zhang J, Joyce BA, Fernandez VM, Taylor AG, Lees AK. Epioptic Workshop, Erice, June 1996.
34 See for example, Krishnamurthy M, Lorke A, Petroff PM. *Surf. Sci.* 1994; **304**: L493.
35 Cullis AG, Robbins DJ, Barnett SJ, Pidduck AJ. *J. Vac. Sci. Technol. A,* 1994; **12**: 1924.
36 Patrikarakos DG, Shukla N, Pemble ME. *J. Crystal Growth* 1997; **170**: 215.
37 Nishizawa J, Kurabayashi T, Abe H, Nozoe A. *Surf. Sci.* 1987; **185**: 249.
38 Pemble ME, Shukla N, Turner AR *et al. Faraday Trans.* 1995; **91**: 3627 (special surface science edition).

39 Frederick BG, Leibsle FM, Lee MB, Kitching KJ, Richardson NV (in preparation).

40 Blanchet GB, Estrup PJ, Stiles PJ. *Phys. Rev. Lett.* 1980; **44**: 171.

41 Mochan WL, Barbera PJ, Borenstein Y, Adjeddine A. *Physica A* 1994; **207**: 334.

42 Hoffmann P, Rose KC, Fernandez V, Bradshaw AM, Richter W. *Phys. Rev. Lett.* 1995; **75**: 2039.

43 Jacob W, Dose V, Kolac U, Fauster T, Goldmann A. *Z. Phys. B Condensed Matter* 1986; **63**: 459.

44 Dempsey DG, Kleinmann L. *Phys. Rev. B* 1977; **16**: 5356.

45 Urbach LE, Percival KL, Hicks JM, Plummer EW, Dai H-L. *Phys. Rev. B* 1992; **45**: 3769.

10 Plasma Chemistry at Polymer Surfaces

S.H. WHEALE and J.P.S. BADYAL

Department of Chemistry, Science Laboratories, University of Durham, Durham DH1 3LE, UK

1 Introduction

Since the early 1970s there has been an enormous expansion in the use of plasmas [1]. This technology is of vital importance to a number of the world's largest manufacturing industries — microelectronics, steel, aerospace, automotive, biomedical, toxic waste management, etc. [2]. However, the behaviour of these electrical discharges is poorly understood at the molecular level due to the inherent complexity of this form of matter. In order to develop commercially viable industrial processes, an understanding of the fundamental processes is imperative.

Non-equilibrium plasma treatment of polymeric materials can be used to improve the surface properties — optical reflection, adhesion, friction coefficient, surface energy, permeability, biocompatibility, etc. A systematic overview is presented of the current state of knowledge related to the interaction of non-isothermal plasmas with solid polymer surfaces.

2 Low-pressure non-equilibrium plasmas

The term *plasma* was first used in 1929 by Langmuir to describe ionised gases [3]. A plasma can be considered as a partially or fully ionised gaseous state of matter which contains atoms and/or molecules in ground and excited states, ions of either polarity, electrons and electromagnetic radiation [4]. All of these species can be regarded as potential reactants [5]. This quasi-neutral gas comprising charged and neutral particles usually behaves in a collective manner [6]. It is normally regarded as being an electrically conducting medium; although there is no overall charge imbalance, local perturbations from neutrality can occur.

2.1 Discharge theory

Plasmas can be produced using a variety of means: electric fields, heating, laser radiation and chemical processes. When a gas is subjected to an electric field, randomly occurring free electrons originating from cosmic rays or background radioactivity [7] become accelerated and undergo elastic and inelastic collisions; the latter cause ionisation of the gas, together with the formation of secondary electrons [4]. This process leads to a cascade effect and the production of ions, atoms, metastables, free radicals and electromagnetic radiation.

2.2 Equilibrium and non-equilibrium plasmas

Plasmas can exist in three forms: *complete thermodynamic equilibrium*, where the temperatures of all the species are equal (stars, explosions, etc.); *local thermodynamic equilibrium*, where everything except the photons is at the same temperature (electric arcs, plasma jets, etc.); and *non-equilibrium*, where the electron temperature (*c.* 10 000 K) far exceeds the temperature of the bulk gas (300–500 K, as in a glow discharge). Equilibrium plasmas are useful for inorganic chemical synthesis [8], nuclear fusion [4], metallurgy [4], etc.

Non-equilibrium plasmas open up non-thermally activated reaction pathways. In the absence of external magnetic fields, their degree of ionisation is low (10^{-5}–10^{-1}) so the gas consists of mainly neutrals at ambient temperature [9,10]. In this case, average electron energies can span 1–30 eV; however, only the electrons contained within the high-energy tail are capable of causing ionisation [11]. Non-equilibrium plasmas provide low-temperature processing environments under which thermodynamically unfavoured reactions are able to proceed. Such cold plasmas have found application in low-temperature materials processing — amorphous silicon deposition [12], plasma polymerisation [13], etching [14], restoration of archaeological artefacts [15], polymer surface modification [16], etc.

2.3 *Generation of low-pressure non-equilibrium plasmas*

Non-equilibrium plasmas can be generated at atmospheric and low (10^{-4}–10 mbar) pressures. In the case of low-pressure electrical discharges, there are three main components to a plasma processing system.

1 *Source of electrical power.* Electrical power frequencies spanning the d.c. to microwave range are used at power levels ranging from 1 to 5000 W.

2 *Coupling mechanism.* The electrical power source can be resistively, capacitively or inductively coupled to the gas under investigation. A matching network is usually necessary to ensure that there is efficient power dissipation.

3 *Plasma environment.* The reactor geometry and other variables, such as type of gas, pressure, flow rate, power level, processing time, must all be taken into consideration. Normally, the chamber is evacuated to a pressure well below its operating pressure, and then feed gas is introduced, followed by ignition of the electrical discharge.

Variation of the electrical discharge parameters (gas flow rate, gas pressure, reactor geometry, substrate temperature, frequency or intensity of power, etc.) can have a direct influence upon the plasma characteristics (electron density, electron energy distribution, gas density, residence time, etc.)

3 Fundamental processes

A variety of different reaction pathways are potentially viable at the surface of a polymer substrate immersed in a non-polymer forming electrical discharge (Fig. 10.1).

3.1 *Neutrals*

Neutral species in the form of atoms, radicals, molecules and metastables can all take part in reactions at the plasma–solid interface [5].

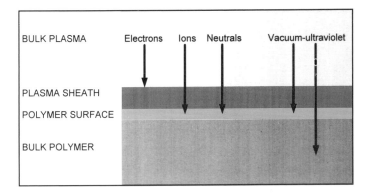

Figure 10.1. Physicochemical processes occurring at the plasma–polymer interface.

3.2 *Electrons*

During plasma modification, electrons can participate in electron capture processes at the surface [5].

3.3 *Ion bombardment*

Any surface in contact with a plasma gains a negative charge due to the constituent electrons being more mobile than the ions. This leads to the build-up of a space-charge layer at the plasma–solid interface, which is known as the *plasma sheath*. The resultant electric field repels electrons away from the surface and accelerates incident ions towards itself. For floating substrates and chemically reactive plasmas, ion bombardment does not play a significant role. However, in the case of noble gases, the absence of chemical interactions makes ion-induced surface modification a major reaction pathway [17].

3.4 *Electromagnetic radiation*

The main source of electromagnetic radiation within an electrical discharge is the relaxation of metastables [18]. Vacuum-ultraviolet (VUV) irradiation of the substrate can lead to the formation of free radicals, followed by their reaction with incident plasma species [19]. All organic polymers exhibit a strong absorbance below 160 nm [20,21], hence penetration of VUV radiation into the subsurface results in chain scission and cross-linking [22]. By placing a VUV transparent window between the electrical discharge and the polymer substrate, it is possible to block out all of the reactive species apart from the VUV component [23]; such experiments have demonstrated the important role played by VUV radiation during plasma modification. However, the extent of treatment is critically dependent upon the absorption characteristics of the gas [24].

4 Polymer behaviour

The individual constituents of a plasma (electrons, ions, neutrals and photons) are capable of interacting with an underlying polymer substrate in either an isolated or synergistic fashion to result in cleaning, etching, cross-linking, activation, and functionalisation [25]. Such changes can be used to alter the surface energy, polymer molecular weight distribution, chemical composition and topography at a polymer surface.

4.1   *Chain scission*

Chain scission can lead to the formation of chains of low molecular weight [26,27] which can be washed off with solvent [28]. If this material is incompatible with the substrate then it tends to form droplets on the polymer surface [29].

4.2   *Chain mobility*

The movement of chain segments at the surface of some polymers can lead to dynamic behaviour during or following plasma treatment [30]. This may result in migration of polymer chains into the subsurface [31,32], desorption [29,33], or solubilsation into an adjacent medium [31]. Polymer tacticity can also play an important role during plasma modification since it can influence chain mobility [34].

4.3   *Cross-linking*

Reactive and noble gas plasmas have been shown to be capable of inducing cross-linking down to depths of 3 µm as a result of direct and radiative transfer mechanisms [19]. The direct component consists mainly of ions and metastables interacting with the outermost layers, whereas radiative VUV radiation penetrates into the subsurface and bulk [22]. Typically, a lower level of cross-linking is found for chemically reactive plasmas than for inert gas plasmas [35]. The structural nature of the polymer under investigation can be an important factor — for instance, polyethylene tends to undergo cross-linking whereas polypropylene prefers chain scission [36]. Typically, a chemically modified top layer covers underlying cross-linked material; the density of the latter drops with increasing depths [35]. The cross-linked layer can limit polymer chain mobility and thereby provide stability to the overlying treated layer [35]; this can improve its heat resistance, frictional behaviour, cohesive strength, and form a diffusion barrier layer.

4.4   *Surface activation*

Polymer surface activation occurs as a result of the impingement of plasma species. Free radicals are created which can further react with incident plasma species; alternatively they can participate in chemical reactions upon termination of the plasma — surface grafting, cross-linking, oxidation upon exposure, etc.

4.5   *Synthon approach*

Structure–behaviour relationships can prove to be a reliable guide for predicting and designing interfacial behaviour. Clearly the type of gas and polymer employed during plasma modification governs what kind of treated surface is generated. For instance, in the case of oxygen plasma removal of polymer, strong linkages (e.g., aromatic and polar functional groups) inhibit etching, whereas weak bonds within the chain (e.g., C–C bonds) can enhance degradation [37]. This rationale can be extended to plasma modification of polymer surfaces, where the type of functionality generated at the substrate can be correlated to the structure of the parent polymer [38–42].

5   Characterisation methods

Surface analysis has been widely used to help understand the changes taking place at polymer surfaces during plasma modification. Many techniques have been employed, including X-ray and ultraviolet photoelectron spectroscopy (XPS and UPS), secondary ion mass spectrometry (SIMS), ion scattering spectroscopy (ISS), electron microscopy, scanning probe microscopies, contact angle measurements and infrared spectroscopy. Each technique has its inherent advantages and disadvantages, therefore it is vital that a multi-technique approach is used. The effectiveness of some of the more informative analytical techniques is outlined below.

5.1   *X-ray photoelectron spectroscopy*

X-ray photoelectron spectroscopy is a surface-sensitive analytical technique. XPS is based upon the photoejection of a single electron during X-ray irradiation of the substrate. By suitable selection of the X-ray source, photoelectrons can be generated which originate from less than 5 nm below the surface. Each atom in the periodic table has its own characteristic core energy levels, and hence identification of the elements present at the surface is possible (apart from hydrogen) [43]. Signal intensities provide information concerning elemental concentration [44]. For an individual atom, different chemical environments can give rise to a shift in the core-level binding energy. For instance, different types of oxidised carbon environment (e.g., alcohol versus ester) can be distinguished [38]. Various standard polymers and reference compounds are normally used to assist in the identification of unknown surface species created during plasma modification [45]. Sometimes it is not possible unambiguously to identify functional groups present at a plasma-treated polymer surface (e.g., alcohol versus ether); in this case, derivatisation of the surface using labelling reagents can help to resolve these issues [46,47].

Valence band XPS (XPS-VB) spectra provide a qualitative insight into changes occurring in molecular structure during plasma treatment (e.g., cross-linking [19,48, 49]). It can also be helpful for differentiating between groups with the same core level shift (e.g., phenyl versus methyl [45]). Another attribute of this technique is that chemical changes within specific functional groups along the polymer backbone can be followed [19,48,50].

5.2   *Near edge X-ray absorption fine-structure (NEXAFS) spectroscopy*

NEXAFS spectroscopy is used to provide molecular specific information by following the excitation of electrons from core levels into unoccupied molecular orbitals [51]. This technique has been successfully used to determine the orientation of chemical bonds at plasma-treated surfaces [52] and also to follow variations in conjugation [53]. The major drawback of this method is that interpretation is not always so straightforward.

5.3   Secondary ion mass spectrometry

SIMS uses mass spectrometry to study secondary ions emitted from a surface during bombardment by energetic primary particles [54]. A high primary particle beam current leads to *dynamic* SIMS, whereas a low beam current provides *static* SIMS (SSIMS). The former is destructive and often used for depth profiling, while SSIMS can provide structural chemical information and high surface sensitivity. SSIMS has been extensively used to study polymer surfaces in a qualitative manner [55]; it is molecular-specific, can easily differentiate between polymers [56], and is able to identify branching, or unsaturation [57]. The signal-to-noise ratio is high, which provides a sampling depth of one to two atomic layers.

5.4   Electron microscopy

Investigation of the physical changes imparted during plasma treatment has in the past been mainly restricted to scanning electron microscopy (SEM) studies. The intensity of low-energy secondary electrons emitted from a specimen surface during the rastering of the substrate by a concentrated beam of electrons forms the basis of SEM. This technique has shown that the physical nature of a polymer surface can undergo change during plasma treatment [58–60]. Surface roughening has been correlated to better adhesive behaviour as a result of mechanical interlocking [60]. However, in a number of studies, SEM has not been sufficiently sensitive to pick up structural rearrangements which may have taken place at the substrate surface [19]. Another major drawback of SEM is that it usually requires insulating samples to be coated with a conductive layer which can lead to the deformation of soft samples (e.g., polymers); this can also mask any plasma-induced surface modification. Furthermore, SEM probes the specimen with a high-energy electron beam, which can damage the polymer surface during analysis.

5.5   Atomic force microscopy

The relatively recent invention of atomic force microscopy (AFM) overcomes the above-mentioned limitations of SEM. AFM works by scanning a very sharp tip, attached to a lightly sprung cantilever, across the sample surface, while keeping the repulsive force between the probe and the surface constant. Nanometre resolution of non-conducting substrates can routinely be achieved using AFM, without the need for any additional sample preparation. A number of studies have shown that significant changes in surface morphology can occur during plasma treatment [50,61]. Some of the obtained structures are stable [74], while others can be washed off with solvent [27,62].

5.6   Contact angle measurements

The contact angle is taken as the angle between the solid surface and the liquid gas tangent [63]; it is dependent upon the chemical and topographical nature of the surface

[61]. If the surface is hydrophilic the contact angle of a water droplet will be small, while for a hydrophobic surface the angle is large. This technique is very surface-sensitive. The variation of contact angle with ageing time has been attributed to macromolecular rearrangement occurring at plasma-treated polymer surfaces; sometimes this can be misleading, since some modified layers dissolve into the liquid employed for the contact angle measurements [47,64–67].

5.7 Infrared spectroscopy

Infrared spectroscopy is normally not sensitive enough to detect chemical modification at the surfaces of plasma-treated polymer surfaces. However, extensive treatment can lead to changes in the characteristic vibrational features of the substrate [68].

6 Influence of gas

The chemical nature of the feed gas employed during plasma treatment can have a strong influence upon the surface modification of polymers.

6.1 Noble gases

Noble gas plasma treatment of polymer surfaces creates free-radical centres [69–71] which subsequently participate in hydrogen abstraction, cross-linking, or reaction with foreign molecules (air, polymerisable monomers, etc. [72]); see Tables 10.1–10.3. Modelling studies have shown that comparable levels of surface modification can be achieved using low-energy noble gas ion beams [17,73]; this is consistent with there being an absence of any chemical interactions between the glow discharge and the surface.

Table 10.1 Summary of helium plasma treatments.

Polymer	XPS	SEM	Contact angle	Adhesion	Other
Polyethylene	[77]	[77]	[77,87]	[87,90]	Friction: [77]
					Weight loss: [113]
Polypropylene	[50,114]	[113]			AFM: [50]
					Electrical conductivity: [50,114]
					Weight loss: [113]
Polystyrene	[77]	[77]	[77,87]	[87,90]	Friction: [77]
Polyethylene terephthalate			[87]	[87,90]	Weight loss: [113]
Polycarbonate			[87]	[87]	
Polymethylmethacrylate	[50]				AFM: [50]
					Electrical conductivity: [50]
Polysulfone	[75]		[87]	[87]	AFM: [75]
Polyethersulfone	[74]				AFM: [74]
Polytetrafluoroethylene	[96]				AFM: [96]
Polyvinylfluoride			[87]	[87]	
Polyvinylidene fluoride				[90]	
Nylon				[90]	Weight loss: [113]
Polyvinylchloride	[77]	[77]	[77]		Friction: [77]
Silicone rubber	[77]	[77]	[77]		Friction: [77]

Table 10.2 Summary of neon plasma treatments.

Polymer	XPS	Other
Polypropylene	[114]	Electrical conductivity: [114]
Polysulfone	[75]	AFM: [75]
Polyethersulfone	[74]	AFM: [74]
Polytetrafluoroethylene	[96]	AFM: [96]

Topographical changes also occur during noble gas plasma treatment; their physical appearance and stability strongly depend upon the nature of the polymer substrate under investigation and which inert gas is being used [74–76]. For instance, He plasma treatment of polyethersulfone generates cone-like features at the surface [74].

Inert gas plasma treatments can be used to lower the frictional resistance of a polymer substrate (this can be important for biological applications) [77]; however, subsequent biaxial orientation of such treated surfaces has been shown to result in the formation of ultra-fine protrusions, which impart good slip behaviour (this can be critical during film processing on an industrial scale) [78]. Also inert gas plasma modification can improve the electrical conductivity of a polymer surface [50].

6.2 Nitrogen

Nitrogen plasma treatment of polymers introduces mainly primary amine groups at the surface along with cross-linking [29,31,79,80]; however, the rate of surface reaction is found to be slower compared to corresponding oxygen plasma treatments [19]

Table 10.3 Summary of argon plasma treatments.

Polymer	XPS	FT-IR	SEM	Contact angle	Adhesion	Other
Polyethylene	[19,113,115]	[35]	[19]	[35,116]	[116]	DSC: [35] Elemental analysis: [35] XPS-VB: [19] XRD: [35]
Polypropylene	[50,73,114]					AFM: [50] Electrical conductivity: [50,114] XPS-VB: [73]
Polystyrene	[115]					ESR: [69]
Polyethylene terephthalate	[78,115]		[76]			SIMS: [78] TEM: [78]
Polymethylmethacrylate	[50]					AFM: [50] Electrical conductivity: [50] ESR: [70]
Polysulfone	[75]					AFM: [75]
Polyethersulfone	[74]					AFM: [74]
Polyimide	[107]	[107]	[107]			Electron temperature: [107]
Nylon	[115]					
Polytetrafluoroethylene	[71,72,96, 113,115, 117]	[117]	[72,117,118]			AFM: [96] ESR: [71] SIMS: [117]
Silicone rubber			[116]	[116]		

Table 10.4 Summary of nitrogen plasma treatments.

Polymer	XPS	FT-IR	SEM	Contact angle	Adhesion	Other
Polyethylene	[19,31,79, 113,119]		[19]	[119,120]		ESR: [113] Langmuir probe: [120] XPS-VB: [19]
Polypropylene	[29,73, 104,114]	[29]	[29,113]	[104]	[104,121]	AFM: [104,114] Electrical conductivity: [114] SIMS: [29,121] Weight loss: [113] XPS-VB: [73]
Polystyrene	[79]					
Polyethylene terephthalate	[80]	[61]	[61]	[61]		
Polymethylmethacrylate	[50]					AFM: [50] Electrical conductivity: [50]
PEEK	[33]					
Polyimide	[107]	[107]		[107]		Electron temperature: [107]
Polytetrafluoroethylene	[96,113, 122,123]			[118,123]		AFM: [96]

(Table 10.4). In the case of polypropylene, a preferential reaction of the syndiotactic phase is observed [29]. Amine functionalised polymer chains can possess a dynamic mobility which allows them to be transported into the subsurface with the aid of a wetting permeable solvent, and they can also subsequently be pulled back out towards the surface by using an aqueous protonation medium [31]. Some nitrogen plasma treated polymer surfaces slowly oxidise upon exposure to air to form transient oxygenated moieties which gradually disappear with storage time [33].

6.3 Hydrogen

Hydrogen plasma reduction of polytetrafluoroethylene (PTFE) surfaces results in the formation of a layer of defluorinated material 2 nm thick; this is accompanied by a corresponding drop in its contact angle with water, thereby making the substrate more suitable for bonding [81,82] (Table 10.5). The predominant reaction pathway is considered to be fluorine abstraction by atomic hydrogen from the PTFE surface to form HF (this is highly favourable due to its exothermic nature); the free radical centre left behind either undergoes reaction with subsequent incident hydrogen atoms or participates in cross-linking at the surface [83]. Similarly, other types of polymer surfaces are also able to undergo hydrogen plasma reduction [74].

6.4 Ammonia

Ammonia plasma treatment results in the incorporation of amino groups at polymer surfaces [84] (Table 10.6). These functionalities can give rise to an improvement in adhesion and blood compatibility of the substrate [60,84].

Table 10.5 Summary of hydrogen plasma treatments.

Polymer	XPS	FT-IR	Contact angle	Other
Polyethylene	[82,119]	[35]	[35,119,120]	DSC: [35]
				Elemental analysis: [35]
				ESR: [113]
				Langmuir probe: [120]
				XRD: [35]
Polypropylene	[50,114]			AFM: [50]
				Electrical conductivity: [50,114]
Polymethylmethacrylate	[50]			AFM: [50]
				Electrical conductivity: [50]
Polyethersulfone	[74]			AFM: [74]
Polysulfone	[75]			AFM: [75]
Polytetrafluoroethylene	[81,96,124,125]		[81,125]	AFM: [96]
Polyvinylidenefluoride	[124]			

6.5   *Oxidation*

Plasma oxidation can be used to remove contaminants present on a polymer surface [85]. It can also lead to oxygen incorporation (various groups can be formed, including alcohols, ethers, esters and acids), which in turn gives rise to improved bondability of the substrate [86–88] or a change in its dielectric performance [89]. Care needs to be exercised, since an over-treatment of the polymer substrate can produce a weak boundary layer as a result of extensive chain scission, which can have a detrimental effect on the adhesive performance of the treated surface [90].

6.5.1   OXYGEN PLASMAS

Oxygen plasma treatment of polymer surfaces comprises degradation of the substrate and reaction with ions, atoms, ozone, metastables of atomic and molecular oxygen, electrons, and a broad electromagnetic spectrum (Table 10.7). In the case of floating low-pressure radio frequency (RF) oxygen plasmas, the concentration of oxygen atoms is of the order of 1% (trace amounts of water can increase this value [91]), the ozone concentration is less than 0.02%, the electron density is close to 4×10^{14} m^{-3}, with an electron temperature of approximately 4 eV, the thermal energy of the neutral oxygen

Table 10.6 Summary of ammonia plasma treatments.

Polymer	XPS	FT-IR	SEM	Contact angle	Adhesion	Other
Polyethylene	[60]	[60]			[60]	
Polypropylene						Radioactive labelling: [84]
Polyethylene terephthalate		[61]	[61]	[61]		
Polyvinylchloride						Radioactive labelling: [84]
Polytetrafluoroethylene	[68,118, 126,127]	[68,118]	[68,118]	[68,118, 126,127]	[68,118]	Radioactive labelling: [84]
Polycarbonate						Radioactive labelling: [84]
Polymethylmethacrylate						Radioactive labelling: [84]

Table 10.7 Summary of oxygen plasma treatments.

Polymer	XPS	FT-IR	SEM	Contact angle	Adhesion	Other
Polyethylene	[19,36,39, 48,64,82, 86,92,95, 106,119, 128–132]	[59,133]	[19,59, 128]	[35,36,64, 87,94,116, 119,120, 133]	[36,64,86, 87,90,94, 116]	DSC: [35] Elemental analysis: [35] GPC: [59] Langmuir probe: [120] Mass spectrometry: [59] SIMS: [36,95] XPS-VB: [19,48] XRD: [35]
Polypropylene	[36,38,50, 52,59,64, 65,67,88, 114]	[59,133]	[59]	[36,64,65, 67,87,94, 133]	[36,64,87, 94]	AFM: [50,52] Electrical conductivity: [50,114] GPC: [59] Mass spectrometry: [59] NEXAFS: [52] SIMS: [36,65] XPS-VB: [52,88]
Polystyrene	[23,28,38, 39,41,48, 64–66,95]	[28]		[64–66, 87,94]	[64,87,90, 94]	Emission spectroscopy: [66] GPC: [66] SIMS: [65,95] XPS-VB: [99]
PEEK	[39,99,103, 129]					SIMS: [99]
Polymethylmethacrylate	[50,59,95]	[59]	[59]	[94]	[94]	AFM: [50]: Electrical conductivity: [50] GPC: [59] Mass spectrometry: [59] SIMS: [95]
Polyethylene terephthalate	[59,64]	[59,61]	[59,61]	[61,64,87, 94]	[64,87,90, 94]	GPC: [59] Mass spectrometry: [59]
Polycarbonate	[64]			[64,87,94]	[64,87,94]	
Polyimide	[59,89, 107,129]	[59,107]	[59]	[94,107, 129]	[94]	Electron temperature [107] GPC: [59] Mass spectrometry: [59]
Polycaprolactone		[133]		[133]		
Polytetrafluoroethylene	[58,59,96, 117,126, 134,135]	[59]	[58,59, 117]	[58,94, 117,126]	[94]	AFM: [96] GPC: [59] Mass spectrometry: [59] SIMS: [117]
Polyvinylfluoride	[59]	[59]	[59]	[94]	[90,94]	GPC: [59] Mass spectrometry: [59]
Polyvinylidenefluoride	[132]			[87]	[87,90]	
Nylon				[94]	[90,94]	
Polysulfone	[75]			[87]	[87]	AFM: [75]
Polyethersulfone	[74]					AFM: [74]
Silicone rubber				[116]	[116]	

atoms is 0.1 eV, and the emission spectrum is dominated by three intense atomic oxygen lines at 130.2 and 130.5 nm and 130.6 [92]. Electron impact dissociation of molecular oxygen is considered to be the predominant reaction pathway for the formation of atomic oxygen [93].

The activation energy for the reaction of atomic oxygen with a polymer substrate is significantly lower than the activation energy required for the diffusion of molecular oxygen into the bulk polymer [59], therefore treatment tends to be localised at the surface. Modelling studies have shown that the reaction probability of ground-state molecular oxygen with a polymer substrate in the presence of VUV radiation emitted by an oxygen plasma is low [24], hence it is the attack of atomic oxygen at VUV-activated surface sites which gives rise to oxygenation [24]. Surface modification quickly reaches a steady state in terms of chemical composition [38]; this can be attributed to an overall balance between oxygen incorporation and evolution of volatile reaction products (H_2O, CO, CO_2, oligomers, etc. [94]). For a fixed set of experimental parameters, the relative importance of these competing surface processes is dependent upon the type of polymer under investigation. All unstabilised polymers degrade upon exposure to an O_2 plasma, but the rates of oxidation are dependent upon their structure. Functional groups which readily react with oxygen without causing extensive chain scission (e.g., phenyl rings [41,95] help to generate a more oxidised surface (e.g., carboxylate groups [39,41]), while structural units which are highly susceptible to cleavage will encourage ablation, depolymerisation and the unveiling of fresh polymer [39,95]. This leads to a whole range of chemical functionalities being generated at a polymer surface during oxygen plasma treatment (oxidised groups, cross-linked centres, unsaturated bonds, etc.) [52]. Most polymers tend to experience an increase in oxygen content at the surface during oxygen plasma treatment; two notable exceptions are PTFE, for which there is no change (although there is significant surface roughening [96]), and polymethyl methacrylate (PMMA), which suffers a loss in surface oxygen content [97]. The plasma operating conditions can also influence the surface chemistry: for instance, oxygen incorporation is greater at low powers in the case of polyethylene, whereas polystyrene does not display this behaviour [48]. High power densities [59], long treatment times [58], and heating above the polymer glass transition temperature (T_g) [98] all lead to extensive roughening of the polymer substrate as a result of etching.

Most oxygen plasma-treated surfaces are found to undergo ageing effects, leading to hydrophobic recovery combined with a decrease in bondability [64]. Any oxidised material of low molecular weight present on plasma-treated polymer surfaces may be washed off with a solvent [99]. Such low molecular weight oxidised material can be formed via the attachment of atomic oxygen to free radical sites (these may have been created either by hydrogen abstraction or VUV activation); this is followed by chain scission [27,40]:

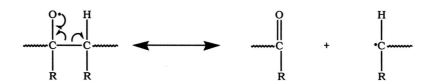

Where R = H or CH_3

Scheme 10.1 Generation of low molecular weight oxidised material.

Table 10.8 Summary of air plasma treatments.

Polymer	XPS	FT-IR	SEM	Contact angle	Adhesion	Other
Polyethylene	[60,136]	[60]			[60,136]	
Polypropylene	[104]			[104]	[104]	AFM: [104]
Polyethylene terephthalate	[136]	[61]	[61]	[61]	[136]	
PEEK	[103]					
Polytetrafluoroethylene	[68,137]	[68]	[68]	[68,137]	[68,137]	

Extended ageing of oxygen plasma modified polyethylene results in almost a complete loss of oxygenated functionalities from the surface (probably due to desorption of low molecular weight oxidised material) to leave behind a highly cross-linked layer [48].

Metal-containing polymers (polysilanes, polysiloxanes, etc.) quickly form an etch-resistant inorganic oxide layer [37,100,101] during plasma oxidation. Ageing effects have also been observed for oxygen plasma treated polysiloxane films; this is primarily attributed to cracking of the overlayer oxide film as a consequence of its markedly different mechanical properties [37] and density [102] relative to the underlying polymer substrate.

6.5.2 AIR

Air and oxygen plasma treated polymer surfaces generally tend to have similar chemical compositions [103,104] but differing levels of cross-linking; this is to be expected on the basis of each plasma emitting its own unique radiative VUV component (Table 10.8). Air plasmas are found to take longer than oxygen plasmas to achieve the same level of surface modification.

6.5.3 WATER

Water plasma treatment is normally found to produce higher levels of oxygenation than corresponding oxygen plasma treatments [105] (Table 10.9).

6.5.4 CARBON DIOXIDE

CO_2 plasma treatment also leads to a greater level of oxygen incorporation than oxygen plasma treatment [106,107] (Table 10.10). The difference in molecular structure between O_2 and CO_2 feed gases is reflected in a much higher concentration of carbonate groups being generated at the surface for the latter. In the case of polypropylene, extensive breakdown of the polymer surface into low molecular weight oxidised oligomers is reported [26].

Table 10.9 Summary of water plasma treatments.

Polymer	XPS	FT-IR	Contact angle	Other
Polystyrene	[105]			
Polymethylmethacrylate	[138]	[138]		ISS: [138]
Polytetrafluorethylene	[127]	[127]		

Table 10.10 Summary of carbon dioxide plasma treatments.

Polymer	XPS	FT-IR	SEM	Contact angle	Other
Polyethylene	[47]	[47]		[47]	
Polypropylene		[26]	[26]		NMR: [26]
Polystyrene	[23]				
Polyethylene terephthalate		[61]	[61]	[61]	

6.5.5   OTHER GASES

Compared to oxygen plasma treatment, NO and NO_2 glow discharges also give rise to higher levels of surface oxidation, along with more hydrophilic surfaces [107].

6.6   *Fluorination*

Dissociation of CF_4 in an RF electrical discharge can be summarised as follows [108]:

$$CF_x \quad -e \rightarrow \qquad F + CF_{x-1}$$
(where $x = 4$ to 1)

Fluorine atoms are reported to be the most chemically reactive species during CF_4 plasma treatment of polymer surfaces [40], (Table 10.11). Ions, electrons and electronically excited species play a relatively minor role [109]. Permeation of atomic fluorine deep into the subsurface, followed by reaction with the substrate, is consistent with this viewpoint [109,110]. Extended Huckel molecular orbital calculations have been used to explain why fluorine atoms react via hydrogen abstraction in the case of saturated hydrocarbon polymers, while addition to double bonds is the preferred reaction

Table 10.11 Summary of fluorination (CF_4) plasma treatment.

Polymer	XPS	SEM	Contact angle	Other
Polyethylene	[40,42,110, 129*,139†]	[110]	[110]	Emission spectroscopy: [110, 139†] RBS: [110] SIMS: [110]
Polypropylene	[42,140]		[140]	
Polystyrene	[41,42†]			
Polyisoprene	[40,42]			
Polyethylene terephthalate	[32,42, 109,141]	[109]	[32,109]	
Polyetheretherketone (PEEK)	[42,103]			
Polycarbonate	[112]		[112]	Emission spectroscopy: [112] SIMS: [112]
Polyethersulfone	[74]			AFM: [74]
Polyimide	[40]			

* Based on F_2 plasma.
† Based on CF_4 and SF_6 plasmas.

pathway for polymers containing unsaturated centres, thereby yielding a greater level of surface fluorination for the latter [40,42].

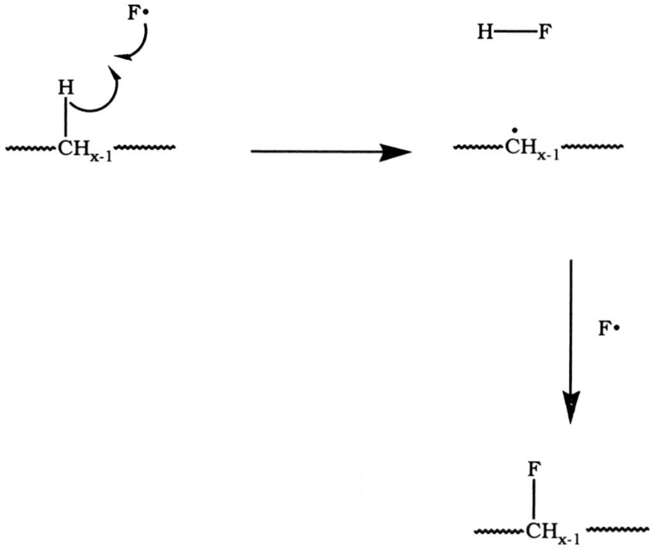

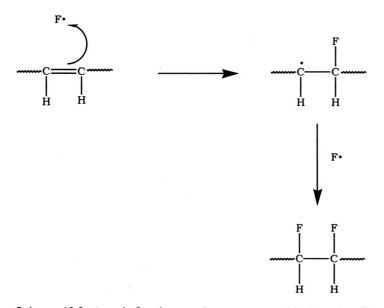

Scheme 10.2. Atomic fluorine attack at saturated hydrocarbon.

Scheme 10.3. Atomic fluorine attack on unsaturated hydrocarbon functionalities.

This structure–behaviour relationship has been utilised to introduce differing levels of fluorination along the length of a polymer chain consisting of both aliphatic and aromatic segments [111].

CF$_4$ glow discharge modification of polymer surfaces can be used to improve their water repellence [112]. In the case of polyethylene terephthalate and nylon, surface hydrophobicity introduced during CF$_4$ plasma treatment deteriorates upon immersion in water [32]; this behaviour has been attributed to the rotational and diffusional migration of fluorinated moieties into the polymer subsurface. A way round this is first to deposit a highly immobile cross-linked plasma polymer layer, which is subsequently fluorinated using a CF_4 plasma treatment [32].

7 Conclusions

Plasma modification of polymer surfaces can be used to improve their wettability, adhesion, permeability, gas barrier, electrical conductivity, abrasion resistance, biocompatibility, etc. A detailed understanding of structure–behaviour relationships at the molecular level allows the industrial end-user to optimise the macroscopic behaviour of such plasma-treated polymer surfaces.

8 References

1 Coburn JW. *IEEE Trans. Plasma Sci.* 1991; **19**: 6.
2 Boenig HV. *Fundamentals of Plasma Chemistry and Technology.* Lancaster: Technomic, 1988.
3 Langmuir I. *Phys. Rev.* 1929; **33**: 954.
4 Von Engel A. *Electric Plasmas: Their Nature and Uses.* London: Taylor & Francis, 1983.
5 Grill A. *Cold Plasma in Materials Fabrication: From Fundamentals to Applications.* New York: IEEE Press, 1994.
6 Chen FF. *Introduction to Plasma Physics.* London: Plenum Press, 1974.
7 Penning FM. *Electrical Discharges in Gases.* Holland: Philips Technical Library, 1957.
8 Baddour RF, Timmins RS (eds). *The Applications of Plasmas to Chemical Processing.* Cambridge, Mass.: MIT Press, 1967.
9 Chapman B. *Glow Discharge Processes.* New York: Wiley-Interscience, 1980.
10 Rossnagel SM. In Vosen JL, Wernen K (eds) *Thin Film Processes II.* London Academic Press, 1991; Chapter II-1.
11 Reif R, Kern W. In Vosen JL, Wernen K (eds) *Thin Film Processes II.* London: Academic Press, 1991; Chapter IV-1.
12 Veprek S, Sarrot F-A, Rambert S, Taglauer E. *J. Vac. Sci. Technol. A* 1989; **7**: 2614.
13 Biederman H, Osada Y. *Plasma Chem. Polym. Adv. Polymer Sci.* 1990; **95**: 57.
14 Flamm DL, Donnelly VM, Ibbotson DE. *J. Vac. Sci. Technol. B* 1983; **1**: 23.
15 Veprek S, Eckmann C, Elmer JT. *Plasma Chem. Plasma Processing* 1988; **8**: 445.
16 Liston EM. *J. Adhesion* 1989; **30**: 199.
17 Wells RK, Ryan ME, Badyal JPS. *J. Phys. Chem.* 1993; **97**: 12879.
18 Clark DT, Dilks A, Shuttleworth D. In Feast WJ, Clark DT (eds) *Polymer Surfaces.* Bath: Wiley, 1978; Chapter 9.
19 Gerenser LJ. *J. Adhesion Sci.* 1987; **1**: 303.
20 Partridge RH. *J. Chem. Phys.* 1966; **45**: 1685.
21 Partridge RH. *J. Chem. Phys.* 1968; **49**: 3656.
22 Clark DT, Dilks A. *J. Polym. Sci. A: Polym. Chem.* 1978; **16**: 911.
23 Shard AG, Badyal JPS. *J. Phys. Chem.* 1991; **95**: 9436.
24 Hopkins J, Wheale SH, Badyal JPS. *J. Phys. Chem.* 1996; **10**: 14062.
25 Mayoux C. *IEEE Trans. Dielectrics and Electrical Insulation* 1994; **1**: 785.
26 Ponicin-Epaillard F, Chevet B, Brosse J-C. *Eur. Polym. J.* 1990; **26**: 333.
27 Hopkins J, Boyd RD, Badyal JPS *J. Phys. Chem.* 1996; **100**: 6755.
28 Onyiriuka EC, Hersh LS, Hertl W. *J. Colloid and Interface Sci.* 1991; **144**: 98.
29 Poncin-Epaillard F, Chevet B, Brosse J-C. *Makromol. Chem.* 1991; **192**: 1589.
30 Andrade JD. *Polymer Surface Dynamics.* New York: Plenum, 1988.
31 Everhart DS, Reilley CN. *Surf. Interface Anal.* 1981; **3**: 126.

32 Yasuda T, Yoshida K, Okuno T. *J. Polym. Sci. B: Polym. Phys.* 1988; **26**: 2061.

33 Jama C, Dessaux O, Goudman P, Gengembre L, Grimblot J. *Surf. Interface Anal.* 1992; **18**: 751.

34 Hook TJ, Gardella JA, Salvati L. *J. Mater Res.* 1987; **2**: 117.

35 Yao Y, Liu X, Zhu Y. *J. Appl. Polym. Sci.* 1993; **48**: 57.

36 Morra M, Occhiello E, Gila L, Garbassi F. *J. Adhesion* 1990; **33**: 77.

37 Taylor GN, Wolf TM. *Polym. Eng. Sci.* 1980; **20**: 1087.

38 Clark DT, Dilks A. *J. Polym. Sci. A Polym. Chem.* 1979; **17**: 957.

39 Shard AG, Badyal JPS. *Macromolecules* 1992; **25**: 2053.

40 Cain SR, Egitto FD, Emmi F. *J. Vac. Sci. Technol. A* 1987; **5**: 1578.

41 Tepermeister I, Sawin HH. *J. Vac. Sci. Technol. A* 1992; **10**: 3149.

42 Hopkins J, Badyal JPS. *J. Phys. Chem.* 1995; **99**: 4261.

43 Wagner CD, Riggs WM, Davis LE, Moulder JF, Muilenberg GE. *Handbook of X-ray Photoelectron Spectroscopy*. Minnesota Perkin-Elmer Corp., 1979.

44 Briggs D, Seah MP (eds). *Practical Surface Analysis by Auger and X-ray Photoelectron Spectroscopy*. Chichester: Wiley, 1983.

45 Beamson G, Briggs D. *High Resolution XPS of Organic Polymers, The Scientific ESCA 300 Database*. Chichester: Wiley, 1992.

46 Tasker S, Backson SCE, Richards RW, Badyal JPS. *Polymer* 1994; **35**: 4717.

47 Dickie RA, Hammond JS, de Vries JE, Holubka JW. *Anal. Chem.* 1982; **54**: 2045.

48 Wells RK, Badyal JPS, Drummond IW, Robinson KS, Street FJ. *J. Adhesion Sci. Technol.* 1993; **7**: 1129.

49 Foerch R, Beamson G, Briggs D. *Surf. Interface Anal.* 1991; **17**: 842.

50 Collaud M, Groenig P, Nowak S, Schlapbach L. *J. Adhesion Sci.* 1994; **8**: 1115.

51 Stohr, J, Outka DA. *Phys. Rev. B* 1987; **36**: 7891.

52 Gross T, Lippitz A, Unger WES, Friedrich JF, Woll C. *Polymer* 1994; **35**: 5590.

53 Ryan ME, Hynes AM, Wheale SH, Hardacre C, Ormerod RM, Badyal JPS. *Chem. Mater.* 1996; **8**: 916.

54 Brown A, Vickerman JC. *Surf. Interface Anal.* 1986; **8**: 75.

55 Briggs D, Brown A, Vickerman JC. *Handbook of Static Secondary Ion Mass Spectrometry*. Chichester: Wiley, 1989.

56 Briggs D. *Surf. Interface Anal.* 1982; **4**: 151.

57 Briggs D. *Surf. Interface Anal.* 1990; **15**: 734.

58 Morra M, Occhiello E, Garbassi F. *Langmuir* 1989; **5**: 872.

59 Whitaker AF, Jang BZ. *J. Appl. Polym. Sci.* 1993; **48**: 1341.

60 Mercx FPM. *Polymer* 1994; **35**: 2098.

61 Wrobel AM, Kryszewski M, Rakowski W, Okoniewski M, Kubacki Z. *Polymer* 1978; **19**: 908.

62 Strobel M, Duntatov C, Strobel JM, Lyons CS, Perron SJ, Morgen MC. *J. Adhesion Sci. Technol.* 1989; **3**: 321.

63 Comyn J. *Int. J. Adhesion Adhesives* 1992; **12**: 145.

64 Morra M, Occhiello E, Garbassi F. In Mittal KL (ed.) *Metallised Plastics 2*. New York: Plenum Press, 1991; 363.

65 Occhiello E, Morra M, Garbassi F, Johnson D, Humphrey P. *Appl. Surf. Sci.* 1991; **47**: 235.

66 Occhiello E, Morra M, Cinquina P, Garbassi F. *Polymer* 1992; **33**: 3007.

67 Garbassi F, Morra M, Occhiello E, Barino L, Scordamaglia R. *Surf. Interface Anal.* 1989; **14**: 585.

68 Collins GCS, Lowe AC, Nicholas D. *Eur. J. Polym. Sci.* 1973; **9**: 1173.

69 Kuzuya M, Noguchi A, Ito H, Kondo S-I, Noda N. *J. Polym. Sci. A: Polym. Chem.* 1991; **29**: 1.

70 Kuzuya M, Noguchi A, Ishikawa M *et al.* *J. Phys. Chem.* 1991; **95**: 2398.

71 Momose Y, Ogino TM, Okazaki S, Hirayama M. *J. Vac. Sci. Technol. A* 1992; **10**: 229.

72 Tan KI, Woon LL, Wong HK, Kang ET, Neoh KG. *Macromolecules* 1993; **26**: 2832.

73 Collaud M, Nowak S, Kuttel OM, Groning P, Schalpbach L. *Appl. Surf. Sci.* 1993; **72**: 19.

74 Hopkins J, Badyal JPS. *Macromolecules* 1994; **27**: 5498.

75 Hopkins J, Badyal JPS. *J. Polym. Sci. A: Polym. Chem.* 1996; **34**: 1385.

76 Hseih Y-L, Timm DA, Wu M. *J. Appl. Polym. Sci.* 1989; **38**: 1719.

77 Triolo P, Andrade JD. *J. Biomed. Mater. Res.* 1983; **17**: 129.

78 Nakayama Y, Soeda F, Ishitani A, Ikegami T. *Polym. Eng. Sci.* 1991; **31**: 812.

79 Foerch R, McIntyre NS, Sodhi RNS, Hunter DH. *J. Appl. Polym. Sci.* 1990; **40**: 1903.

80 Mutel B, Dessaux O, Goudman P, Gengenbre L, Grimblot J. *Surf. Interface Anal.* 1993; **20**: 283.

81 Yamada Y, Yamada T, Tasaka S, Inagaki N. *Macromolecules* 1996; **29**: 4331.

82 Clark DT, Wilson R. *J. Polym. Sci. A: Polym. Chem.* 1987; **25**: 2643.

83 Gallaher TN, DeVore TC, Carter III RO, Anderson C. *Appl. Spectrosc.* 1990; **34**: 408.

84 Hollahan JR, Stafford BB, Falb RD, Payne ST. *J. Appl. Polym. Sci.* 1969; **13**: 807.

85 Ladizesky NH, Ward IM. *J. Mater. Sci.* 1983; **18**: 533.

86 Plawky U, Londshien M, Michaeli W. *Acta Polymer.* 1996; **47**: 112.

87 Wade, Jr. WL, Mammone RJ, Binder M. *J. Appl. Polym. Sci.* 1991; **43**: 1589.

88 Burkstrand JM. *J. Vac. Sci. Technol.* 1978; **15**: 223.

89 Wu S-H, Denton DD, De Souza-Machado R. *J. Vac. Sci. Technol. A* 1993; **11**: 291.

90 Hall JR, Westerdahl CAL, Bodnar MJ, Levi DW. *J. Appl. Polym. Sci.* 1972; **16**: 1463.

91 Herron JT, Schiff HI. *Can. J. Chem.* 1958; **36**: 1159.

92 Whitaker AF, Jang BZ. *SAMPE J.* 1994; **30**: 30.

93 Friedel P, Gourier S. *J. Phys. Chem. Solids* 1983; **44**: 353.

94 Hansen RH, Pascale JV, De Benidictis T, Rentzepis PM. *J. Polym. Sci. A: Polym. Chem.* 1963; **3**: 2205.

95 Lianos L, Parrat D, Hoc TQ, Duc TM. *J. Vac. Sci. Technol. A* 1994; **12**: 2491.

96 Ryan ME, Badyal JPS. *Macromolecules* 1995; **28**: 1377.

97 Groning P, Collaud M, Dietler G, Schlapbach L. *J. Appl. Phys.* 1994; **76**: 887.

98 Joubert O, Paniez P, Pons M, Pelletier J. *J. Appl. Phys.* 1991; **70**: 977.

99 Pawson DJ, Ameen AP, Short RD, Denison P, Jones FR. *Surf. Interface Anal.* 1991; **18**: 13.

100 Jurgensen CW, Rammelsberg A. *J. Vac. Sci. Technol. A* 1989; **7**: 3317.

101 Fonseca JLC, Barker CP, Badyal JPS. *Macromolecules* 1995; **28**: 6112.

102 Packirsamy S, Schwam D, Litt MH. *J. Mater. Sci.* 1995; **30**: 308.

103 Onyiriuka EC. *J. Vac. Sci. Technol. A* 1993; **11**: 2941.

104 O'Kell SO, Henshaw T, Farrow G, Aindow M, Jones C. *Surf. Interface Anal.* 1995; **23**: 319.

105 Evans JF, Gibson JH, Moulder JF, Hammond JS, Goretzki H. *Fresenius Z. Anal. Chem.* 1984; **319**: 841.

106 Shard AG, Badyal JPS. *Polym. Commun.* 1991; **32**: 217.

107 Inagaki N, Tasaka S, Hibi K. *J. Polym. Sci. A: Polym. Chem.* 1992; **30**: 1425.

108 Truesdale EA, Smolinsky G. *J. Appl. Phys.* 1979; **50**: 6594.

109 Yasuda T, Okuno T, Miyama M, Yasuda Y. *J. Polym. Sci. A: Polym. Chem.* 1994; **32**: 1829.

110 Khairallah Y, Khonsari-Arefi F, Amouroux J. *Thin Solid Films* 1994; **241**: 295.

111 Emmi F, Egitto FD, Matienzo LJ. *J. Vac. Sci. Technol. A* 1991; **9**: 786.

112 Occhiello E, Morra M, Garbassi F. *Angew. Makromol. Chem.* 1989; **173**: 183.

113 Yasuda H. *J. Macromol. Sci. Chem. A* 1976; **10**: 383.

114 Collaud Couen M, Groenig P, Dielter G, Schalpbach L. *J. Appl. Phys.* 1995; **77**: 5695.

115 Yasuda H, Marsh HC, Brandt S, Reilley CN. *J. Polym. Sci. A: Polym. Chem.* 1977; **15**: 991.

116 Sowell RR, DeLollis NJ, Gregory HJ, Montoya O. *J. Adhesion* 1972; **4**: 15.

117 Morra M, Occhiello E, Garbassi F. *Surf. Interface Anal.* 1990; **16**: 412.

118 Fessehaie MG, McClain S, Barton CL, Swei GS, Suib SL. *Langmuir* 1993; **9**: 3077.

119 Hollander A, Behnisch J, Zimmermann H. *J. Appl. Polym. Sci.* 1993; **49**: 1857.

120 Behnisch J, Hollander A, Zimmermann H. *J. Appl. Polym. Sci.* 1993; **49**: 117.

121 Arefi F, Andre V, Montazer-Rahmati P, Amouroux J. *Pure Appl. Chem.* 1992; **64**: 715.

122 Golub MA, Lopata ES, Finney LS. *Langmuir* 1993; **9**: 2240.

123 Kusabiraki M. *Jpn. J. Appl. Phys.* 1990; **29**: 2809.

124 Clark DT, Hutton DR. *J. Polym. Sci. A: Polym. Chem.* 1987; **25**: 2643.

125 Yamada Y, Yamada T, Tasaka S, Inagaki N. *Macromolecules* 1996; **29**: 4331.

126 Badey JP, Urbaczewski-Espuche E, Sage D, Duc TM, Chabert B. *Polymer* 1994; **35**: 2472.

127 Xie X, Gengenbach TR, Griesser HJ. In Mittal KL (ed.) *Contact Angle, Wettability and Adhesion.* Utrecht, Netherlands: VSP, 1993; 509.

128 Foerch R, McIntyre NS, Hunter DH. *J. Polym. Sci. A: Polym. Chem.* 1990; **28**: 193.

129 Emmi F, Egitto FD, Matienzo LJ. *J. Vac. Sci. Technol. A* 1991; **9**: 786.

130 Clark DT, Dilks A. *J. Polym. Sci. A: Polym. Chem.* 1979; **17**: 957.

131 Munro HS, Beer H. *Polym. Commun.* 1986; **27**: 79.

132 Golub MA, Cormia RD. *Polymer* 1989; **30**: 1576.

133 Normand F, Granier A, Leprince P, Marec J, Shi MK, Clouet F. *Plasma Chem. Processing* 1995; **15**: 173.

134 Wydeven T, Golub MA, Lerner NR. *J. Appl. Polym. Sci.* 1989; **37**: 3343.

135 Golub MA, Wydeven T, Cormia RD. *Polymer* 1989; **30**: 1571.

136 Saphieha S, Cerny J, Klemberg-Sapieha JE, Martinu L. *J. Adhesion* 1993; **42**: 91.

137 Kasemura T, Ozawa S, Hattori K. *J. Adhesion* 1990; **33**: 33.

138 Hook TJ, Gardella Jr JA, Salvati LS. *J. Mater. Res.* 1987; **2**: 117.

139 Khairallah Y, Khonsari-Arefi F, Amouroux J. *Pure Appl. Chem.* 1994; **66**: 1353.

140 Strobel M, Corn S, Lyons CS, Korba GA. *J. Polym. Sci. A: Polym. Chem.* 1985; **23**: 1125.

141 Krentsel E, Fusselman S, Yasuda H, Yasuda T, Miyama M. *J. Polym. Sci. A: Polym. Chem.* 1994; **32**: 1839.

11 Metal–Organic Chemical Vapour Deposition for the Preparation of New Materials and Interfaces

J.O. WILLIAMS, N. MAUNG and A.C. WRIGHT

Advanced Materials Research Laboratory, Multidisciplinary Research and Innovation Centre, North East Wales Institute, Plas Coch, Mold Road, Wrexham LL11 2AW, UK

1 Introduction

Metal–organic chemical vapour deposition (MOCVD) — also called metal–organic vapour phase epitaxy (MOVPE) when there is an explicit epitaxial relationship between the substrate and the thin film — has since its discovery by Manesevit and co-workers [1] in the mid-1960s become an accepted technique for the preparation of high-quality (in terms of structure, composition and purity) materials. The technique has been reviewed on a number of occasions, particularly from the point of view of theory and experimental details [2,3], the preparation of high-specification electronic device structures based on III–V and II–VI binary, ternary and quaternary structures [4], and precursor chemistry [5,6]. The approach taken in this review is, however, different from that taken by other reviewers in that we recognise that the wide range of materials prepared by MOCVD have large surface-to-volume ratios and constitute a range of fascinating interfacial structures prepared either accidentally or by design. In the majority of cases, structure and compositions have been evaluated by post-growth techniques such as X-ray diffraction [7], transmission electron microscopy [8], electrical measurements [9] and photoluminescence [10]. However, recently and in selected cases, *in situ* optical monitoring techniques are increasingly being used to probe the growing surface. We summarise reflectance anisotropy spectroscopy (RAS) work on the ZnSe–GaAs system in Section 5 and the use of the technique to understand the structure of the growing surface. Such an understanding will allow MOCVD to be used for the preparation of materials and structures that may lead to new technologies, particularly at the nanometric (quantum) level, and capitalising on particular surface-sensitive properties (dimensionality). We do not cover in any detail the field of conventional III–V materials, which has become by far the main area for application of MOCVD [4], but prefer to include a treatment of metals (Section 2), oxides and superconductors (Section 3), wide-bandgap nitride semiconductors (Section 4), wide-bandgap II–VI compound semiconductors (Section 5) and novel III–VI compound semiconductors (Section 6).

2 Metals

Although the preparation of metals using organometallic precursors is sometimes termed chemical vapour deposition (CVD), there are sufficient similarities between this and MOCVD that it can be included in this review. We take the example of metallic copper and evaluate the various surfaces and interfaces that can be prepared when Cu(II) precursors are reduced by dihydrogen under MOCVD-like conditions. From the device point of view, copper has distinct advantages over W and Al metal layers, having

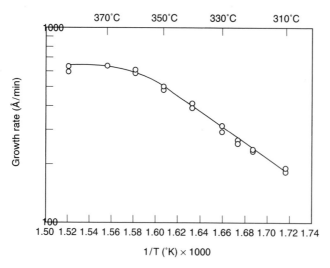

Figure 11.1. Plot of growth rates against reciprocal temperature for CVD copper films deposited at 310–385°C, 0.042 Cu(hfac)$_2$ mole fraction, and 10 torr total H$_2$ pressure, From [14].

lower resistivity than W and improved electromigration resistance over Al and also an increased resistance to stress-induced voltage, believed to result from copper's higher melting point than Al. Copper is, therefore, the preferred choice for metallisation in many cases of very large-scale integration (VLSI) technology. For such applications the crystalline form and interfacial properties are of crucial importance. The most common precursor for copper is Cu(hfac)$_2$ (hfac = hexafluoroacetonyl acetate) [11], closely followed by (hfac)Cu(L), where L is an appropriate neutral ligand. Plots of growth rate against reciprocal temperature (see Fig. 11.1) are reminiscent of those found in more traditional MOCVD, reflecting a surface-controlled process at lower temperatures changing to a diffusion-controlled process at higher temperatures. The advantage of using H$_2$ as a carrier gas in controlling the copper deposition rate is clearly shown in Fig. 11.2 [12]. Temperature and carrier gas concentration are the major factors that

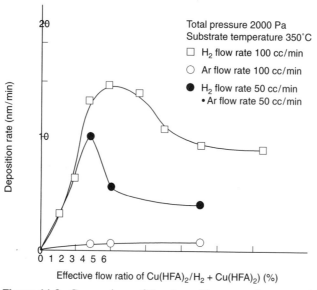

Figure 11.2. Comparison of the deposition rate of copper with various carrier gases. From [14].

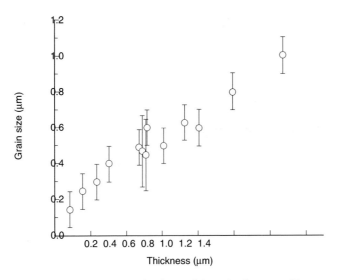

Figure 11.3. Average grain sizes of deposited copper films as a function of film thickness. From [14].

control film purity as measured using Auger electron spectroscopy (AES). Substrate selectivity has been reported for copper [13]. Deposition has been shown to occur selectively on several metals and at least one metal silicide, but not on SiO_2 or Si_3N_4. This gives rise to selective deposition and is important in device manufacture. Grain size increases linearly with increasing film thickness (see Fig. 11.3) and can be correlated with the growth rate. Furthermore, films grown under the reduced-pressure conditions frequently used to induce surface-reaction limited conditions are smooth and the crystallites well connected [14]. Thinner films are believed to be discontinuous based on resistivity measurements.

3 Oxides and superconductors

In 1986 Bednorz and Müller [15] discovered that partial substitution of lanthanum in La_2CuO_4 by divalent alkaline earth elements resulted in the preparation of superconducting material with a T_c of about 30 K. The metal substitution introduces an excess of holes into the host matrix, which subsequently act as charge carriers (p-type superconductor). Since then, various other copper oxide based materials have been investigated which show the onset of superconductivity at high temperatures ($10 \leqslant T_c \leqslant 134$ K). The most popular of these has been $YBa_2Cu_3O_{7-x}$ (also known as YBCO and 123) which was found to have a T_c of about 90 K [16]. YBCO has been the most studied superconducting material because of its excellent properties: ease of preparation as a single phase, high critical temperature ($T_c = 95$ K), high critical current density ($J_c = 10^6$ A cm^{-2} at 77 K), and the absence of highly toxic substances in its preparation. All of the high-temperature superconductor (HTS) materials contain CuO_2 planes which play a crucial role in determining the electrical and superconducting properties. The various HTS compounds simply differ according to whether the CuO_2 sheets are separated by atoms in the rocksalt or fluorite structure, layers of metal atoms, or various combinations of all three [17,18]. The transport properties were also found to depend strongly on the microstructure of the film (defects, grain boundaries, etc.). Shortly after the discovery of HTS the preparation of and investigation of thin films started. The major motivations for this work were, first, the large number of possible applications for thin films of HTS materials such as superconducting quantum interference devices (SQUIDs), bolometers and microwave detectors; and second, the suggestion that the two-dimensional nature of these compounds and the two-dimensional character of thin films might enable fruitful investigations of the fundamental properties of HTS materials. Another reason became clear later, when it was found that the values of J_c in thin films were several orders of magnitude higher than in the bulk material. The reason for this is that the deposition of YBCO layers by MOCVD leads to the preparation of highly textured films, enhancing the values of J_c that are attainable. In 1988, the first films of YBCO were prepared by MOCVD [19]. Since then, thin films of most of the materials with high T_c values have been grown by this technique. Higher T_c values have been achieved with Bi- and Tl-based superconductors, but the coexistence of several phases makes both the development and application of these compounds more difficult. More recently, Hg-based and mercury cuprate superconductors have attracted the attention of the scientific community because of the potential performance advantages of these superconducting compounds [20].

HTS compounds present a particularly challenging set of properties that make the preparation of high-quality, reproducible thin films potentially problematic. They are very difficult to grow under even the most ideal circumstances. A major contributor to the problems that arise is the complex stoichiometry of the materials. All of the HTS compounds are composed of a minimum of four elements that must be deposited within close limits of the correct ratio in order to obtain optimised films. Most of the elements are highly reactive, leading to the possibility of unwanted chemical reactions that can lead to deviations from chemical stoichiometry and purity in the films. Since the crystal structure of the materials is highly anisotropic, as are the resultant electrical and superconducting properties, crystallographic alignment in the films is critical. These difficulties notwithstanding, most of the HTS materials with high T_c values have been grown by this technique [21–28] and CVD preparation of high-T_c superconducting films has been reported using thermal decomposition, plasma excitation, and photochemical excitation of organometallic compounds and metal β-diketonate complexes. Several problems remain, however, one of which is the inadequate purity and chemical stability of commercially available source materials.

The special features of MOCVD — high growth rates, large area deposition and coating of complex shapes — make it a technique of choice for industrial applications, and this technique must be regarded as a promising candidate for high-volume thin-film production. The quality of a YBCO thin film may be evaluated by the same two characteristic properties as are used for bulk superconducting materials: T_c and J_c. Matsuno et al. [31] have reported the highest values in a YBCO thin film prepared by MOCVD ($T_c = 92$ K and $J_c = 6.3 \times 10^6$ A cm^{-2} at 77 K). Surface resistance is another important characteristic of the films to be taken into account for microwave applications.

Substrates play a critical role in epitaxial growth of HTS, and the most important properties of substrates as regards MOCVD are: (i) minimal lattice mismatch between substrate and film in order to avoid mechanical stresses and defects at the interface, because the superconducting current is highly anisotropic and its maximum value is realised in the (a, b) plane, implying that the a- and b-axes of rhombohedral and orthorhombic substrates should be as close as possible to the values of 3.8 Å or 5.4 Å; (ii) chemical inertness to avoid a reaction between the substrate and film or interdiffusion; (iii) no phase transitions; (iv) good physical property match for thin film applications, e.g. similar thermal expansion coefficients, small dielectric constants and dielectric losses. Chemical compatibility is especially difficult to achieve for these materials since ideally there should be no chemical reaction between the film and the substrate.

The necessary matching of the film and substrate with respect to lattice parameters, atomic positions, crystallographic orientation, etc., are the critical issues as regards controlled epitaxial growth, although the precise nature of the device application has a bearing on the choice for any particular application. All superconducting devices based on Josephson junctions, hybrid superconductor–semiconductor devices such as the semiconductor-coupled Josephson junction, microwave devices such as filters and delay lines, as well as current-carrying applications, require slightly different optimum characteristics from the substrate. Perovskite-structure oxides ($SrTiO_3$, $LaGaO_3$, $NdGaO_3$, etc.), oxides with a crystal structure not based on the perovskite cell (MgO, sapphire (Al_2O_3), yttria-stabilised ZrO_2 (YSZ) and semiconductors (Si and GaAs) as well as metals (Ag, Au, Cu and Hastealloy (Ni–Cr–Mo alloy)) have all been used as substrate materials [32].

Since interdiffusion was found to be a severe problem for high-quality HTS films on various substrates, a number of different approaches have been tried, the deposition of a chemically stable buffer or barrier layer between the substrate and the HTS film being one of the most common. T_c and J_c values strongly depend on the average composition and microstructure of the film, which can be controlled by the experimental conditions. In accordance with data from the literature, Busch *et al.* [33] and Leskela *et al.* [29] suggest that the critical current density decreases with increasing film thickness (above 100 nm). However, the reason for this behaviour is not clear and may be correlated with the loss of epitaxy described above.

The effect of film composition on thin film properties such as T_c and J_c has been studied by many authors. Zhao and Norris [34] investigated the electric and magnetic properties as a function of Ba/Y and Cu/Y atomic ratio. In copper-rich films it was found that CuO precipitates were formed which could act as oxygen sources. Interestingly for Y-rich compositions, Y_2O_3 precipitates have been observed in YBCO films which could act as effective sites for magnetic flux pinning, enhancing the value of J_c in the presence of high magnetic fields.

There is increasing interest in the chemistry of perovskite-phase mixed-metal oxide materials with the general formula ABO_3, such as $PbTiO_3$ (PT), $PbZrO_3$ (PZ) and $PbZr_xTi_{1-x}O_3$ (PZT), because they crystallise in polar space groups, giving rise to a variety of interesting properties. These properties include ferroelectric, piezoelectric and pyroelectric behaviour, which can be utilised in applications such as optoelectronic sensors, light detectors and mechanical transducers. Because the physical properties associated with the perovskite phase are directly related to the intrinsic crystal chemistry of these materials, control over composition, homogeneity and crystallisation behaviour is extremely important. For many applications it is necessary to synthesise pure, homogeneous crystalline materials, with the correct metal-atom stoichiometry. In order for these materials to be incorporated into existing solid-state devices, e.g. as thin films, low crystallisation temperatures (below 400°C) are often desirable. However, a number of the technologically important ABO_3 phases are polymorphic and can crystallise with the non-polar pyrochlore structure, which does not exhibit ferroelectric, pyroelectric or piezoelectric properties. Crystallisation of the pyrochlore phase in Pb-containing ABO_3 materials, rather than the perovskite phase, is generally associated with Pb deficiency, which means that control of composition and stoichiometry is extremely important. Many applications of these ferroelectric ceramic materials in electronics require the use of relatively thin films [34], particularly applications including multi-layer ceramic capacitors, electro-optic devices including optical memories and various pyroelectric and piezoelectric transducers used in thermal detectors. Consequently, there has been much interest in the deposition of thin films of lead titanate, lead zirconate titanate and barium titanate, among others, by a variety of methods, including physical vapour deposition [35] (e.g., sputtering) and more recently MOCVD [36].

MOCVD studies on oxide materials are as illustrated by the ZrO_2 system [37]. One of the primary disadvantages of MOCVD in the growth of ternary and more complex oxides is that unless the metal-organic precursor materials are particularly well chosen, it may prove difficult to achieve the required cation stoichiometry in the deposited film. Thus it may be necessary to conduct extensive investigations into the deposition of simple binary oxides containing the cations of choice, before deposition of more complex materials can be optimised. The study of the deposition of ZrO_2 and TiO_2 is a first step along the route to the deposition of complex oxides such as $PbTiO_3$, $PbZrO_3$ and $Pb(Ti,Zr)O_3$, which are employed in a wide variety of electronic and electro-optic devices. While the deposition of TiO_2 by MOCVD has been well studied, there have been relatively few attempts to deposit ZrO_2 thin films via this route. Most reported studies have used β-diketonates as precursors, although the metal alkoxides zirconium tetraisopropoxide and zirconium tetra-*tert*-butoxide (ZTB) have also been used in a limited number of investigations. ZTB decomposes more readily than the iso-propoxide, and is therefore potentially better suited as a precursor for MOCVD growth.

Many novel variations of the basic CVD process have been developed to eliminate some of the practical problems encountered in using this technique for the fabrication of oxide films. Specific difficulties include the high toxicity of some precursors, the incorporation of impurities (e.g., carbon) into the film, the need to control the stoichiometry, and the need to reduce parasitic gas-phase and surface reactions which often occur at some stage in the process. Furthermore, the need for high temperatures during typical CVD processes can cause interdiffusion at the various interfaces, thus preventing the formation of sharp interfaces required in some devices.

Some of these problems may be alleviated if a single-source precursor could be developed with a sufficiently strong Zn–O bond to withstand the conditions necessary to remove the substituent groups on the Zn and O atoms. Although many zinc compounds form Lewis acid–base adducts with oxygen compounds, the strength of the resulting Zn–O dative covalent bond is quite weak and unsuitable for the growth of these materials.

4   Wide-bandgap nitride semiconductors

Materials based on gallium nitride and its alloys with aluminium and indium have been recently developed as epitaxial thin films by MOCVD to the point where short-wavelength (blue or green) optoelectronic devices are now commercially available [38,39]. These commercial devices are, at the time of writing, exclusively produced by MOCVD, with material made by molecular beam epitaxy (MBE) currently falling behind in terms of device quality. While only light-emitting diodes (LED) operating in the green to violet range are presently available, albeit with very high intensity, laser operation has now been demonstrated by the same company that manufactures the LEDs [40].

Other device applications for these materials include transistors for high-power, high-temperature operation [65], rugged field emission sources for flat panel displays [66] and ultraviolet detectors [68].

Until recently, the only commercially available blue LEDs were made from the indirect-gap material silicon carbide (SiC), which have poor (12 mCd) output compared with the GaN based devices (1000 mCd). Thus GaN-based nitrides are now serious competitors to the MBE-grown ZnSe-based II–VI materials currently being developed world-wide for use in blue laser diodes.

The GaN-based nitrides are all the more remarkable considering the very high (10^{10} cm^{-2}) defect densities known to exist in the commercial LEDs [41] as revealed by cross-sectional transmission electron microscopy (TEM) (Fig. 11.4). In conventional III–V optoelectronic materials based on GaAs, such defect densities would yield useless material, with defect concentrations of 10^4 cm^{-2} or less being the norm. The high levels of defects in the nitrides appear to result from the use of highly mismatched (14%) sapphire substrates. C-plane sapphire, having a surface mesh with a hexagonal arrangement of atoms, has been used in the commercial LEDs to produce GaN grown in the hexagonal wurtzite form, rather than the cubic sphalerite phase. A recent study

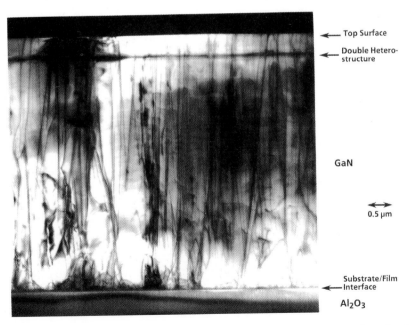

Figure 11.4. Dark-field two-beam TEM micrograph showing the defect distribution along a typical GaN commercial device. The dislocation density is in the range 2–10×10^{10} cm^{-2}. From [41].

indicates that, like the II—VI materials described in the following section, the low levels of additional elements introduced to control the electrical conductivity of GaN also affect to a small degree the defect density within the grown layer [77].

It should be pointed out at this stage that the bonding of both the cubic and hexagonal is tetrahedral and that the structural differences between the two only arise from the stacking sequence of the metal–nitrogen bilayers. The hexagonal structure arises from the stacking sequence ABABAB along the [0001] direction. The usual notation for this structure is 2H, which describes the double periodicity of the bilayers in the hexagonal form. The cubic form has a stacking sequence of the form ABCABC along the [111] direction and denoted in the short-form notation by 3C. A useful way of relating the two structures is to think of the hexagonal form as a twinned variant of the cubic phase with the twin planes being adjacent to each other. In the cubic form, one atom's bonds are rotated 60° with respect to its nearest neighbours. Across the 'twin' planes in the hexagonal form, the Ga–N bond does not rotate. Figure 11.5 shows the stacking sequences for the two polytypes of GaN.

The large mismatch leads to columnar growth, with the films containing low-angle grain boundaries leading to a mosaic structure. This can be seen in the plan-view transmission electron micrograph of Fig. 11.6. These boundaries arise from the small degree of rotation of the grains with respect to each other. However, parallel cleaved laser facets in the laser diode structures have been obtained by growth on the (1120) A-plane and exploiting cleavage of the (1102) r-plane [42]. The cleaved facets of these

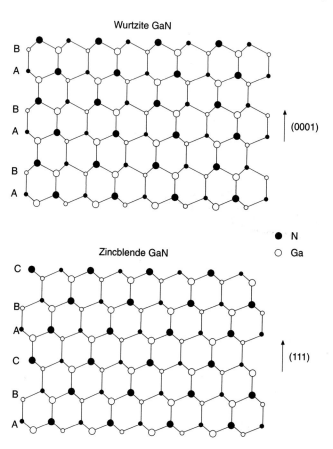

Figure 11.5. Crystal structures of the two common nitride polytypes showing the stacking sequence for the cubic (zincblende) and hexagonal (wurtzite) forms. From [182].

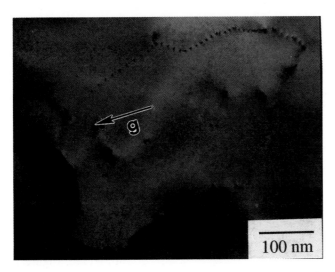

Figure 11.6. Plan-view bright-field TEM micrograph, g = (2110), taken near the top surface of a GaN film. The dislocation lines appear as short segments, with the majority of them forming well-defined low-angle grain boundaries. From [183].

nitride layers have been seen to be highly stepped due to the rotation of the layer over the substrate of several degrees arising from the large mismatch. Figure 11.7 shows a schematic drawing of the situation.

All nitride film growth uses a two-step process with an initial thin buffer layer grown at low temperature (c. 500°C) before growth of the main epilayers at a higher temperature (c. 1000–1100°C). The buffer layer is not amorphous as sometimes stated, but is actually nanocrystalline, as evidenced from high-resolution transmission electron microscopy (HRTEM). This initially formed layer then undergoes recrystallisation/reorientation by solid phase epitaxy and forms the columnar grains described above. The resulting island coalescence achieves defect-free regions separated by the low-angle grain boundaries. Thus the buffer layer effectively provides a large number of nucleation sites for growth of the main epilayer.

Further important aspects of III–V nitride layer epitaxy are those of polarity and interfacial steps. Like other tetrahedrally bonded compound semiconductors, GaN, both in its hexagonal and cubic forms, is polar. For example, when hexagonal GaN is grown on to basal (0001) plane sapphire, the first layer of atoms deposited could be

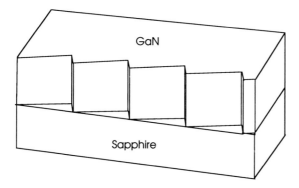

Figure 11.7. Schematic diagram of stepped cleavage of GaN thin film on a sapphire substrate arising from a rotation of a few degrees about the surface normal.

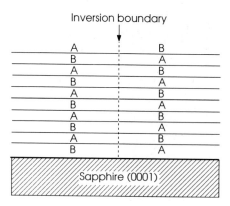

Figure 11.8. Schematic diagram of inversion boundary arising from GaN growth on sapphire basal plane. Letters A and B refer to the stacking sequence of bilayers seen in Figure 11.5.

either Ga or N. Whichever occurs will determine the orientation of the polar axis, ABABABAB or BABABABA, where A and B represent the stacking sequence of the bilayers growing normal to the substrate. If both types occur in different regions of the layer, then domain boundaries will arise where these regions join, as shown in the schematic diagram of Fig. 11.8. Such antiphase domain behaviour appears to have been observed in practice when hexagonal AlN is grown on to (0001)-oriented sapphire [63] (Fig. 11.9). This study reported the crystallography of AlN films on both the usual (0001) and also the (1012) r-plane of sapphire, and showed how the two structures were arranged at the plane of the interface. Figure 11.10 shows the idealised interfacial mesh for AlN on (0001) sapphire.

The polarity determines the nature of the exposed surface — either gallium — or nitrogen-terminated. The same type of argument applies for the cubic form when grown with the [111]-axis normal to any substrate, although the stacking sequence is of the ABCABC type. The polarity of the layers can be determined by the preference for a given bond, as shown when GaN is grown on to basal plane 2H SiC. The carbon face of the substrate prefers to bond directly to nitrogen rather than gallium but the silicon face will bond to gallium first, thus allowing selectivity of polarity. However, the carbon face of 2H SiC is known to be difficult to clean for MOCVD growth and although hydrogen plasma cleaning has been successfully applied to oxide removal from the silicon-terminated face without the need for *in situ* high-temperature reduction [56], this is not easily applied within the MOCVD environment.

Steps on the substrate surface can also introduce defects within the growing layer. If the step height is not the same as the interplanar spacings of the layer, then stacking fault related defects are expected to occur. This does in fact apply to the growth of hexagonal GaN on the basal plane of sapphire. The elimination of surface steps by careful substrate orientation and polishing should therefore have some effect on defect densities.

For growth of GaN or AlN on (0001) sapphire, the need for metal–oxygen bonding at the interface dictates a 30° in-plane rotation of the nitride lattice with respect to that of the sapphire [63] giving an orientation relation of:

$(0001)_{AlN} \parallel (0001)_{sapphire}; [01\overline{1}0]_{AlN} \parallel [1\overline{2}10]_{sapphire}; [1\overline{2}10]_{AlN} \parallel [1\overline{1}00]_{sapphire}$

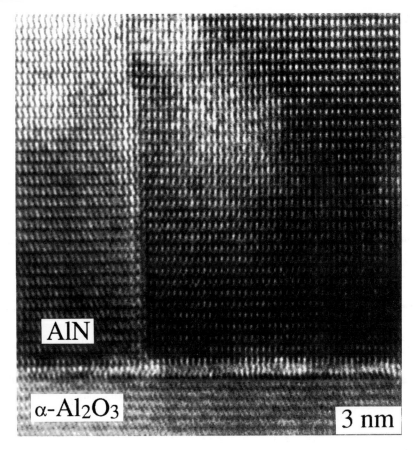

Figure 11.9. High-resolution lattice image of the AlN–sapphire(0001) interface, showing inversion boundary running normal to the plane of the interface. From [184].

For the case of growth on the r-plane of sapphire; the surface mesh of atoms is more complex but the orientation relation appears to be:

$(11\bar{2}0)_{AlN} \parallel (10\bar{1}2)_{sapphire}$; $[0001]_{AlN} \parallel [10\bar{1}1]_{sapphire}$; $[1\bar{1}00]_{AlN} \parallel [\bar{1}2\bar{1}0]_{sapphire}$

The interfacial mismatch varies from 2.9% in the $[0001]_{AlN}$ direction to 13.3% in the $[1\bar{1}00]_{AlN}$ direction. The crystal quality of AlN films grown on to the r-plane was

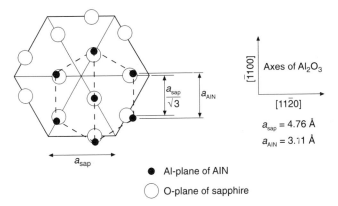

Figure 11.10. Schematic representation of an idealised in-plane atomic arrangement in the case of an (0001)AlN film grown on to (0001)sapphire. From [184].

observed to be inferior to those grown on to the basal plane.

Alternative substrates to sapphire are thus currently being sought. It has been shown that GaAs can be used with lower growth temperatures which would enable easy cleaving for laser diode fabrication. Other substrates tried include SiC (conductive but very expensive), ZnO (limited availability and poor temperature stability), spinel (non-conductive) and GaN. The latter would be ideal from the point of view of lattice match but for the difficulty in manufacture (currently made by a high-pressure process because of the high dissociation pressure of GaN). Homoepitaxial growth of GaN has been reported [62] with threading dislocation densities of 10^8 cm^{-2}. This study also revealed the presence of highly conductive gallium droplets within the substrate, making electrical measurements on the grown layer impossible. A further observation from this work is that the surface morphology and layer luminescence depended on whether the layer was grown on to the gallium side or the nitrogen side of the (hexagonal) substrates; smoother films with sharper emission lines being obtained from the nitrogen face of the substrate [67]. Although (111)-oriented spinel can produce material of good quality [61] and with a common cleavage plane, the demonstration of parallel cleaved facets when grown on to A-plane sapphire, coupled with sapphire's ready availability and cheapness, means that spinel has little to offer in this application.

Cubic substrates are better suited if the sphalerite form of GaN is required, but recent work using MBE as the growth technique has indicated that close control of the surface stoichiometry is also required — at least for MBE growth on GaAs(100) substrates — if pure sphalerite epitaxial films must be formed [44]. Under nitrogen-rich conditions, grains of hexagonal material start to be formed, and a roughening of the surface morphology occurs. In contrast, gallium-rich conditions lead to the formation of hexagonal material also containing crystallites of cubic GaN formed via nucleation within gallium droplets on the surface. Electron diffraction was used to monitor the surface reconstructions during these studies of MBE GaN growth, which showed that under a nitrogen-only flux the (001)GaN surface is unreconstructed while an impinging gallium-only flux produces a (2×2) reconstruction at 0.5 monolayer coverage and a centred c(2×2) reconstruction at 1.0 monolayer coverage.

The conclusions drawn from the MBE studies cannot of course be assumed to apply to the MOCVD growth of GaN-related material. However, it does seem to point towards the need for a similar *in situ* monitoring (and thus control) of the MOCVD growth of GaN. The surface structure-sensitive technique of RAS should therefore be of considerable use for accurate growth control; very little work has been done thus far on the surface structures and surface processes of MOCVD nitride growth.

The possibilities for further control over which of the two polytypes of GaN are formed are extended by changing the orientation of the substrate, as has been shown in the case of GaN on GaAs, where both (111)$_A$ and (111)$_B$ orientations were used to select, via the hexagonal symmetry of the (111)GaAs surface, the hexagonal form. The standard trimethylgallium (TMG)–ammonia recipe was used in this work [180] and with (111)$_A$ substrates a changeover from cubic to hexagonal form occurred at temperatures above around 800°C. Additionally, it was found that growth on (111)$_B$ resulted in poorer-quality films that separated from the substrate fairly easily. This temperature-mediated selection of the two forms opens up the possibility of bandgap engineering for carrier confinement, as the cubic form has a smaller bandgap than the hexagonal form. It was also found that the transition from cubic (grown first) to hexagonal occurred with a sharp interface, i.e. easily, but that the obverse experiment of growing cubic material on to a hexagonal layer took place more gradually and resulted in a highly faulted structure. The reasons for this difference are not well understood at present.

The MOCVD process used both by Nichia [43] and many other research workers makes use of TMG and ammonia (NH_3) as the principal source materials. These materials react very rapidly (0.2 s) to form an adduct even when cold, which appears to be detrimental to the growth process. This interaction is between a Lewis acid (TMG) and a Lewis base (ammonia) thus:

$$(CH_3)_3Ga + NH_3 \Leftrightarrow (CH_3)_3Ga{:}NH_3$$

Subsequent adduct-derived reactions have been suggested to lead to a gas-phase depletion of the growth nutrients, resulting in a degradation of growth uniformity and efficiency. Recent experimental work [76] with deuterated sources at elevated temperatures using *in situ* mass spectroscopy has indicated that fast elimination of methane follows on from the initial adduct formation:

$$(CH_3)_3Ga : NH_3 \Rightarrow (CH_3)_2Ga : NH_2 + CH_4$$

This product may undergo further reactions, including oligomerisation. These studies indicated that TMG no longer exists as a distinct chemical species within the reactor, and that the predominant species within the MOCVD environment is the methane-eliminated species $(Ga(CH_3)_2{:}NH_2)x$, with further reactions of this species in the gas phase through further decomposition or further oligomerisation leading to larger molecules and eventually particles. While the mass spectroscopy work was not able specifically to identify the mass of the oligomers, the most dominant entity in the MOVPE reactor environment was deemed to be $(Ga(CH_3)_2 : NH_2)x$ with $x = 3$, which has a ring-like structure.

As a result Nichia developed a novel cross-flow reactor which, although it is described as improving interactions of the precursors with the substrate, may also avoid mixing of these gases until the substrate is reached [43]. Other similar reactor designs have appeared in the literature [73], including a complex 'showerhead' design [75].

MOCVD of GaN uses quite high growth temperatures (1000–1100°C), which make stringent requirements on the stability of the substrates. Such high temperatures are required due to the inefficient cracking of ammonia at lower temperatures, although more recent work suggests that adequate growth at temperatures around 850°C is possible. The poor cracking efficiency of ammonia thus leads to very high V/III ratios of up to 20 000 : 1. Other areas of concern are the purity (dryness) of the ammonia currently available and its very corrosive nature.

The first step in GaN material growth appears to be the preparation of a low-temperature buffer layer of around 50 nm grown at 550°C. The structural quality of this layer, as evidenced by TEM, shows a polycrystalline structure which is subsequently recrystallised when the next higher-temperature growth step is performed. This is analogous to the growth of GaAs on Si and avoids the columnar growth typical when higher temperatures are used to initiate the GaN deposit.

Doped material is attained by using magnesium for p-type conduction and silicon for n-type. The magnesium source used is bis-cyclopentadienyl magnesium, while for silicon doping silane is used. The natural tendency to lose nitrogen from the lattice leads to n-type material, and p-type conduction can only be attained if the devices are given a post-growth anneal in an N_2 ambient. This latter step is necessary to allow hydrogen incorporated into the lattice to diffuse out of the material, which otherwise compensates for the magnesium. Annealing temperatures greater than 800°C are required to outdiffuse hydrogen completely from GaN and AlN, and temperatures above 600°C for InN films. Before the introduction of an anneal process step, low-energy electron irradiation was employed to activate the magnesium dopant; this is now unnecessary, although still in use [58]. Because of the relatively large binding energy of the magnesium acceptors, activation ratios of only 10^{-2} are generally achieved, thus to obtain high doping levels (10^{18} cm^{-3}) large amounts of magnesium must be incorporated (10^{20} cm^{-3}).

Gallium nitride films of very high resistivity (above 10^6 Ω cm) have been prepared by a plasma-assisted MOCVD approach [70]. A microwave-generated nitrogen plasma at a pressure of 25 Pa was used, together with a separate TMG–hydrogen flux, to grow films of GaN at 0.15–0.2 μm h^{-1} on (0001)-oriented sapphire. Secondary ion mass spectrometry (SIMS) measurement revealed very high levels (10^{19}–10^{20} cm^{-3}) of carbon and hydrogen (c. 1 : 1 ratio) in the films, and this was interpreted to form a C–H complex, with the carbon acting as a compensated p-type acceptor. Experiments to remove the incorporated hydrogen by thermal annealing did not alter the resistivity and thus material in this state may have potential for semi-insulating layers within GaN-based devices.

The tendency to lose nitrogen is greater in the indium-containing alloys and it has been found necessary to clad such layers (normally used to create a pseudomorphic well region) with thin aluminium gallium nitride to prevent dissociation of the indium-rich regions (or even indium droplets) during growth of the p-type region [40]. In addition, a much higher over-pressure of trimethylindium than of trimethylgallium seems to be required to achieve reasonable indium incorporation at the high temperatures required to obtain good luminescence. A thermodynamic analysis of the growth of $In_xGa_{1-x}N$ indicates that the indium content is controlled by competition between incorporation and desorption kinetics, with lower temperatures (c. 500°C) resulting in a linear relation between the gas-phase composition and the level of indium incorporated [71]. At higher temperature (800°C), this relation becomes very nonlinear and much higher indium gas-phase levels are required to obtain the same indium concentration within the layer. Higher growth rates also lead to greater incorporation of indium as the adsorbed species becomes grown over before it has time to desorb. Further studies have shown that a reduced reactor pressure also results in a greater level of indium incorporation and that the optimum temperature for InGaN alloys is considerably lower than that for GaN growth [72]: 700°C for InGaN and 1000–1050°C for GaN.

Low-temperature atomic layer epitaxy (ALE) was employed to reduce the relative flow rate of the indium source for a given indium incorporation [64]. Up to 27 atomic per cent of indium incorportion was reported. No hydrogen was present in the carrier, which was nitrogen. It was also determined that ALE enabled a lower temperature to be used for cracking ammonia than for conventional MOCVD. This effect was attributed to the ALE process facilitating a more active surface, one covered with Ga/In adatoms or their monomethyl-metal radicals. ALE should also permit a higher surface diffusion than MOCVD, thus enabling lower growth temperatures. Evidence for this view is supported by the low peak width (6 arcmin) in X-ray diffraction rocking curves obtained for the indium-rich alloy.

Other problems with the indium-containing active layers include mismatch with the cladding GaN material, which appears to be responsible for the much lower optical power output for devices in the orange regions where greater indium concentrations are used. This greater mismatch results in misfit dislocations which presumably are responsible for this power reduction, but further work is required to understand the detailed effects of defects on the optoelectronic properties, as it is generally assumed that dislocations are benign in the III–V nitrides. The first double heterostructure diodes used thick (100 nm) InGaN active layers, which dislocated due to both lattice and thermal mismatch. These appear to have been replaced by thin (2 nm) pseudomorphic InGaN layers to realise devices with output powers of over 4 mW in the blue at 20 mA drive current [59]. The microstructure of these devices and the quality of the interfaces are as yet unknown.

Material based on the analogous AlN is normally of very high resistivity, but recent work using MOCVD appears to show that when high partial pressures of propane are used (0.5×10^{-3} torr), large levels of carbon incorporation result and the resistivity of this p-type material drops to only $0.1 \, \Omega \, cm$ [45]. Activation energies of only 33 meV were measured with hole concentrations of the order of $10^{18} \, cm^{-3}$. This work appears to have been performed in a UHV-related apparatus rather than a conventional MOCVD system, however.

Aluminium nitride (AlN) is, however, sometimes used as the initial buffer layer prior to growth of GaN. This thin (30 nm) layer is grown on the sapphire substrate at low temperatures (500–600°C) prior to growth of GaN at 1000–1100°C, and enables better structural quality than by direct growth without any low-temperature buffer layer. Recently, the trend has been to use GaN instead as the low-temperature buffer layer which can also be induced by direct nitridation of the sapphire. This is achieved by a short exposure of the substrate to ammonia at 1050°C [46]. The length of the exposure has a direct effect on both the extent of the mosaic columnar structure and the density of threading dislocations within the GaN layer, with longer (300 s) exposures resulting in higher dislocation densities (in excess of $10^{10} \, cm^{-3}$), narrow X-ray line widths (002) and low electron mobilities. Conversely, short (60 s) exposures lead to low dislocation densities ($c. \, 70 \times 10^8 \, cm^{-3}$) and better photoluminescence quality.

Work on sapphire substrates has shown that the initially formed GaN buffer (or nucleation) layer grown directly (i.e. not via nitridation of the sapphire surface) at 600°C is in fact cubic not hexagonal, but that a transformation to the hexagonal form occurs on increasing the temperature to 1050°C [47]. This is also accompanied by grain growth of this initial nucleation layer from 33 nm to 77 nm sized islands but no appreciable decrease in the stacking fault density or mosaicity. Clearly, optimisation of the initial growth conditions for epitaxial GaN and its alloys is important, particularly with the highly mismatched sapphire substrates.

The two-step approach to nitride growth on sapphire has been studied in detail, where the main guide to structural quality was the surface morphology [60]. Both H_2 and N_2 carrier gases were investigated and, in summary, better (flatter, more mirror-like) layers were produced on (0001) sapphire at a high H_2 partial pressure (and a high total pressure), a high growth temperature and a low growth rate. These studies also showed that thermal dissociation of GaN at the high growth temperatures generally used is suppressed by the NH_3 in the growth ambient.

There is also current research to develop alternative precursor chemicals to the usual TMG and ammonia recipe. This appears to have focused on single-source precursors rather than alternatives to ammonia as a more easily cracked source of nitrogen. One idea is to replace the gallium–carbon bond with a gallium–nitrogen bond as in the azides. However, binary azides such as $[Ga(N_3)_3]_2$ are not volatile and are also explosive. By replacing one azide group by one NMe_2 group, a liquid and volatile source for gallium nitride is obtained [48]:

$$N_3\!-\!Ga\!-\!NMe_2 \quad \xrightarrow[T > 400°C]{\text{LPMOCVD}} \quad GaN + 5/2\,N_2 + \diagup\!\!\!\diagdown NMe_2$$

Further substitutions of the azide groups can be made and this may also be achieved with the aluminium and indium compounds. The growth temperatures used were typically 300°C (InN) and 500°C (GaN), which are much lower than with the TMG and ammonia route (1000–1100°C). No ammonia at all was used to deposit GaN, InN and AlN, which were of either a polycrystalline structure or highly preferred orientation.

Other alternative sources of nitrogen to ammonia which may enable lower growth temperatures include hydrazine, H_2N_2, which is more reactive [57]. However, hydrazine and its dimethyl derivative (1,1-dimethylhydrazine) are toxic and highly inflammable. Indeed, special precautions must be taken in the handling of these substances to exclude materials and impurities which can explosively decompose them. Nevertheless, good-quality GaN can be produced with hydrazine at growth temperatures of 900°C with far lower hydrazine partial pressures than would be required with ammonia. Another source of nitrogen with similar characteristics to hydrazine is hydrogen azide, HN_3. This material decomposes at around 300°C to form N_2 and a metastable HN (nitrene) species which is a ready source of active nitrogen. While hydrogen azide is also very toxic and potentially explosive (in the liquid state), good-quality epitaxial GaN has been produced [79].

A variety of amine-related compounds [74] has also been investigated in a plasma-assisted MOCVD reactor, which found that ammonia still produced the best GaN material from the quality of the photoluminescence. In general, alkylamines are poor sources of nitrogen for the growth of nitrides, due to their poor decomposition efficiency which is directly related to the strength of the C–N bond.

Other compounds with direct metal to nitrogen bonds are those such as hexakis(dimethylamido)dialuminium, $Al_2(NMe_2)_6$, namely:

When used together with ammonia, amorphous films of AlN were deposited at temperatures as low as 100°C with growth rates as high as 9 μm h^{-1} [49]. These films contained up to 30 atomic per cent of hydrogen, and films deposited at 300°C were polycrystalline, with still 26 atomic per cent hydrogen incorporated. Similar results were obtained with the gallium analogue. The key to understanding these particular types of precursor is that they react in the gas phase with the ammonia via a transamination reaction. When used without the ammonia, no deposition occurs until much higher temperatures are reached and the resultant films are easily wiped from the substrates. Large concentrations of carbon within the layers have been reported in the absence of ammonia [55] and the same investigators reported the use of plasma-assisted MOCVD of GaN and AlN. These precursors are not particularly volatile compared to TMG, being solids that sublime under reduced pressure (10^{-2} torr at 70–80°C). Other, related precursor types incorporating a direct gallium–nitrogen bond have been investigated [50]. At the time of writing, high-quality epitaxial material has yet to be demonstrated with any of the 'single'-source precursors.

A third route to the synthesis of III–V nitride layers is a hybrid of the organometallic and hydride approaches [51]. It is known that the hydride route can give high growth rates, and the basic idea is to form gaseous gallium chloride *in situ* by reaction of TMG and HCl prior to interaction with the substrate. The nitrogen source used is still ammonia, as with the usual route to GaN. The growth temperature used was 850°C, a considerable reduction on the usual 1000–1100°C, and to avoid the reaction of both ammonia and HCl with the GaAs substrates used, a thin (30 nm optimum) buffer layer of GaN was grown first at the reduced temperature of 500°C. The growth rates were 3 μm h^{-1} and with the GaAs substrates used it was claimed that the films were purely cubic in habit. Cross-sectional TEM of these films showed that the initial buffer layer was polycrystalline after growth at 500°C, but after ramping the temperature to 850°C this transformed to a highly faulted epitaxial film of generally cubic habit [52].

Direct use of the hydride/chloride approach has realised growth rates of the order of 30–70 μm h^{-1} using sapphire substrates with a ZnO buffer layer [53]. This raises the possibility of growing very thick cubic GaN films on GaAs which can then be etched off to provide free-standing GaN substrates for the MOCVD growth of nitride layers. Further investigations of the hydride route to GaN have been made using both (100) and (111) orientations of the GaAs substrate to select the cubic and hexagonal forms, respectively [54].

One final interesting precursor has been reported [78] in which both gallium and nitrogen are present in the correct stoichiometric ratio but without organic groups:

This precursor decomposes readily at 500–700°C in the absence of any carrier, to yield GaN, GaCl$_3$ and N$_2$. While GaN thin films on sapphire could be demonstrated at growth temperatures as low as 550°C, substantial amounts of incorporated chlorine were found with a very poor-quality layer. Growth temperatures of 650–700°C resulted in epitaxial material with growth rates of 7–350 nm min^{-1} with no detectable chlorine.

5   Wide-bandgap II–VI materials

At the time of writing, a serious competitor for the nitride alloys described above are those materials based on Zn–Cd–Mg–Se–S, better known as the II–VI compound semiconductors. The electronic bandgap of these materials varies from 4.4 eV for MgS to 1.7 eV for CdSe, and both LEDs and laser diodes operating in the green and blue-green part of the visible spectrum have been reported by many research groups. True blue (c. 470 nm wavelength) emission has only been achieved with pulsed operation thus far. Unlike the nitride alloys, however, II–VI alloys are not commercially available in these devices. The main reason for this appears to be the short lifetimes of the devices made so far. Blue-green laser diodes have been reported with lifetimes of 101 h [80], but this is far short of the lifetimes of 10 000 h typically required. The race to make bright blue LEDs commercially viable appears to have been won by the nitrides, although the outome for blue laser diodes still remains open.

The II–VI alloys have been fabricated both by MBE and MOCVD. A review of the physics and device science developed from the MBE grown materials has recently been published [81]. In contrast to the nitrides, the best material to date has been produced by MBE, and the reasons for this are discussed below, as are the short lifetimes of the devices. Much of the known data on wide-bandgap II–VI material growth presented below is thus from the MBE studies rather than MOCVD.

For laser diodes, two basic alloy structures have been used, and both of these incorporate quantum wells as part of the design [81]. The structures have usually been either wells of $Zn_{1-x}Cd_xSe$ with $ZnSe_{1-y}S_y$ barriers or a ZnSe well with $Zn_{1-x}Mg_xSe_{1-y}S_y$ barriers. Recently, the two seem to have been amalgamated, thus requiring a growth system with five sources plus two dopants, rather more than with the nitrides. A typical laser diode structure is shown in the schematic drawing of Fig. 11.11. The magnesium addition results in a device structure with better optical confinement (waveguiding) by means of its lower refractive index while still maintaining a lattice match with the substrate. This quaternary system has been reported to give rise to periodic composition variations due to the alloy clustering effects at the growing surface [82]. A device with a graded refractive index separate confinement heterostructure (GRINSCH) has been described by the Sony group [115] which gave superior turn-on characteristics. Such graded interfaces are more easily achieved via MOCVD than with MBE as gas flows are easier to modulate/regulate than via temperature changes in the effusion cells of an MBE system.

Almost all of the material grown to date, whether by MBE or MOCVD, has been on GaAs(100) substrates, and this brings us to the first of the material problems of II–VI alloys. There is clearly a need to lattice-match to the GaAs, and this can be attained with either $ZnSe_{1-y}S_y$ or, to attain greater optical and electron confinement, $Zn_{1-x}Mg_xSe_{1-y}S_y$. While this material may be lattice-matched and strain-free at the growth temperature, the differences in thermal expansion coefficient between the II–VIs and GaAs are sufficient to create strains in the layer at room temperature. High strain levels are also

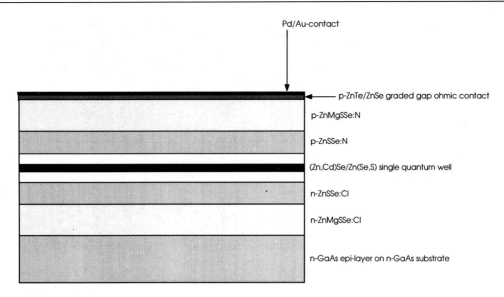

Pd/Au-contact

p-ZnTe/ZnSe graded gap ohmic contact

p-ZnMgSSe:N

p-ZnSSe:N

(Zn,Cd)Se/Zn(Se,S) single quantum well

n-ZnSSe:Cl

n-ZnMgSSe:Cl

n-GaAs epi-layer on n-GaAs substrate

Figure 11.11. Typical layer make-up of II–VI laser diode. The latest devices have material with a graded index surrounding the quantum well region.

created in the pseudomorphic quantum well regions because lattice matching is not possible here.

Like conventional III–V alloys, the luminescent output of II–VI materials is affected by the presence of extended defects such as threading dislocations and stacking faults, which appear to be very easy to introduce into the grown layer either during growth or during cool-down from growth temperature. The defect levels have generally been too high to enable long device lifetimes and, although the very latest laser diode devices from the Sony group appear to have very low initial dislocation levels (10^3 cm^{-2}), they still fail after relatively short times. Plan-view transmission electron micrographs of failed laser diodes correlated the non-emitting regions of the device with patches of dislocation networks originating from the quantum well [112]. These dislocation networks appeared to be nucleated at threading dislocations originating at stacking faults which were bounded by Frank (sessile) partial dislocations with Burger's vector $(a/2)\langle 110 \rangle$ lying in the (001) junction plane. These Frank partials have been, via *in situ* electron beam heating of TEM specimens, observed with Burger's vectors of $(a/3)\langle 111 \rangle$ subsequently dissociating into Shockley partial dislocations with $b = (a/6)\langle 211 \rangle$ and a 60° type perfect dislocation with $b = (a/2)\langle 011 \rangle$ [98]. Figure 11.12, taken from a failed device, shows a plan-view TEM image of a pair of triangular-shaped tracking faults that originate at the II–VI/III–V interface, with dislocation networks in the plane of the device quantum well.

It is possible that point defects, created by the intense optical and electrical power levels reached during laser action, agglomerate to form stacking faults which then act as sources for dislocation generation, multiplication and thus device failure. It is also known that non-radiative recombination transfers approximately 2.5 eV to the lattice [81], sufficient energy to power such defect reactions given that the energy to form a Schottky defect (a cation–anion vacancy pair) in CdS is 2.16 eV [100]. Recent work by the Sony group also suggests that point defects are now the limiting factor in device lifetime [115], although they conclude that the optical field intensity does not accelerate defect generation. This should be compared with the fact that in gallium nitride the

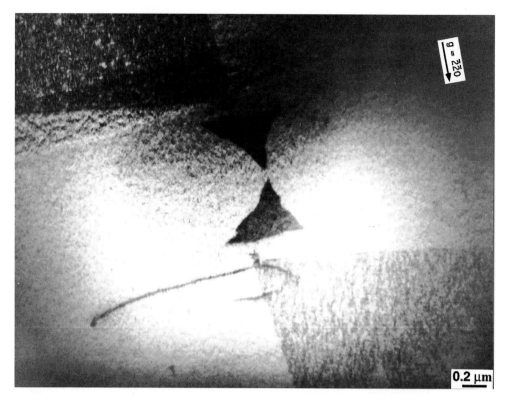

Figure 11.12. Plan-view TEM image showing two patches of dislocation networks which have nucleated at threading dislocations which in turn originate at a pair of triangular-shaped stacking faults.

bond energy is several times that of the bandgap and thus defect generation from photon absorption is not expected to occur.

A further feature of growing II–VI alloys on GaAs is both the valence mismatch at the interface and the tendency for an interfacial phase to form. This phase appears to be a III–VI compound, Ga_2Se_3, which, while also cubic with tetrahedral bonding, has one-third of the metal sites vacant. This phase is prone to form in the MOCVD growth of ZnSe when the substrate is exposed first to the group VI element, selenium, and thus can be prevented from forming by starting the growth cycle with zinc. The presence of this phase in MOCVD-grown ZnSe at the interface with GaAs(100) substrates has been characterised by TEM [96] using lattice imaging together with computer simulation on cross-sections. Figure 11.3 shows a cross-sectional TEM image of interfacial Ga_2Se_3. The presence of cubic gallium selenide at the interface has been attributed to the origin of the stacking faults seen in MBE-grown ZnSe [91,99] due to the known high levels of vacancies in this phase [137]. Such a concentrated source of vacancies could feasibly lead to the creation of intrinsic type faults and their growth by vacancy-assisted climb of the bounding partial dislocations. Interfacial cubic gallium selenide is covered in more detail in Section 6.

The influence of the precise initial growth conditions on the structure and optoelectronic properties of ZnSe on GaAs(100) have been extensively studied using MBE by a number of research groups in an attempt to reduce the defect density in the layers. In fact, far more surface science has been conducted in the II–VI materials than in the

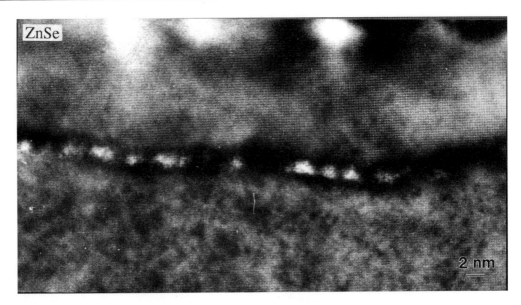

Figure 11.13. Cross-sectional TEM image of the ZnSe–GaAs interface, indicating the formation of Ga_2Se_3 (bright regions) at the interface. From [112].

nitrides. For MBE growth at least, the consensus is that two-dimensional growth from the beginning is required to attain low threading dislocation densities [84]. The 2D growth mode is obtained by first creating an arsenic-rich GaAs surface with the (2×4) reconstruction and then exposing this surface to a zinc flux for a period of 1 min [84]. The threading dislocation densities were then 6×10^6 cm^{-2}. The worst conditions of growth appear to be a gallium-stabilised GaAs(100) surface first exposed to a selenium flux which produced threading dislocation densities of 5×10^9 cm^{-2}, three orders of magnitude higher. Material grown with a zinc-initiated first step is believed to induce the formation of a zinc–arsenic compound at the interface, possibly Zn_3As_2, with the same tetrahedral structure as GaAs and ZnSe [91,99], although no firm evidence (in the form of modelling of TEM image contrast or high spatial resolution microanalysis) for such an interfacial phase has been presented thus far. An arsenic-stabilised GaAs surface simultaneously exposed to both zinc and selenium has a threading defect density of 3×10^8 cm^{-2}.

Photoluminescence of the layer produced under the Zn- and Se-initiated conditions showed that the Y^0 peak was absent in material grown under the Zn-initiated 2D growth mode [84]. The 3D growth conditions resulted in an extensive network of interfacial misfit dislocations of the perfect 60° type and strong Y^0 emission, while 2D growth conditions appear to induce the formation of 30° Shockley partials (at the interface) and attendant stacking faults rising up along {111} planes through the layer. It is also interesting to note that mismatch dislocations lying at the II–VI/III–V interface have also been cited as contributing to actual luminescence [83]. Thus the presence of the Y^0 emission appears strongly correlated with perfect dislocations lying at the plane of the ZnSe–GaAs interface. These results should be compared with those obtained by cathodoluminesence (CL) imaging in which the Y^0 peak was observed as broken lines of intensity aligned along the [110] direction for 2D growth and as random spots in 3D growth [181]. However, no TEM was performed to correlate these observations with the CL data. The stacking faults are generally aligned along [0$\bar{1}$1] rather than [011] directions, indicating that strain relief in the layer is anisotropic. The microstructure of the films prepared under both 2D and 3D growth conditions as seen in plan-view TEM is shown in Fig. 11.14. The stacking fault asymmetry has also been observed in MOCVD-grown ZnSe and discussed in terms of the differences in dislocation mobilities between the α and β dislocation types (either zinc or selenium termination at the dislocation core) [97]. The defect structures developed in ZnSe, by either MBE or MOCVD, also appear to be heavily influenced by doping [97,98]. p-type doping of MBE-grown ZnSe, using nitrogen, resulted in a greater overall defect density, while n-type doping of MOCVD ZnSe resulted in a reduction in stacking fault density [97]. Doping is expected to affect the mobilities of α and β types of dislocation differently and

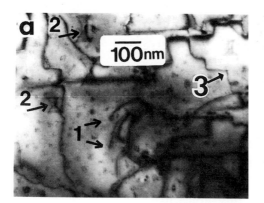
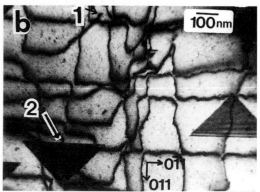

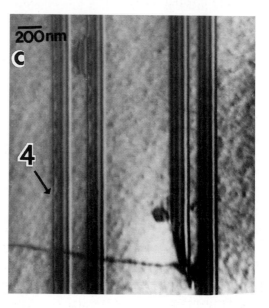

Figure 11.14. Plan-view TEM images of ZnSe growth under: (a) 3D conditions via selenium-exposed Ga-stabilised GaAs (4 × 1) surface; (b) intermediate 2D conditions via zinc- and selenium-exposed As-stabilised GaAs (2 × 1) surface; (c) 2D conditions via zinc-exposed As-stabilised GaAs (2 × 1) surface. Note development of stacking faults in full 2D growth mode. Arrows marked 1 indicate threading segments of dislocations, arrows marked 2 indicate triangular stacking faults, arrow marked 3 indicates a zig-zag misfit dislocation and arrow 4 indicates a Shockley partial misfit dislocation. From [84].

so changes in the defect structure of the layers should occur [97].

The reasons for this difference appear to be in the creation of faults as islands coalesce during 3D growth, the strained and thus dilated (the layer would normally be in compression) nature of the island edges somehow leading to fault formation as they grow outwards and meet. These islands have also be observed and studied by scanning tunnelling microscopy (STM) [85,86]. The selenium-exposed arsenic-rich GaAs(100) (2×4) surface adopts a selenium-terminated (2×1) reconstruction which is rather unreactive and thus ZnSe nucleation is difficult on this surface. Under these conditions, nucleation is seen to take place at the edges of steps rather than on the terraces. There is also a marked tendency for the ZnSe to grow as elongated islands aligned with the rows of selenium dimers. This STM study also investigated the smoothness of the GaAs(100) surface for both the oxide-desorbed and grown GaAs buffer layer cases. The oxide-desorbed surface was seen to be much rougher than when a GaAs buffer layer was grown and residual matter, possibly oxide, was also seen in the former case. It is worthwhile noting that passivating the GaAs(100) surface with group VI species such as sulfur via chemical pretreatments has become fashionable [87] and that in the light of these observations this may not now be an optimum way of preparing the GaAs(100) substrate for MOCVD. When the arsenic-stabilised (2×4) GaAs(100) was exposed simultaneously to Zn and Se, i.e. ZnSe growth from the beginning, a highly disordered but more uniformly covered surface resulted. The Zn-initiated growth was not investigated in the STM studies. Figure 11.15 shows STM images of the growth of ZnSe on GaAs(001) under the conditions described above.

The best structural quality appears to have been achieved when GaAs buffer layers have first been created. Defect densities as low as $10^4 \, \text{cm}^{-2}$ have been reported [88] when GaAs buffers with a zinc-initiated growth are employed. Even lower etch pit densities (EPDs) — low $10^3 \, \text{cm}^{-2}$ — have very recently been found to occur if an additional thin ZnSe (35 Å) migration-enhanced epitaxy (MEE) step follows the zinc exposure [113]. The MEE step is used to attain a better 2D growth mode and avoid any

(a)   (b)

Figure 11.15. Scanning tunnelling microscope images of: (a) discrete islands of ZnSe formed on GaAs(001)–Se (2×1) surface; (b) disordered 2D surface structure of ZnSe on GaAs(001) (2×4) surface. From [86].

tendency for 'islanding' to occur. While the preparation of GaAs buffers is possible in MBE, even in a single-chamber system, it is far more difficult to implement in MOCVD, as the addition of Ga and As sources to the reaction vessel would give rise to the risk of non-intentionally doping the ZnSe via reactor wall contamination.

Further work on the relationship between the structure of the initial surface prior to growth of the ZnSe layer and the defect density has been reported [185]. Following the creation of a GaAs epilayer with a (2×4) As-stabilised surface and zinc termination, thin ZnSe pseudomorphic films were subsequently grown with different Se/Zn flux ratios ranging from 0.3 to 10. A minimum Schottky stacking fault density of $10^4\,\text{cm}^{-2}$ was found at a flux ratio of 2. This flux ratio was observed by RHEED to yield a ZnSe surface reconstruction containing both (2×1) and $c(2 \times 2)$ corresponding to a surface stoichiometry with equal coverage for both Zn and Se. Growth under conditions other than this resulted in far higher stacking fault densities — in excess of $10^8\,\text{cm}^{-2}$ at Se/Zn flux ratios of 0.3. Flux ratios higher than 2 also resulted in a larger stacking fault density, but saturating at around 10^5–$10^6\,\text{cm}^{-2}$. Additionally, the stacking fault orientation was found to change from lying on $\{111\}_A$ planes under zinc-rich conditions to $\{111\}_B$ under Se-rich conditions. The core structure of the dislocations bounding these faults also changes from zinc-terminated to selenium-terminated, in line with the changes in fault orientation and thus surface stoichiometry. These results indicate that very close control of the surface stoichiometry is required for fault-free ZnSe growth. The asymmetric distribution of fault density with flux ratio can be explained in terms of the greater tendency for selenium to desorb than for zinc and of faults arising from stacking errors due to an excess of one species over another.

A way of attaining the required 1 : 1 Se/Zn surface stoichiometry has been shown [186] where RHEED was used to monitor the surface reconstructions on ZnSe. An ALE-like approach with a time delay between selenium and zinc exposure allowed an initially fully covered Se-terminated ZnSe surface with the (2×1) structure partially to desorb selenium under conditions in which no zinc flux was present. An optimum time delay of around 30 s was found to yield the lowest stacking fault density (and also the lowest Y^0 photoluminescence emission intensity). This work used ZnSe substrates, and the lowest stacking fault levels attained were limited by those present in the substrate itself, of $6 \times 10^4\,\text{cm}^{-2}$. While this work gave no real proof of the actual surface stoichiometry occurring, it gives additional credence to the premise that careful control of the surface stoichiometry is essential for ZnSe growth if defects are to be eliminated and that methods of attaining this in MOCVD growth will need to be developed.

An indication that an arsenic-stabilised GaAs surface in MOVPE could be attained in MOVPE reactors has been given by recent work in MBE which used an atomic hydrogen plasma source to clean the GaAs surface of oxide prior to growth [114]. While conventional preheating (580°C) of chemically cleaned and etched GaAs gave a Ga-rich (4×2) surface (which roughened if heating became too long), the atomic hydrogen cleaning step (360°C) resulted in an As-terminated (2×4) reconstruction. While etching of the (normally fused silica) reactor walls must be avoided, it may be possible to use radio frequency induction to induce a hydrogen plasma at reduced pressure to perform a similar oxide-removal step in MOVPE.

The single report in the literature on the effect of initial growth conditions for MOCVD-prepared ZnSe on defect densities has yielded results somewhat at variance with the MBE data [89]. This study used tertiarybutylarsine (TBAs) to pre-treat and obtain an As-rich GaAs(100) surface. In line with the MBE work previously reported, this surface was then exposed to either Zn and Se source flows prior to ZnSe growth. Briefly, the results found were that there was an optimum exposure of the GaAs to TBAs above and below which material with a higher EPD was formed if Zn and Se were then introduced simultaneously to this As-stabilised surface. A much higher EPD was found to occur if this surface was first exposed to zinc (c. 2×10^7 cm^{-2}). Further recent work using MBE has indicated that the GaAs(100) surface, stabilised with the c(4 × 4) As-rich reconstruction, induces both arsenic precipitates and very high Shockley partial dislocation densities [91]. The c(4 × 4) As-rich GaAs(100) surface has a double arsenic layer as opposed to the single arsenic layer of the (2 × 4) GaAs surface reconstruction, and this fact does correlate with the observation that an excess TBAs flow during the MOCVD pre-treatment of GaAs prior to ZnSe growth leads to higher defect densities. However, this is not consistent with the very low EPD results of the MBE work described above, when MEE was used, as the starting GaAs surface in that case had the c(4 × 4) stoichiometry. When the MEE approach was used with a (2 × 4) reconstructed GaAs surface somewhat higher (though still very low at 10^3 cm^{-2}) EPDs were found. Clearly, for MOCVD of ZnSe, the situation is more complex and will require further investigation. It is also now known, for instance, that exposure of GaAs(100) to a dimethylzinc flux in a hydrogen carrier will etch this substrate [90].

Very recent work has in fact begun to address the nature of the surface before and during growth of ZnSe of GaAs(100), using *in situ* optical reflectance techniques [92,93]. RAS measures the difference in reflectivity of light oriented in two orthogonal polarised states over a wide spectral range (200–800 nm) [94]. Light of energy above the material bandgap gives difference spectra that are surface-sensitive and serve as fingerprints to a particular surface structure. The anisotropy in reflectance arises from the fact that the surface atoms on (100) tend to form pairs or dimers which then order into arrays. Clearly, the reflective contribution from each dimer depends on its orientation with respect to the electric field vector for a given polarisation of incident light and the different ordering patterns (reconstructions) of these dimers appear to give characteristic spectra.

In principle, RAS spectra obtained from a surfaces with known surface reconstruction, i.e. obtained via RHEED under ultrahigh vacuum (UHV) conditions, should correspond directly to the same surface structure under MOCVD growth conditions where similar spectra are observed. However, other factors affect the RAS spectra, notably surface roughness [95], and a phenomenological model has been presented to interpret the reflectance anisotropy in terms of parameters related to surface roughness or anisotropy for InAs growth on GaAs. Thus there is no guarantee at present that similar spectra obtained from both UHV MBE growth and from MOCVD do in fact refer to the same surface structure. A recent study of the MOCVD growth, of ZnSe on GaAs(100) [92] using RAS showed that initial exposure of the surface to selenium resulted in an ever-increasing reflectance anisotropy during growth, and a phenomenological model for this has been put forward in terms of anisotropic surface roughness [93]. This surface anisotropy has then been directly observed by cross-sectional TEM [92], where a one-dimensional grooving of the ZnSe surface had taken place and this surface morphology developed very early on in the growth process. The grooving was seen to be highly faceted but not corresponding to exact {111} planes, thus showing that the sides of the grooves had a very high density of steps. Growth initiated with zinc was observed to develop a random roughness with no obvious anisotropy in either the RAS spectra or in TEM cross-sections.

Reflectance anisotropy has also been shown to arise from the buried ZnSe–GaAs interface, as well as from the free surface of the layer, and these two contributions can be analysed separately by processing the RAS spectra [187]. This MBE work indicated that the Se-terminated (2×1) structure persists in some form even after overgrowth by ZnSe. Cross-sectional TEM then showed that cubic Ga_2Se_3 was likely to be present at the ZnSe–GaAs interface, although peaks in the RAS spectra could not be directly correlated with the known electronic structure of Ga_2Se_3. The inference of this work is that cubic gallium selenide forms at the interface under these growth conditions, with its associated structural vacancies. As with previous work from the same research group, analysis of TEM image contrast indicated that a Zn–As interfacial layer formed when a Zn-exposed $c(4 \times 4)$ GaAs surface was used for ZnSe growth. Processed RAS spectra indicated that this initial surface does not allow the formation of any Ga–Se related structure. By contrast, a Zn-exposed (2×4) GaAs surface overgrown with ZnSe showed signs of Ga–Se bond formation in processed RAS data. RAS spectra were not observed to change when the (2×4) arsenic-terminated GaAs surface was exposed to a zinc flux and also no changes in surface reconstruction were seen in the RHEED patterns obtained from this surface. This indicates that the surface zinc atoms do not interact greatly with the surface As-dimers and that RAS cannot therefore be used as a sensitive monitor of zinc coverage in this case. This is in contrast to when the double arsenic layer terminated $c(4 \times 4)$ GaAs surface was exposed to a zinc flux where the loss of top-layer arsenic dimers was interpreted from changes in the RAS spectra and RHEED patterns which now showed the subsequent formation of a (1×4) reconstruction. All of this work was performed using pseudomorphic ZnSe layers — thicknesses below which dislocations form as the presence of dislocation arrays may influence further reflectance anisotropy.

RAS has been used to obtain a database of spectra for MBE-grown ZnSe on GaAs under the three surface structures known for ZnSe [102]: the zinc-rich $c(2 \times 2)$ structure, which exhibits no dimer, features the selenium-rich (2×1) dimer structure; and the trimer $c(2 \times 2)$ structure under very selenium-rich conditions. While these surface structures cannot be guaranteed to be found under MOCVD conditions, this work was able to deduce from observations of the time-resolved decay of the (2×1) Se-rich surface to the $c(2 \times 2)$ Zn-rich configuration, an activation energy for selenium desorption of 0.7 eV. The $c(2 \times 2)$ Zn-rich surface does not appear to contain dimers and gives a flat RAS spectrum.

In summary, it is clear that the nature of the GaAs(100) surface and its pre-treatment is crucial if this substrate is to be used for the growth of II–VI wide-bandgap materials. It is also clear that, judging from the literature, even in MBE-grown material, sufficiently high levels of extended defects persist in the layers to reduce device lifetimes severely. In the case of MOCVD of ZnSe-related materials, the prospects for control of the microstructure are limited by the lack of surface control — difficulty in forming an arsenic-rich surface of the correct stoichiometry and the tendency for the background levels of group VI elements present within the reactors to 'poison' the surface. A way forward may be to move towards the use of ZnSe single-crystal substrates which have recently been prepared using seeded vapour transport; LEDs grown by MBE on these substrates demonstrated lifetimes of up to 1000 h at room temperature [101].

Two other issues that affect the prospects for II–VI optoelectronics are those of doping and electrical contacts. Obviously, in order to fabricate II–VI optoelectronic devices, such as blue LEDs and laser diodes based on ZnSe, it is necessary to be able to control n- and p-type doping from background levels (below 10^{15} cm^{-3}) up to and exceeding 10^{18} cm^{-3}. Materials such as ZnSe and ZnS are by nature n-type, and highly conductive material can be produced with the addition of donors such as chlorine and iodine. This can be readily achieved, and where the triethyl derivatives of Al, Ga and In, or alkyl halides such as ethyl iodide [116], n-buthyl iodide [116] and n-butyl chloride [117] have all been used to prepare n-type ZnSe [118].

p-type conduction, however, has been difficult to obtain in ZnSe and related alloys. Acceptors such as lithium tend to be incorporated as interstitials rather than substitutionally and their diffusion rate through the lattice is high, thus precluding reproducible doping and hence the fabrication of stable devices. Lithium can be readily introduced into the layers using volatile metal–organic sources such as t-BuLi [119], CpLi [120] and LiN(Ph$_3$)$_2$ [120]. Other potential acceptors, such as arsenic and phosphorus, tend to produce deep levels. To date, the only successful acceptor with a reasonably shallow state (110 meV) is nitrogen with free hole concentrations up to 10^{18} cm^{-3} having been achieved [121]. In MBE, incorporation of nitrogen has been achieved using radio-frequency excitation to produce a molecular beam of nitrogen atoms which have a much greater sticking coefficient than molecular nitrogen. Despite actual nitrogen incorporation levels of the order of 10^{19} cm^{-3}, the level of activated nitrogen in the lattice is of the order of no more than 10^{18} cm^{-3}, probably due in part to compensation but also because the acceptor state is not particularly shallow.

However, the incorporation of electrically active N in ZnSe by MOCVD has proved very difficult. The use of NH$_3$ has been unsuccessful [116]. The high decomposition temperature of NH$_3$ makes it incompatible with the requirements for low-temperature ZnSe growth. Although pre-cracking of NH$_3$ and plasma assistance [133] have also been investigated, the doping efficiency remains low.

A large number of primary and secondary alkylamines, RNH$_2$ and R$_2$NH, have therefore been investigated as alternatives to NH$_3$ [122]. Unfortunately there is generally a large difference between the level of incorporated and electrically active N in the ZnSe layers. It has been shown that this arises from the unwanted incorporation of hydrogen, leading to (N–H) species which do not act as acceptors [123,124]. In order to avoid hydrogen passivation of N acceptors, nitrogen sources which do not contain the (N–H) bond, such as trialkylamines, have been investigated [117]. However, it is unlikely that the mechanism of amine pyrolysis is so simple (such 'rules of thumb' are unlikely to be of any real use in these multi-component systems) and reactions between the products of the decomposition reactions and the dihydrogen ambient will inevitably complicate the dopant kinetics. The use of N$_2$ in plasma-assisted MOCVD has also been investigated without any real success [125]. Various other more complex amines such as triallylamine [117], ethylazide [131], trimethylsilylazide [126] and directly bonded single-source precursors (direct Zn–N bond) di(trimethylsilyl)amido zinc [126] have also been investigated, but progress is hampered by a lack of understanding of the basic growth chemistry. Using purely thermal MOCVD, triallyamine has achieved hole concentrations in the low 10^{17} cm^{-3} range [117].

The work on ethylazide [131] did indicate that the nature of the ZnSe surface greatly affected the degree to which nitrogen was incorporated, with a zinc-stabilised surface enhancing the nitrogen incorporation. This was performed using an ALE-like approach, with a cyclic alternate supply sequence of diethylzinc, ethylazide and diethylselenium. Nevertheless, annealing of the grown sample under an N$_2$ atmosphere was required to remove hydrogen that had been incorporated into the ZnSe layer to achieve hole concentrations of 5×10^{17} cm^{-3}. Unless carefully done, such post-growth annealing is likely to result in evaporation of selenium and zinc from the surface layers of the material.

Hydrogen azide (HN$_3$) is a potential source of nitrogen for p-type doping in MOCVD ZnSe but, as pointed out above on nitride growth, it is very toxic and potentially explosive in the liquid state. However, it has been shown [132] that hydrogen azide can be made *in situ* by the decomposition of ethylazide (C$_2$H$_5$N$_3$) into HN$_3$ and ethylene. Hydrogen azide then decomposes further into nitrene (HN) and dinitrogen gas. The nitrene group may then strongly interact with the surface but also lead to the incorporation of hydrogen, thus passivating the acceptor state.

For successful p-type doping it is also important to employ low growth temperatures (ideally no higher than 350°C) so that the conductivity of the ZnSe layers is not compromised by the generation of intrinsic defects. This means that the Se precursor must be carefully selected. Although H$_2$Se does allow growth at low temperatrues (250–350°C), it is believed to generate large numbers of hydrogen radicals on pyrolysis [127] and a high VI/II ratio is generally required for good layer morphology, making the substitution of Se by N very difficult. The recently developed precursor *t*-Bu$_2$Se has the desired low-temperature growth characteristics and has subsequently been used in combination with Et$_2$Zn [128], and Me$_2$ZnNEt$_3$ [112] in n-doping studies, although as yet the level of progress achieved compared to MBE is disappointing. Very recently Pohl *et al.* [129] have investigated several zinc amides and the corresponding amines. Doping efficiency appears, however, to be too low for device applications.

Some success has been attained with photo-assisted epitaxy to make working LED structures, via the enhancement of the sticking probability of nitrogen from a variety of nitrogen-bearing precursors [103], via reduction of the growth temperature, possible with this technique. So far, photo-assisted growth has enabled high acceptor doping with reasonable activation levels above 10^{17} cm^{-3} [130].

Some work has been done to investigate the effect of substrate orientation on the level of acceptor concentration in ZnSe [132]. Although this work was not MOCVD but MBE, the results do show that misorientation of the GaAs substrate has a significant effect on the nitrogen incorporation level. Nominally exact (100) orientations had hole concentrations only a fifth or a quarter of the size of substrates misoriented by 4° towards either (110) or (010). Additionally, the polar (211)$_A$ surface was investigated. This surface has more zinc atom dangling bonds (to which nitrogen would be expected to bond as an acceptor) available than selenium atom dangling bonds. More than an order of magnitude's difference was observed in hole concentration between (211)$_A$ and (211)$_B$, although the absolute hole concentrations were down in the 10^{15} cm^{-3} rather than the 10^{16}–10^{17} cm^{-3} range of the (100)-oriented cases.

Electrical contacts to wide-bandgap II–VIs is another problem area, particularly for p-type material which still exhibits Schottky type contact behaviour (rectifiers) even when high work-function metals such as gold are used, owing to the deep-lying valence band of the ZnSe. Thus, graded ZnSe–ZnTe contact layers have been employed to give ohmic contacts in which a superlattice of ZnSe–ZnTe was grown, with a gradually changing Te content from zero to 100% [104]. Zinc telluride is intrinsically p-type but, grown, adjacent to ZnSe, yields a 1 eV barrier for holes — hence the need for a graded contact. It appears that controlling the Te level in MBE was sufficiently difficult to require this 'pseudo-grading' scheme and that while MOCVD could provide a true graded layer to ZnSe, it would require a further source material (tellurium) to be added to the growth system. Thus a MOCVD system, taking on board all the advances developed by MBE, would require six main sources plus two dopants — a complex system indeed.

MOCVD of II–VI materials like ZnSe was originally performed with hydride group VI sources such as H_2Se and H_2S [105]. The best-quality material appears to have been grown at temperatures around 300°C, which is fairly low by III–V standards. However, the zinc source typically used, dimethylzinc, reacts in the gas phase with the hydride to form an adduct which appears to be detrimental to wafer uniformity due to gas-phase depletion of the nutrients. In addition, the actual growth mechanism at the surface is unknown but may involve interaction of the adduct with the hot surface or ZnSe species formed in the gas phase. Extended interaction in the gas phase is now known to lead to the ZnSe particle formation by successive elimination of methyl and hydrogen fragments yielding free methane and a polymer-like species which eventually becomes particulate [106]. The hydride itself does not crack to form selenium species until temperatures well above 600°C are reached. This is not an option for II–VI growth, as evaporation of both selenium and zinc from the growth surface would occur.

Ways to prevent adduct formation have involved making an adduct between dimethylzinc (DMZ) and triethylamine (TEA) to block the gas-phase interaction with the hydride. However, the DMZ–TEA adduct is known to be unstable and dissociates easily in the heated zone of the reactor, thus while the 'adduct' may work as a means of preventing reaction with the hydride, exactly how this happens is not clear. Alternatively, the addition of pyridine also prevents polymer formation by a similar blocking mechanism.

Alternative group VI precursors have been much in vogue of late, both to reduce or eliminate entirely the gas-phase pre-reaction and to eliminate the possibility of incorporation of atomic hydrogen into the crystalline lattice. Interstitial hydrogen is known to act as a compensator for nitrogen in ZnSe and thus must be eliminated if nitrogen doping in MOCVD of ZnSe-related alloys is to be successful. Thus those organoselenium precursors such as ditertiarybutylselenium which do not incorporate a selenium–hydrogen bond have found favour. Recent work has indicated that the breakdown pathway for this precursor leads in any case to the formation of a Se–H species, and so the advantage is negated [107]. Another factor to consider is the possibility for carbon incorporation. Dimethylselenium cracks in hydrogen to form $CH_3Se\cdot$ and $CH_3\cdot$ radicals. Reaction of $CH_3\cdot$ radicals with the dihydrogen carrier gas leads to the formation of methane but $CH_3Se\cdot$ is thermally stable and could be incorporated into the growing crystal, leading to carbon incorporation. Other more complex group VI precursors may break down in a more favourable way. The growth temperatures of ZnSe for all these organoselenium compounds are considerably higher than those used for the hydride, typically of the order of 400–450°C [112] unless photo-assisted growth is used. The sulfur analogue, ditertiarybutylsulfide, has been found to crack easily at 300°C, and higher temperatures resulted in gas-phase cracking of this source [108].

The growth of magnesium-containing material is also not without problems. The choice of magnesium precursor chemicals is very restricted and the most volatile source is currently bis(cyclopentadienyl)magnesium, which is a solid at room temperature. Condensation in the supply lines to the reactor is a potential problem and the vapour pressure of this source is not accurately known at the time of writing. In addition, severe gas-phase pre-reactions with sulfur sources such as thiols have been reported [109], which must be eliminated if uniform highly crystalline epitaxial layers are to be produced.

Magnesium sulfides and selenides normally adopt the sixfold coordinated rocksalt structure in the bulk form. Epitaxial MgS has recently been reported to form in the fourfold coordinated sphalerite form when grown on to GaAs substrates [110]. However, other reports only find the occurrence of the rocksalt form when grown on to either near-lattice-matched GaP or mismatched GaAs, although a thiol was used as the sulfur source [109] instead of the hydride [110]. This recent work [109] has addressed the issue of the transition between the two structure types, as the composition is changed from MgS (rocksalt cubic) to CdS (cubic sphalerite) but the gas-phase pre-reaction did not permit a full evaluation of this interesting system, as polycrystalline material often resulted. Similar studies on the $Zn_xMg_{1-x}Se$ system [111] have been conducted using Raman spectroscopy as the investigative tool for measuring the phase ranges, and considerable solubility (*c.* 50%) of ZnSe and MgSe was found.

6   Novel III–VI compound semiconductors (Ga₂Se₃)

The interest in III–VI compounds appears to have grown out of the realisation that they can be observed to form at the interface between II–VI compounds (such as ZnSe) and III–V substrates (such as GaAs). For the case of the polar heteroepitaxy of II–VI and III–V zincblende materials [134], it has been shown that an ideal planar geometry is not allowed. The reason for this is that interface charge effects arise when excess electron-rich Ga–Se bonds or electron-deficient As–Zn bonds are formed at the interface. The absence of charge neutrality may lead to the formation of local electrostatic fields and/or trapping states which disturb the electrical nature of the interface and subsequent growth stages, since electric field enhancement of diffusion by many orders of magnitude is possible. On the basis of electrostatic considerations, Harrison *et al.* [135] suggested that polar heteroepitaxial interfaces require at least two transition planes with graded stoichiometry to circumvent these fundamental problems.

Different GaAs(100) surface reconstructions are available during epitaxial growth [134]. The GaAs(100)(2 × 4)–As surface has a large number of accessible As/Ga surface stoichiometries, varying about around 3 : 1. Typical GaAs growth conditions generally lead to the more As-deficient stoichiometries within this range, approaching the desired 1 : 1 stoichiometry. By contrast, the GaAs(100)(4 × 2)–Ga-terminated surface provides an excess of Ga (Ga : As of 3 : 1) which, during typical growth conditions, tends to become exaggerated, thus leading away from the desired stoichiometric surfaces. These conclusions are consistent with observations of 2D nucleation of ZnSe on (2 × 4) GaAs surfaces, and three-dimensional nucleation on the (4×2) surfaces. Intermediate GaAs reconstructions, such as the (4×6) and the (3×1) are of great fundamental importance, since they provide more optimal stoichiometric conditions.

Tu and Khan [136] first suggested the formation of a (Ga, Se) compound layer at the ZnSe–GaAs heteroepitaxial interface. One of the stable phases of (Ga, Se) compounds, Ga_2Se_3, is known to have a structure identical to that suggested by the observations. It has a zincblende structure, and one-third of Ga sites are left vacant [137] (Fig. 11.16). As a result of these vacancies, the lattice parameter of Ga_2Se_3 is about 5% smaller than those of ZnSe and GaAs. One question that needs to be addressed is what effect a high concentration of vacancies will have on the stability of the transition structure, as these vacancies are expected to yield to strains in the structure. Thermodynamic data of III–VI compounds such as heats of formation of Ga_2Se_3 and In_2Te_3 [138], however, show that these compounds are highly stable phases despite the existence of a high concentration of vacancies in their structures.

Since one-third of the Ga sites in Ga_2Se_3 are left vacant, it seems reasonable to suggest that the superstructure at the interface is an ordered arrangement of structural vacancies. HRTEM images and electron diffraction patterns support this model. A model based on structural vacancies has been developed to explain the main features of the experimental results. In the model, only one-half of sites of the first Ga layer of the Ga_2Se_3 epilayer are occupied by atoms, leaving the remainder of the sites as structural vacancies. These vacancies form a $c(2 \times 2)$ superlattice in this layer. Conversely, the second Ga layer in the epilayer is almost fully occupied by atoms. In the third Ga layer, the structural vacancies locally form the $c(2 \times 2)$ ordered structure, which is shifted by half a period from that of the first Ga layer. In Ga layers beyond the third, structural vacancies form a nearly random arrangement. Figure 11.17 shows a TEM cross-section lattice image of the observed superstructure.

The formation of the $c(2 \times 2)$ structure induced by the GaAs substrate surface is believed to be a function of the vacancy concentration in the $c(2 \times 2)$ ordered structure. One-half of all Ga sites are vacant in the ordered structure, significantly greater than the vacancy concentration in Ga_2Se_3. The top Ga atomic plane of the GaAs substrate is, however, completely occupied by Ga atoms, so that each Se atom between these two Ga

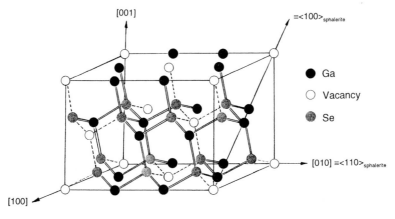

Figure 11.16. Crystal structure of Ga_2Se_3 as described by Lubbers and Luite [137]. The outline of the monoclinic unit cell is shown and its relation to the sphalerite substructure indicated.

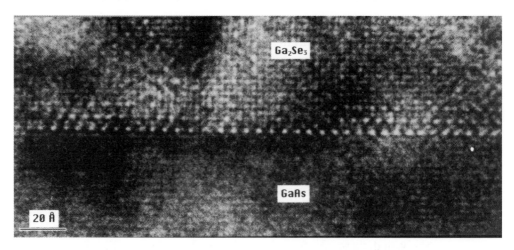

Figure 11.17. High-resolution TEM image of the superstructure of the Ga_2Se_3–GaAs interface. From [143].

atomic planes has three nearest-neighbour Ga atoms, which are close to those of a selenium atom in the Ga_2Se_3.

It has been shown that if the Ga_2Se_3 layer forms directly on the GaAs substrate surface, most of the Se atoms at the interface will have four nearest-neighbour Ga atoms and therefore result in the formation of a large number of electron-rich bonds. This suggests, therefore, that the $c(2 \times 2)$ ordered structure is a transitional structure between GaAs and Ga_2Se_3, created in order to minimise electron-rich bonds at the interface. The high degree of lattice ordering of the $c(2 \times 2)$ structure may be explained by the simple geometrical nature of $c(2 \times 2)$ ordered arrangement in the (100) growth plane.

In the interfaces of the II–VI/III–V heterostructures, formation of thin interfacial layers, consisting mainly of III–VI compounds, have been reported in CdTe–InSb [139], ZnSe–GaAs [140–144], and ZnTe–GaSb [145,146]. Since Ga–Se compounds have been studied the most we will restrict our discussion to these materials. Which selenide can the surface compound be ascribed to? The most likely candidate is Ga_2Se_3, as has been suggested by several researchers [147,148]. The α phase of this compound has a defect zincblende structure, in which one-third of the Ga sites are vacant. The lattice constant is 5.43 Å. The mismatch to the GaAs lattice is only 5% and thus the formation of a coherent film a few atomic layers thick is possible. Another candidate is the layer compound, GaSe [145]. The thickness of a unit layer is 7.97 Å, which is very close to twice the lattice constant of the surface unit cell for (001) GaAs, 4.00 Å. The a-value (the lattice constant in the basal plane) of 3.76 Å is also close to 4.00 Å and the mismatch is 6.4%. Thus, it may be possible that a GaSe-like compound could grow epitaxially on a (001) GaAs surface with the basal plane normal to the surface. It has also been reported that an interfacial Ga_2Se_3 layer can improve the crystallinity of an epitaxially grown ZnSe layer on the GaAs substrate [148]. In addition, III–VI compounds are themselves wide-bandgap semiconductors and have the potential to become materials for 'blue-emitting devices'. For example, the bandgap energy of Ga_2Se_3 is 2.6 eV, and doping control is expected to be much easier than for II–VI compounds, since their bonding is less ionic [149].

A further interesting development is that GaAs surfaces treated with selenium have recently shown dramatically improved properties such as enhanced photoluminescence [150], reduced numbers of bandgap surface states and unpinning of the Fermi level [151]. Reduction of the number of interface states has also been reported for ZnSe–GaAs heterojunctions prepared on a GaAs surface with a selenide layer formed by heating to a high temperature in a selenium ambient [152].

Interest in these III–VI phases is increasing both for control of the interface quality of the II–VI/III–V heterostructures and for exploring the properties of new semiconductor materials. Efforts have been made to grow these III–VI epitaxial layers deliberately on to suitable III–V substrates by a variety of epitaxial growth techniques. These include MBE [153–157], metal-organic molecular beam epitaxy (MOMBE) [149,158,159], MOVPE [160–164] and vapour phase epitaxy (VPE) [165]. The majority of the work has been done on Ga–Se compounds and a variety of selenium sources have been used, including hydrogen selenide [147,160–163], diethyl- (DESe) [20], di-isopropyl- (DIPSe) [161–163] and ditertiarybutylselenide (DtBSe) [161–163], with trimethylgallium (TMGa) as the gallium source. The Ga–Se films are deposited by flowing the gallium and selenium precursors in hydrogen at 1 atm pressure over a substrate heated to between 623 and 873 K [160–164]. Diethylsulfide (DES) and diethyltelluride (DETe) have both been used in conjunction with elemental gallium, to grow Ga_2Se_3 and Ga_2Te_3 respectively, by MOMBE [149,158,159]. Various substrates and orientations have been examined, including GaAs(100) [149,153–159,161–164], GaP(100) [153–156,161–163] and GaP(110) [166] and GaSb(100) [167].

The properties of bulk Ga_2Se_3 [168], In_2Te_3 [169], Ga_2S_3 [169] and Ga_2Te_3 [170] have been reported in several recent papers. The authors all suggest that III–VI compounds have structures based on the sphalerite unit cell (for Ga_2S_3 it is a structure based on the wurtzite unit cell) with two-thirds of cation sites occupied and one-third vacant, whereas the anion sites are fully occupied. Evidence has been presented that, under slow cooling conditions, vacancy ordering develops in these compounds. The compound Ga_2Te_3 has been reported to have an ordered vacancy structure based on the sphalerite unit cell [171]. However, the exact ordering pattern is unknown. When the cooling rate is high, the vacancies are likely to be disordered. There are structural models for vacancy ordering in Ga_2Se_3, suggesting that cation sites are either fully occupied or completely vacant, resulting in supercells three times the size of the sphalerite unit cell along certain directions. As a consequence, the symmetry of the cubic sphalerite is lost, and a monoclinic or orthorhombic structure is formed.

The other (Ga, Se) phase, GaSe, has a layered structure, although different from other metal chalcogenides (of the TX_2 type) such as $MoSe_2$ or $NbSe_2$ [172–174], and layers are bound together by weak van der Waals forces [175]. Thus, although a unit layer of TX_2 compound consists of X (chalcogen)–T (transition metal)–X three-atomic layers, the layered structure can be regarded as a spontaneous superlattice formed by stacking four monatomic sheets in a Se–Ga–Se–Ga sequence [176,177]. On a cleaved face of GaSe, Se atoms are arranged in the sixfold triangular-like TX_2 structure. The Se–Se length on the surface is 0.3755 nm, while the As–As length is 3.998 nm on the top hexagonal layer of the GaAs(111) surface. Therefore the lattice mismatch is as large as 6% in this combination. In spite of the large lattice mismatch, it is believed that GaSe grows epitaxially on chalcogen-treated GaAs(111) surfaces by the van der Waals epitaxy

(VDWE) mechanism [174]. GaSe has strong optical and electrical anisotropy [176,177], and it was reported that the luminescence of the excitons in GaSe exhibited quasi-2D characteristics [178]. Epitaxial GaSe has been successfully grown on GaAs(111), GaAs(110) and GaAs(112) by MBE [50] and on GaAs(100) by MOVPE [161,162]. On the GaAs(111) substrate, the *c*-axis of the GaSe layer was aligned perpendicular to the substrate surface. On the other hand, each unit layer of GaSe was inclined on the GaAs(100) and GaAs(112) substrates when the growth temperatures were higher than 500°C at high growth fluxes of Ga and Se. Thus GaSe films could be grown epitaxially on GaAs(100), since the (100) surface of GaSe has the same symmetry as that of the (111) surface of GaAs. The GaSe layers tend to grow along the {111} plane of the zincblende structure. Figure 11.18 shows a proposed model depicting how hexagonal GaSe grows on the GaAs(001) surface.

It is found that the crystal structure of Ga–Se compounds is a function of both the VI/III ratio and the growth temperature [156]. Vacancy-ordered Ga_2Se_3 is grown at high temperatures and at high VI/III ratios, while the layered GaSe is grown at low growth temperatures and at low VI/III ratios. The existence region or 'growth-parameter space' that is spanned by the Ga–Se compounds is also dependent on the chemical precursor combination used to deposit the epilayer. It should be noted that only limited epitaxial growth was possible with TMGa and H_2Se, due to an extensive room temperature pre-reaction between the precursors. In Ga_2Se_3 layers grown from this combination of precursors there is no tendency towards single domain formation although, and despite the lattice mismatch of 0.76% at room temperature, no threading dislocations appear to be present within the layer. This is very unusual in a tetrahedrally bonded compound semiconductor with this much mismatch, and a ready vacancy-assisted climb mechanism towards the interface very early in the growth cycle has been proposed to explain this behaviour, the large intrinsic vacancy concentration in this material supporting such a mechanism. Unusual tree-like structures in this material have been revealed as arising from plate-like faults lying on {111} planes which have been interpreted as intergrowths of the hexagonal GaSe phase, although this assignment has not been conclusively established.

Other precursor combinations result in very different behaviour; the use of TMGa and DIPSe [161,162] leads to the growth of principally GaSe under normal growth

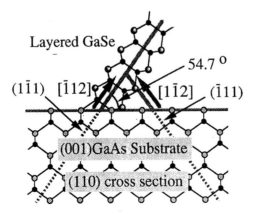

Figure 11.18. Schematic drawing of the proposed interfacial structure of hexagonal GaSe on GaAs(001). From [179].

conditions, while the use of TMGa and DtBSe [161,162] allows the controllable growth of either GaSe or Ga_2Se_3, depending on the precise conditions selected. Indeed, using the latter combination, smooth specular films of Ga_2Se_3 could be grown on to GaP substrates with growth rates c. 1 μm h^{-1}. Although considerable strain is present in such films, along with stacking faults and twins, the structure is epitaxial from interface to surface, with no evidence of polycrystallinity, as found with films of similar thickness produced using the hydride. Additionally, [$\bar{1}$00] diffraction patterns showed that there was a tendency towards the formation of single-domain material. This suggests that the rates of gas-phase and surface reactions determine both the rate of deposition and the specific phase deposited. This is not surprising if one considers that the surface kinetic mechanism is likely to depend upon the surface morphology via the type and density of surface sites for adsorption.

7 Conclusions

In this paper we have demonstrated how a principal route in the preparation of advanced materials, MOCVD, can be used under particular specified conditions to prepare new materials and interfaces. The underlying importance of substrate surface structure is emphasised, and in the case of wide-bandgap II–VI compounds semiconductors we have shown how surface monitoring techniques recently developed at or near atmospheric pressure conditions of growth can contribute to our understanding. In recent years the spectacular success with the GaN and related systems for the fabrication of blue emitting devices without a full understanding of the processes that occur at the substrate surface has justified the empirical approach to MOCVD growth. The technique has also gained considerable success in the preparation of metals, oxides and superconductors to satisfy the demands of device engineers well into the 21st century.

8 References

1 Manesevit HM, Simpson WI. *J. Electrochem. Soc.* 1969; **116**: 1725.
2 Williams JO. In Stradling RA, Klipstein PC (eds) *Growth and Characterisation of Semiconductors*. Bristol: Adam Hilger, 1990; 17.
3 Stringfellow GB. In *Organometallic Vapour Phase Epitaxy*. New York: Academic Press, 1989.
4 Razeghi M. In *The MOCVD Challenge*, Vol. 1. Bristol: Adam Hilger, 1989.
5 Jones AC. *J. Crystal Growth* 1993; **129**: 728.
6 Zanella P, Rossetto G, Brianese N, Ossola F, Porchia N, and Williams JO. *Chem. Mater.* 1991, **3**: 225.
7 Halliwell MAG. *J. Crystal Growth* 1997; **170**: 47.
8 Wright AC, Williams JO. In Miller LS, Mullin JB (eds) *Electronic Materials: from Silicon to Organics*. New York: Plenum Press, 1991.
9 Land DV. *J. Appl. Phys.* 1974; **45**: 3023.
10 Williams JO. In Clark RJH, Hester RE (eds) *Spectroscopy of Advanced Materials*, vol. 19. Chichester: Wiley, 1991.
11 Armitage DN, Dunhill NI, West RH, Williams JO. *J. Crystal Growth* 1990; **108**: 683.
12 Shim HK, Chi KM, Hampden-Smith MJ, Kodas TT, Paffett MF, Farr JD. *Chem. Mater.* 1992; **4**: 788.
13 Awaya N, Arita Y. *Jpn. J. Appl. Phys., Part 1* 1991; **30**: 1813.

14 Kodas TT, Hampden-Smith MJ (eds). In *Chemistry of Metal CVD*. Weinheim: VCH, 1994; 175.

15 Bednorz JG, Muller KA. *Z. Phys. B* 1986; **64**: 189.

16 Wu MK, Ashburn JR, Torng CJ *et al. Phys. Rev. Lett.* 1987; **58**: 908.

17 Adrian FJ, Cowan DO. *Chem. Engg News* 1992; **21 December**: 24.

18 Burdett JK. In Williams AF, Floriani C, Merbach AE (eds) *29th International Conference on Coordination Chemistry*. Lausanne: Helvetica Chimica Acta, 1992; 293.

19 Dahmen K-H, Gerfin T. *Prog. Cryst. Growth Charact.* 1993; **27**: 117.

20 Berry AD, Gaskill DK, Holm RT, Cukauskas EJ, Kaplan R, Henry RL. *Appl. Phys. Lett.* 1988; **52**: 1743.

21 Schilling A, Cantoni M, Guo JD, Ott HR. *Nature* 1993; **363**: 56.

22 Erbil K, Zhang K, Kwak BS, Boyd EP. *SPIE* 1187, 104.

23 Kaul AR. *Mendeleev Chem. J.* 1989; **34**: 60.

24 Schmaderer F, Huber R, Oetzmann H, Wahl G. *Appl. Surf. Sci.* 1990; **46**: 53.

25 Hirai T, Yamane H. *J. Crystal Growth* 1991; **107**: 683.

26 Scheel HJM, Berkowski M, Chabot B. *J. Crystal Growth* 1991; **115**: 19.

27 Schieber M, Han SC, Ariel Y *et al. J. Crystal Growth* 1991; **115**: 31.

28 Yamane H, Hasei M, Kurosawa H, Hirai T. *Jpn. J. Appl. Phys. Part 2* 1991; **30**: L1003.

29 Leskela M, Molsa H, Niinisto L. *Supercond. Sci. Technol.* 1993; **6**: 627.

30 Tonge LM, Richeson DS, Marks TJ *et al. Adv. Chem. Ser.* 1990; **226**: 351.

31 Matsuno S, Uchikawa F, Yoshizaki K. *Jpn. J. Appl. Phys.* 1990; **29**: L947.

32 Phillips JM. *J. Appl. Phys.* 1996; **79**: 1829.

33 Busch H, Fink A, Müller A. *J. Appl. Phys.* 1991; **70**: 2449.

34 Zhao J, Norris P. *Thin Solid Films* 1991; **206**: 122.

35 Chen ZX, Derking A. *J. Mater. Chem.* 1993; **3**: 1137.

36 Gome-Aleixandre C, Sanchez O, Albella JM, Santiso J, Figueras A. *Adv. Mater.* 1995; 7: 111.

37 Gould BJ, Povey IM, Pemble ME, Flavell WR. *J. Mater. Chem.* 1994; **4**: 1815.

38 *Compound Semicond. Mag.* 1, No. 1, 1995. Special issue on short-wavelength light emitting diodes.

39 Nakamura S, Senoh M, Iwasa N, Nagahama S. *Jpn. J. Appl. Phys. Lett.* 1995; **34**: L797.

40 Nakamura S, Senoh M, Nagahama S *et al. Jpn. J. Appl. Phys.* 1996; **35**: L74.

41 Lester SD, Ponce F, Craford MG, Steigerwald DA. *Appl. Phys. Lett.* 1995; **66**: 1249.

42 Nakamura S, Senoh M, Nagahama S *et al. Jpn. J. Appl. Phys.* 1996; **35**: L217.

43 Nakamura S. *Jpn. J. Appl. Phys.* 1991; **30**: 1620.

44 Brandt O, Yang H, Jenichen B, Suzuki Y, Daweritz L, Ploog KH. *Phys. Rev. B*, 1995; **52**: R2253.

45 Wongchotigul K, Chen N, Zhang DP, Tang X, Spencer MG. *Materials Letters* 1996; **26** (4–5): 223.

46 Keller S, Keller BP, Wu Y-F *et al. Appl. Phys. Lett.* 1996; **68**: 1525.

47 Wu XH *et al. Appl. Phys. Lett.* 1996; **68**: 1371.

48 Fischer RA, Miehr A, Ambacher O, Metzger T, Born E. *J. Crystal Growth* 1997; **170**: 139.

49 Gordon RG, Hoffman DM, Riaz U. *J. Mater. Res.* 1992; 7: 1679.

50 Lakhotia V *et al. Chem. Mat.* 1995; 7: 546.

51 Miura Y, Takahashi N, Koukitu A, Seki H. *Jpn. J. Appl. Phys. Part 2* 1995; **34**: L401.

52 Miura Y, Takahashi N, Koukitu A, Seki H. *Jpn. J. Appl. Phys. Part 1* 1996; **35**: 546.

53 Detchprohm T, Hiramatsu K, Itoh K, Akasaka I. *Jpn. J. Appl. Phys.* 1992; **31**: L1454.

54 Tsuchiya H, Hasegawa F, Okumura H, Yoshida S. *Jpn. J. Appl. Phys. Part 1* 1994; **33**: 6448.

55 Hoffman DM *et al. J. Vac. Sci. Technol. A* 1996; **14**: 306.

56 Lin ME *et al. Appl. Phys. Lett.* 1993; **62**: 702.

57 Fujieda S, Mizuta M, Matsumoto Y. *Jpn. J. Appl. Phys.* 1987; **26**: 2067.

58 Koike M *et al. Appl. Phys. Lett.* 1996; **68**: 1403.

59 Nakamura S, Senoh M, Iwasa N, Nagahama S. *Appl. Phys. Lett.* 1995; **67**: 1868.

60 Sasaki T, Matsuoka T. *J. Appl. Phys.* 1995; **77**: 192.

61 Sun CJ *et al. Appl. Phys. Lett.* 1996; **68**: 1129.

62 Ponce FA *et al. Appl. Phys. Lett.* 1996; **68**: 917.

63 Dovidenko K, Oktyabrsky S, Narayan J, Razegi M. *J. Appl. Phys.* 1996; **79**: 2439.

64 Boutros KS *et al. Appl. Phys. Lett.* 1995; **67**: 1856.

65 Morcoç H, Strite S, Gao GB, Lin ME, Sverdlov B, Burns M. *J. Appl. Phys.* 1994; **76**: 1363.

66 Kapolnec D, Underwood RD, Keller BP, Keller S, DenBaars SP, Mishra UK. *J. Crystal Growth* 1997; **170**: 340.

67 Pakula K, Baranowski JM, Stephiewski R *et al.* ICMOCVD-8 conference, Cardiff, June 1996.

68 Kahn MA, Bhattari A, Kuznia JN, Olsen DT. *Appl. Phys. Lett.* 1993; **63**: 1214.

69 McCarron KT, Kline GR, Martin JT, Lakin KM. *Proc. IEEE Ultrasonics Symp.* 1988; **3**: 673.

70 Sato M. *Appl. Phys. Lett.* 1996; **68**: 935.

71 Mihopoulos TG, Simka H, Jensen KF. ICMOCVD-8 conference, Cardiff, June 1996.

72 Scholz F, Härle V, Steuber F *et al. J. Crystal Growth* 1997;**170**: 321.

73 Nishida K, Haneda S, Hara K, Munekata H, Kukimoto H. *J. Crystal Growth* 1997; **170**: 312.

74 Beaumont B, Vaille M, Boufaden T, el Jani B, Gibart P. *J. Crystal Growth* 1997; **170**: 316.

75 Van der Stricht W, Moerman I, Demeester P, Crawley JA, Thrush EJ. *J. Crystal Growth* 1997; **170**: 344.

76 Thon A, Kuech TF. ICMOCVD-8 conference, Cardiff, June 1996. (See also Thon A, Kuech TF, *Appl. Phys. Lett.* 1996; **69**: 55.)

77 Ruvimov S, Liliental-Weber Z, Suski T *et al. Appl. Phys. Lett.* 1996; **69**: 990.

78 McMurran J, Todd M, Kouvetakis J, Smith DJ. *Appl. Phys. Lett.* 1996; **69**: 203.

79 Flowers MC, Jonathon NGH, Laurie AB, Morris A, Parker GJ. *J. Mater. Chem.* 1992; **2**: 365.

80 Taniguchi S, Hino T, Itoh S *et al. Electron. Lett.* 1996; **32**: 552.

81 Nurmikko AV, Gunshor RL. *Solid State Phys.* 1995; **49**: 205.

82 Kuo LH, Salamanca-Riba L, Wu BJ *et al. Appl. Phys. Lett.* 1994; **65**: 1230.

83 Williams GM, Cullis AG, Prior K, Simpson J, Cavenett BC, Adams SJA. Inst. Phys. Conf. Ser. No. 134: Section II. Pages 671–674.

84 Guha S, Munekata H, Chang LL. *J. Appl. Phys. Lett.* 1993; **73**: 2294.

85 Li D, Pashley MD. *J. Vac. Sci. Technol. B* 1994; **12**: 2547.

86 Pashley MD, Li D. *Mater. Sci. Engg B* 1995; **30**: 73.

87 Yablonovitch E, Sandroff CJ, Bhat R, Gmitter T. *Appl. Phys. Lett.* 1987; **51**: 439.

88 Wu BJ, Haugen GM, DePuydt JM, Kuo LH, Salamanca-Riba L. *Appl. Phys. Lett.* 1996; **68**: 2828.

89 Bourret-Courchesne ED. *Appl. Phys. Lett.* 1996; **68**: 1675.

90 Wright AC, Gnoth D, Ng TL, Poole IB, Maung N. *Appl. Surf. Sci.* (submitted).

91 Kuo LH, Kimura K, Yasuda T *et al. Appl. Phys. Lett.* 1996; **68**: 2413.

92 Gnoth D, Evans DA, Poole IB *et al. J. Crystal Growth* 1997; **170**: 198.

93 Kastner MJ, Hahn B, Blumberg R, Sossna E, Duschlo R, Gebhardt W. *J. Crystal Growth* 1997; **170**: 188.

94 Aspnes DE. *J. Vac. Sci. Technol. B* 1985; **3**: 1502.

95 Scholz SM, Muller AB, Richter W *et al. J. Vac. Sci. Technol. B* 1992; **10**: 1710.

96 Wright AC, Williams JO. *J. Crystal Growth* 1991; **114**: 99.

97 Batstone JL, Steeds JW, Wright PJ. *Phil. Mag. A* 1992; **66**: 609.

98 Kuo LH, Salamanca-Riba L, DuPuydt JM, Cheng H, Qiu J. *J. Electron. Mater.* 1994; **23**: 275.

99 Kuo LH, Salamanca-Riba L, Wu BJ *et al. J. Vac. Sci. Technol. B* 1995; **13**: 1694.

100 Greenwood NN. *Ionic Crystals, Lattice Defects and Non-stoichiometry.* xxxx: Butterworths, 1968.

101 Eason DB, Yu Z, Hughes WC *et al. J. Vac. Sci. Technol. B* 1995; **13**: 1566.

102 Zettler J-T, Stahrenberg K, Richter W, Wenisch H, Jobst B, Hommel D. *J. Vac. Sci. Technol. B* 1996; **14**: 2757.

103 Fujita S, Asano T, Maehara K, Fujita S. *Appl. Surf. Sci.* 1994; **79–80**: 270.

104 Fan Y, Han J, He L *et al. Appl. Phys. Lett.* 1992; **61**: 3160.

105 Wright PJ, Cockayne B. *J. Crystal Growth* 1982; **59**: 148.

106 Foster D, Picket N, Cole-Hamilton DJ. *J. Crystal Growth* 1997; **170**: 476.

107 Fan GH, Maung N, Ng TL *et al. J. Crystal Growth* 1996; **170**: 485.

108 Obinata T, Uesugi K, Ikuo G, Suemune I, Machida H, Shimoyama N. *Jpn. J. Appl. Phys.* 1995; **34**: 4143.

109 Poole IB, Ng TL, Maung N, Williams JO, Wright AC. *J. Crystal Growth* (submitted).

110 Konczewicz L, Bigenwald P, Cloitre T *et al. J. Crystal Growth* 1996; **159**: 117.

111 Huang D, Jin C, Wang D, Lui X, Wang J, Wang X. *Appl. Phys. Lett.* 1995; **67**: 3611.

112 Hua GC, Otsuka N, Grillo DC *et al. Appl. Phys. Lett.* 1994; **65**: 1331.

113 Chu CC, Ng TB, Han J *et al. Appl. Phys. Lett.* 1996; **69**: 602.

114 Yu Z, Buczkowski SL, Giles NC, Miles TH. *Appl. Phys. Lett.* 1996; **69**: 82.

115 Ishibashi A. *J. Crystal Growth* 1996; **159**: 555.

116 Shibata N, Ohki A, Katsui A. *J. Crystal Growth* 1988; **93**: 703.

117 Stanzl H, Wolf K, Hahn B, Gebhardt W. *J. Crystal Growth* 1994; **145**: 918.

118 Stutius W. *J. Crystal Growth* 1982; **59**: 1.

119 Kukimoto H. In McGill TC *et al.* (eds) *Growth and Optical Properties of Wide Gap II-VI Low Dimensional Semiconductors.* New York: Plenum, 1989; 119.

120 Yoshikawa A, Muto S, Yamaga S, Kasai H. *J. Crystal Growth* 1988; **93**: 697.

121 Qiu J, DePuydt JM, Cheng H, Haase MA. *Appl. Phys. Lett.* 1991; **59**: 2992.

122 Jones AC. *J. Crystal Growth* 1993; **129**: 728.

123 Kamata A, Mitsuhashi H, Fujita H. *Appl. Phys. Lett.* 1993; **63**: 3353.

124 Wolk A, Ager JW III, Duxstad KJ *et al. Appl. Phys. Lett.* 1993; **63**: 2756.

125 Taudt W, Hermous J, Schneider A, Woitok J, Geurts J, Heuken, M. *Adv. Mat. Opt. Elec.* 1994; **3**: 203.

126 Taudt W, Lampe S, Sauerlander F *et al. J. Crystal Growth* 1996 (accepted for publication).

127 Jones AC. *J. Crystal Growth* 1994; **145**: 505.

128 Taudt W, Wachtendorf B, Beccard R *et al. J. Crystal Growth* 1994; **145**: 582.

129 Pohl UW, Freitag S, Gottfriedsen J, Richter W, Schumann H. *J. Crystal Growth* (in press).

130 Fujita S, Asano T, Maehara K, Toiyo T, Fujita S. *J. Crystal Growth* 1994; **138**: 737.

131 Inoue K, Yanashima K, Takahashi T *et al. J. Crystal Growth* 1996; **159**: 130.

132 Kamata A. *J. Crystal Growth* 1994; **145**: 557.

133 Huh JS, Patnak S, Jensen KF. *J. Electron. Mater.* 1992; **22**: 509.

134 Farrell HH, Tamargo MC, de Miguel JL, Turco FS, Hwang DM, Nahory RE. *J. Appl. Phys.* 1991; **69**: 7021.

135 Harrison WA, Kraut EA, Waldrop JR, Grant RW. *Phys. Rev. B* 1978; **6**: 4402.

136 Tu DW, Kahn A. *J. Vac. Sci. Technol. A* 1985; **3**: 922.

137 Lubbers D, Leute V. *J. Solid State Chem.* 1982; **43**: 339.

138 Mills KC. *Thermodynamic Data for Inorganic Sulphides, Selenides and Tellurides.* Washington: Butterworth, 1974.

139 Zahn DRT, Mackey KJ, Williams RH, Munder H, Geurts J, Richter W. *Appl. Phys. Lett.* 1987; **50**: 742.

140 Wright AC, Williams JO. *J. Crystal Growth* 1991; **114**: 99.

141 Wright AC, Williams JO, Krost A, Richter W, Zahn DRT. *J. Crystal Growth* 1992; **121**: 111.

142 Williams JO, Wright AC, Yates HM. *J. Crystal Growth* 1992; **117**: 441.

143 Li D, Nakamura Y, Otsuka N, Qiu J, Kobayashi M, Gunshor RL. *J. Vac. Sci. Technol. B* 1991; **9**: 2167.

144 Qiu J, Menke DR, Kobayashi M *et al. Appl. Phys. Lett.* 1991; **58**: 2788.

145 Wilke WG, Seedorf R, Horn K. *J. Crystal Growth* 1990; **101**: 620.

146 Halsall MP, Wolverson D, Davies JJ, Lunn B, Ashenford DA. *Appl. Phys. Lett.* 1992; **60**: 2129.

147 Chambers SA, Sundaram VS. *Appl. Phys. Lett.* 1990; **57**: 2342.

148 Li D, Gonsalves JM, Otsuka N, Qiu J, Kobayashi M, Gunshor RL. *Appl. Phys. Lett.* 1990; **57**: 449.

149 Teraguchi N, Kato F, Konagai M, Takahashi K. *Jpn. J. Appl. Phys.* 1989; **28**: L2134.

150 Sandroff CJ, Hedge MS, Farrow LA, Bhat R, Harbison JP, Chang CC. *Appl. Phys.* 1990; **67**: 586.

151 Takatani S, Kikawa Y, Nakazawa N. *Jpn. J. Appl. Phys.* 1991; **30**: 3763.

152 Qiu J, Qian Q-D, Gunshor RL *et al. Appl. Phys. Lett.* 1990; **56**: 1272.

153 Okamoto T, Konagai M, Kojima N *et al. J. Electron. Mat.* 1993; **22**: 229.

154 Okamoto T, Yamada A, Konagai M, Takahashi K. *J. Crystal Growth* 1994; **138**: 204.

155 Okamoto T, Yamada A, Konagai M, Takahashi K, Suyama N. In Ohdomari I, Oshima M, Hiraki A (eds) *Control of Semiconductor Interfaces.* Elsevier Science BV, 1994.

156 Yamada A, Kojima N, Takahashi K, Okamoto T, Konagai M. MBE growth of Ga$_2$Se$_3$ and GaSe and their optical properties. State-of-the-Art Program on Compound Semiconductors XVIII, Honolulu, Hawaii, 16–21 May 1993.

157 Takatani S, Nakano A, Ogata K, Kikawa T. *Jpn. J. Appl. Phys.* 1992; **31**: L458.

158 Teraguchi N, Kato F, Konagai M, Takahashi K, Nakamura Y, Otsuka N. *Appl. Phys. Lett.* 1991; **59**: 567.

159 Teraguchi N, Kato F, Konagai M, Takahashi K. *J. Electron. Mater.* 1991; **20**: 247.

160 Maung N, Fan GH, Ng TL, Williams JO, Wright AC. *J. Crystal Growth* 1996; **158**: 68.

161 Ng TL, Maung N, Fan GH, Poole IB, Williams JO, Wright AC. Growth of cubic gallium selenide by metal organic vapour phase epitaxy—structural investigation by transmission electron microscopy. European Workshop on MOVPE VI, Ghent, Belgium, 25–28 June 1995.

162 Ng TL, Maung N, Fan GH, Poole IB, Williams JO, Wright AC. *Adv. Mater.* (in press).

163 Maung N, Fan GH, Ng TL, Williams JO, Wright AC. *J. Crystal Growth* (submitted).

164 Maung N, Fan GH, Ng TL, Williams JO, Wright AC (in preparation).

165 Morley S, von der Emde M, Zahn DRT *et al. J. Appl. Phys.* 1996; **79**: 3196.

166 Wolfframm D, Bailey P, Evans DA, Neuhold G, Horn K. *J. Vac. Sci. Technol. A* 1996; **14**: 844.

167 Chou CT, Hutchison JL, Cherns D *et al. J. Appl. Phys.* 1993; **74**: 6566.

168 Palatnik LS, Belova EK. *Izv. Akad. Nauk. SSSR. Neorg. Mater.* 1965; **1**: 1883.

169 Finkman E, Tauc J, Kershaw R, Wood A. *Phys. Rev. B* 1975; **11**: 3785.

170 Guymont M, Tomas A, Guittard M. *Phil. Mag.* 1992; **66**: 133.

171 Newman PC, Cundall JA. *Nature* 1963; **200**: 876.

172 Koma A, Sunouchi K, Miyajima T. *J. Vac. Sci. Technol. B* 1985; **3**: 724.

173 Ueno K, Shimada T, Saiki K, Koma A. *Appl. Phys. Lett.* 1990; **56**: 327.

174 Ueno K, Abe H, Saiki K, Koma A, Oigawa H, Nannichi Y. *Surf. Sci.* 1992; **267**: 43.

175 Parkinson BA, Ohuchi FS, Ueno K, Koma A. *Appl. Phys. Lett.* 1991; **58**: 472.

176 Minder R, Ottaviani G, Canali C. *J. Phys. Chem. Solids* 1976; **37**: 417.

177 Le Toullec R, Piccioli N, Mejatty M, Balkanski N. *Nuova Cimento B* 1997; **38**: 159.

178 Minami F, Hasegawa A, Asaka S, Inoue K. *J. Lumin.* 1990; **45**: 409.

179 Kojima N, Sato K, Yamada A, Konagai M, Takahashi K. *Jpn. J. Appl. Phys.* 1994; **33**: L1482.

180 Yang JW, Kuznia JN, Chen QC *et al. Appl. Phys. Lett.* 1995; **67**: 3759.

181 Lin HT, Rich DH, Wittry DB. *J. Appl. Phys.* 1994; **75**: 8080.

182 Strite S, Lin ME, Morcoç H. *Thin Solid Films* 1993; **231**: 197.

183 Qian W, Skowronski M, De Graef M, Doverspike K, Rowland LB, Gaskill DK. *Appl. Phys. Lett.* 1995; **66**: 1253.

184 Dovidenko K, Oktyabrsky S, Narayan J, Razeghi M. *J. Appl. Phys.* 1996; **79**: 2443.

185 Kuo LH, Kimura K, Miwa S, Yasuda T, Yao T. *Appl. Phys. Lett.* 1996; **69**: 1408.

186 Jeon MH, Calhoun LC, Gila BP, Ludwig MH, Park RM. *Appl. Phys. Lett.* 1996; **69**: 2107.

Index

Italic type indicates illustrations